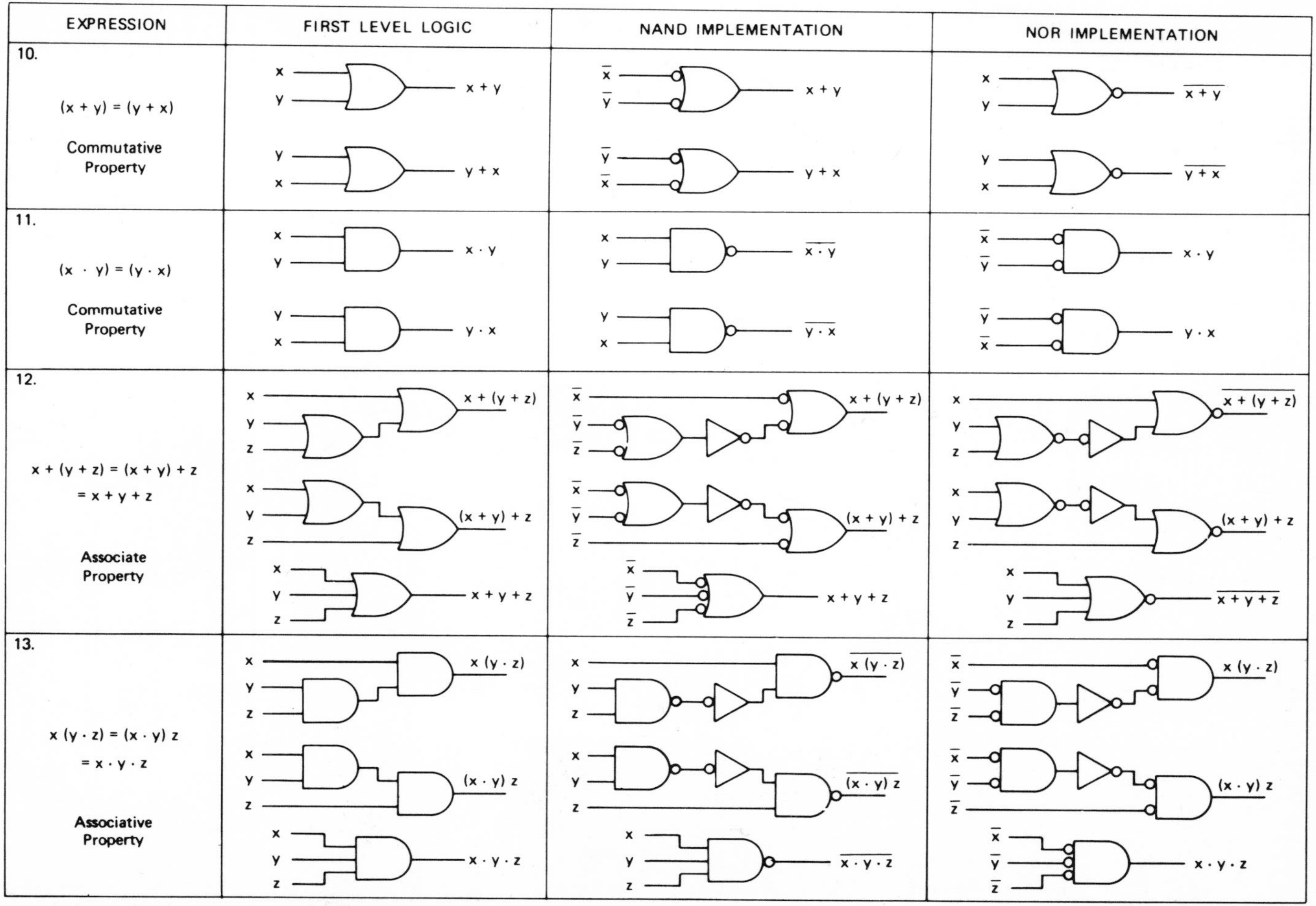

Figure 8-10(b). Implementation of logic expressions.

Second Edition

Digital Computer Circuits and Concepts

Second Edition

Digital Computer Circuits and Concepts

Bill R. Deem
Kenneth Muchow
Anthony Zeppa

Reston Publishing Company, Inc.
A Prentice-Hall Company
Reston, Virginia

Library of Congress Cataloging in Publication Data

Deem, Bill R.
  Digital computer circuits and concepts.

  Includes index.
  1. Electronic digital computers—Circuits.
  2. Electronic data processing.   I. Muchow, Kenneth,
joint author.   II. Zeppa, Anthony, joint author.
III. Title.
TK7888.4.D43   1977   621.3819'58'35   76-49808
ISBN 0-87909-189-4

© 1977 by Reston Publishing Company, Inc.
A Prentice-Hall Company
Reston, Virginia   22090

10   9   8   7

Printed in the United States of America

To Corine, Peggy, and Rose Ann

Contents

Preface

The primary intent of this text is to close the gap that currently exists between the academic theoretical instruction and current industrial practices in digital electronics.

Within the last few years, there have been rapid technological advances in the computer field. The need for clear, comprehensive, and practical information on currently available digital devices has increased just as rapidly.

This text is written primarily for students in technical schools, junior colleges, and community colleges and for electronics technicians currently employed in the field of digital computers. It is written at such a level that it can also be used in high-school electronics programs.

The text will help prepare students for positions in the industry by providing a firm foundation in the basic concepts of logic circuits. To help the student master these basic concepts, the text is illustrated with over 300 drawings, and many practical examples are worked out in detail. Numerous circuits from manufacturers, including practical data, are used throughout the text.

For the practicing digital electronics technician, the text will serve as an excellent reference because modern digital devices and circuits are discussed.

In Chapter 1, the evolution of computers and some applications are discussed. The various number systems needed to understand and work with computer circuits and systems are introduced in Chapter 2. Alphanumeric and interchange codes are also presented here. Chapter 3 covers binary arithmetic, including octal, hexadecimal, and BCD.

Logic, logic functions, logic symbology, and Boolean algebra are presented at a conceptual level in Chapter 4. Gates and flip-flops are discussed in Chapters 5, 6, and 7. Chapter 8 is devoted to the implementation of logic functions. The concepts presented in Chapter 4 are applied in Chapter 8.

Registers, encoders and decoders, counters and timing circuits are covered in Chapters 9 through 12. The arithmetic logic unit (ALU) and its associated circuits are presented in Chapter 13. In Chapter 14 memory units are discussed, including RAMs and ROMs.

Chapter 15, Programming, presents basic programming concepts with applications involving microcomputers. Machine, assembler, and compiler languages are included.

In Chapter 16, microprocessors and microcomputers are described. A practical application of a microcomputer system is presented. Both hardware and software concepts are included.

Many people have contributed in the preparation of the manuscript. We would like especially to express our appreciation to the many companies who were very helpful in making available to us their application notes and circuits. Also a special thanks to the other members of the Electronics Department at San Jose City College who offered their advice and opinions. Finally, we would like to acknowledge the untiring efforts of Corine Deem for typing and preparing the manuscript.

Bill Deem
Ken Muchow
Tony Zeppa

Second Edition

Digital Computer Circuits and Concepts

1

Survey of Digital Computers and Applications

Computing devices were in use long before the advent of modern computers. The use of mechanical means to perform arithmetical calculations has always been one of man's goals. The first counting machine was probably the *counting board,* which consisted of several stiff wires or reeds on a wooden frame. Each wire had ten sliding beans—representing 0 to 9.

Perhaps the earliest known mechanical calculating device was the *abacus,* developed independently by both the Chinese and the Greeks over 4000 years ago. The abacus was the earliest known device that had the capability of denoting a *carry.* The abacus is still used today in China and Japan.

Around 1615 John Napier invented and published tables of logarithms. He devised "Napiers Bones," which were numbered squares used in calculations. William Oughtred used Napier's logarithms in designing the first slide rule in 1621.

An important step in the development of "computing" machines took place in France in 1642: A 19-year-old inventor, Blaise Pascal, having tired of adding long columns of digits, designed and constructed an adding machine. Pascal's machine was composed of a series of gears, each with ten teeth to represent numbers 0 through 9. The gears were turned with a stylus, and when a 9 was reached, it automatically *carried* to the next gear. This same principle is still used in modern electromechanical calculators.

In 1671, Baron Von Leibnitz constructed a machine called a *step reckoner*. Like Pascal's machine, it could add, but it could multiply as well. Twenty years later, Leibnitz improved his machine so that it could also divide and extract roots.

In 1725, Basile Bouchon, a Frenchman, invented punched paper tape for operating cloth-weaving machines. A roll of punched paper was moved past the needles so that only where a hole appeared was the needle allowed to pass through.

In 1801, another Frenchman, J. M. Jacquard, designed a system of binary coded cards for controlling looms. Both the punched paper tape and the punched cards are used today in modern computer systems.

In 1833, Charles Babbage invented a machine called the *analytical engine*. This machine was based on basic principles identical to those used in modern digital computers. Babbage was probably the first man to visualize a general-purpose computer complete with a programming scheme and memory units. He even made use of Jacquard's punched cards for input and output data. Unfortunately, Babbage's computer was never completed. The industry and technology of his time were not sufficiently advanced to produce the components with the necessary tolerances.

In 1854, George Boole, an English mathematician, published *An Investigation of the Laws of Thought on Which Are Founded the Mathematical Theories of Logic and Probabilities*. This work formed the basis of *logical algebra*, in which simple thought processes were represented mathematically. It is important to note here that Boole did not relate his algebra to mechanical or electrical devices.

In 1886, Dr. Herman Hollerith devised a method of recording data on punched cards. Because of the punched card, the 1890 census was completed in less than one-third the time required for the 1880 census. This system is sometimes referred to as the *unit record system* because data is stored as units on cards which can be used repeatedly.

C. E. Shannon, in 1938 at M.I.T., wrote a paper on how an electrical circuit consisting of switches and relays could be represented by mathematical expressions. The basic techniques developed by Shannon are required in the design of all types of switching circuits today.

Howard Aiken designed the first general-purpose computer, the Mark I, a joint venture of Harvard University and IBM. This computer used electromagnetic relays and punched cards.

Among the leading contributors to modern computer technology were an electrical engineer, J. P. Eckert, Jr., and a physicist, J. W. Mauch-ley. In 1945 at the University of Pennsylvania, they completed the Elec-tronic Numerical Integrator and Calculator (ENIAC). The ENIAC used vacuum tubes and was therefore the first all-electronic computer. It was wired to perform a specific sequence of calculations. If a different pro-gram was required, thousands of circuits had to be rewired because the computer was not able to *store* a program.

The ENIAC was a huge machine, weighing over 30 tons. It had 19,000 tubes and hundreds of thousands of resistors, capacitors, and in-ductors. It occupied 15,000 square feet of floor space and consumed almost 200 kilowatts of power. Nevertheless, it could perform 5000 addi-tions per second. (Modern computers can perform millions of additions per second.)

In 1944, Von Neumann began to develop the logical design of a computer capable of a stored program. The program could be changed at will—without the rewiring of the computer.

1.1   MODERN COMPUTERS

The first computers of the late 1940s and early 1950s used relays and vacuum tubes for their logic circuits. Power and tube failures were com-mon. First-generation computers that were commercially available, such as the UNIVAC I, IBM 701, IBM 704, etc., used vacuum tubes. These computers were used essentially for single job operations.

With the advent of the transistor, second-generation computers were designed in the late 1950s. These were general-purpose computers—smaller, faster, and much more reliable. Integrated circuits (ICs) were used in the third-generation computers in the 1960s. Examples of third-generation computers are the IBM 360 series and the UNIVAC 1108.

Fourth-generation computers of the late 1960s to the present utilize solid-state medium-scale integration (MSI) and large-scale integration (LSI). Here, entire processing systems are fabricated on a single silicon chip.

An example of a large computer system is the ILLIAC IV, designed at the University of Illinois and assembled at the Burroughs Corporation in Paoli, Pennsylvania. The ILLIAC IV is actually a series of 64 "slave" computers capable of executing about 200 million instructions per sec-ond. Most computers solve problems by a series of sequential steps (*linear programming*). The ILLIAC IV, however, can perform up to 64 simul-taneous computations (*parallel programming*).

1.2   ANALOG COMPUTERS

There are two general types of computers—analog and digital. An analog device is one whose operation is "analogous" to a mechanical or electrical quantity. It solves problems by using some mathematical model—for example, a mercury thermometer is an analog device; it compares the expansion of a column of mercury with the surrounding temperature.

Analog computers can be used in simulating the performance or characteristics of some future product. For example, the performance of a proposed new jet plane could be simulated by an analog computer long before actual construction begins. All of the design variables—such as wingspan, thrust, and other engineering data—could be fed into the computer, and the output would represent the performance characteristics. Any changes in input variables would immediately affect the output, thereby achieving the desired characteristics.

However, the accuracy of analog computers is limited. The reasons for this are the tolerance limitation of electronic components and our limited ability to read and interpret scales and graphs.

1.3   DIGITAL COMPUTERS

Digital computers operate on the numbers ONE and ZERO—the only digits in the binary number system. The computer can manipulate these digits at extremely high speeds and with great accuracy. High-speed computers can perform the addition cycle in less than 100 nanoseconds. Whereas accuracy in analog computers is limited, the limitation to accuracy in digital computers is the number of digits used. For example, if a number has 32 digits, the binary place accuracy obtainable is equal to 2^{32}.

1.4   WHAT A COMPUTER DOES

A digital computer is an electronic device that consists of input and output devices, arithmetic and control circuits, and a memory. Before the computer can do any work, a program of instructions must be supplied. These instructions are written by a programmer in a language that the computer "understands." This information can be put on punched cards, punched tape, or magnetic tape. A computer operator will sometimes sit at a console and simply type out the instructions. We therefore need to communicate with a computer through some appropriate input device.

The input data is then taken by the computer and translated or coded into some number code or machine language with which the computer can proceed to operate. After all the operations are completed, the coding process is "reversed" at the end, and machine language is translated back into a language we can understand.

Speed is an essential element of input and output devices, just as it is with the internal operation of computers. For example, some of these machines can "read" magnetic tape at a rate of hundreds of inches per second. Thirty thousand printed lines per minute can be reproduced by some output terminals.

Another characteristic of the digital computer is its ability to *store* information. The storage process involves taking the data received as input and "filing" it in a preselected location. This data may be retrieved when called upon, or it may be further processed and returned to storage at another location. The facility of storage is usually an electronic unit or device known as *memory*. The size of the memory may vary from a few thousand to many million binary digits.

Perhaps the biggest advantage of the computer is its ability to manipulate figures and symbols and to perform enormous quantities of arithmetic computations millions of times faster than any human—even one assisted by the slide rule, abacus, or adding machine. Calculations that would require months, and even years, can be done by the computer in a matter of seconds or minutes.

The arithmetic process is essentially one of addition or subtraction. Multiplication and division can be considered as rapid repetition of addition and subtraction, respectively. The arithmetic capacity of the computer is essential in many areas of modern technology. For example, the instantaneous computations involved in space flight make space travel a reality. Weather, census, insurance, etc., all require vast and rapid calculations of an enormous amount of data. The engineering designs of skyscrapers, bridges, and planes are made possible by the arithmetic capacity of the computer.

Another fundamental property of the computer is its ability to make a logical choice. The computer is capable of "comparing" two numbers and "knowing" whether they are equal or if one is greater or less than the other. The ability to compare and make logical choices allows the computer to select the next course of action. For example, suppose a program instructs the computer to add or otherwise calculate until a stated amount is reached and then stop. The computer will instantaneously compare the stated amount with its calculations. When the two are identical, it will stop and go to the next instruction.

1.5 COMPUTER APPLICATION

The following examples of computer applications are not all inclusive. Standard business applications covering such operating procedures as payroll, billing, inventory control, etc., are all familiar. Banks and large companies could not afford to use credit cards without the speed and efficiency of computers. Special types of computers have been developed to perform specialized functions—such as those used in the military, communications, industrial processes, and education. Hospitals have installed computer terminals to monitor patients in intensive care units.

Many school systems have been using computers for years for maintaining records, such as attendance, grades, and other routine matters. Recently, however, computer-aided instruction (CAI) has been introduced. CAI is found in all levels of education, from primary grades through graduate schools. In a sense, the computer becomes the personal tutor for each student.

1.6 THERE IS A COMPUTER IN YOUR FUTURE

It has been predicted that by the year 2000 computers will be as common in the average household as stoves and refrigerators are today. These "microcomputers" will regulate, control, and monitor heating, lighting, and temperature. They may be used to detect and set off alarms to indicate the presence of fire, smoke, or intruders. The homemaker will be able to program her daily routines, such as the selection of menus, special diets, and the starting and stopping of washers and driers and other household appliances. In addition, the computer will be used to help the children with homework and provide storage for personal data, such as health history, tax records, and financial status.

Because of the ever-widening application of computers in all facets of human endeavor, a variety of employment opportunities will continue to exist. These positions will range from the managerial and professional to the clerical. Our interest, however, will focus on those individuals who install, maintain, troubleshoot, and repair computer equipment—the digital technicians and/or the computer technicians.

Number Systems and Codes

Numbers became important to man when he began to ask some basic questions: How much? How many? How far? How big? Number systems have developed as man's needs have increased. For example, certain tribes of aborigines in Australia used a system of numbers from 1 to 5. Since the needs of the aborigine were simple, he used a word meaning "plenty" for quantities greater than 5.

The early Romans, Greeks, and Egyptians used number systems in which symbols were used to represent numerical values. We are all familiar with Roman numerals where V means 5, C means 100, etc. The Egyptians used the symbol ∩ for 10. The symbol ☻ means 100. Zero was not a part of either of these systems. Furthermore, these systems were not "place value" systems; that is, the value for a given symbol never changed, regardless of its place or position in a given number. For example, in the Roman numeral IV, I means 1 and V means 5. When I appears before V, the number is 4 (5 − 1). For the numeral VI, I still has a value of 1, and V a value of 5; but the number has a value of 6 (5 + 1) because of the position of the numeral I. However, V always meant 5. It could never mean 50 or 500. The number 946 would be written as CMXLVI. Imagine performing mathematical operations of addition, subtraction, multiplication, and division with such a numbering system!

Our present number system was devised by Hindu mathematicians about 2000 years ago. Arabs began to use this system around 800 A.D. and introduced it to Europeans about 1200 A.D. We refer to it as the *Arabic number system* because it uses the Arabic symbols 0 through 9. Since there are ten symbols, it is also referred to as the *decimal number system.* The word *decimal* comes from a Latin word meaning "ten." Ten symbols were used because man has ten fingers and he used them for counting purposes. Each symbol is sometimes called a *digit,* the Latin word for "finger." The Arabic number system was revolutionary because it introduced the concepts of *zero* (0) and *place value.*

It is often necessary to store letters and characters in the computer. Like numbers, these are stored in the form of binary digits (bits). Various codes have been introduced to enable the computer to handle data other than numbers. Some of these codes will be discussed in this chapter.

The topics covered in this chapter are:

2.1   Decimal

2.2   Quinary

2.3   Octal

2.4   Hexadecimal

2.5   Binary Number System

2.6   Binary to Octal Conversion

2.7   Binary to Hexadecimal Conversion

2.8   Fractional Numbers

2.9   BCD Codes

2.10 Alphanumeric Codes

2.11 Parity Checking

2.12 Computer Words

2.1   DECIMAL

In counting with the decimal system, one should start with zero. Zero, an important concept indicating the absence of quantity or value, is the digit having the least value. When the count reaches 9, a limit has been reached because 9 is the digit of greatest value. An additional count produces a carry. This carry is equal to 10 and is said to occupy the 10s (10^1) position, whereas the 0 occupies the units (10^0) position. This process continues each time 9 is reached in the units position. When a 9 appears in the 10s position, the next carry to the 10s position will produce a carry to the 100s (10^2) position.

It is important to note that a 1 in the 10s position possesses a value (*weight*) 10 times that of a 1 in the units position. A 1 in the 100s position possesses a value of 10 times that of a 1 in the 10s position. The same is true for each position of greater weight. Because of this relationship, the decimal numbering system is called a *place value system* and its base is 10.

When dealing with number systems having place value, you will find some interesting relationships are useful: (1) the number of digits used in the system is equal to the base; (2) the largest digit is always one less than the base; (3) each position multiplies the value of the digit by the base; and (4) a carry from one position to the next increases its weight base times.

Let's examine a number in the decimal system, for example, the number 687. This number is made up of three digits—6, 8, and 7. The *least significant digit* (LSD) is 7, and its value is 7. The next significant digit is 8 and has a value of 80 (8×10). The 6 is the *most significant digit* (MSD) and has a value of 600 (6×100). The 7 occupies the units position; the 8 occupies the 10s position; and the 6 is in the 100s position. If we were to use scientific notation, the number could be written:

$$(6 \times 10^2) + (8 \times 10^1) + (7 \times 10^0)$$

That is, there are six 100s, eight 10s, and seven 1s.

2.2   QUINARY

Keeping the above relationships in mind, let's use the *quinary* (base 5) number system as another example of a place value system. Quinary was used in some early computers.

There are 5 digits (0, 1, 2, 3, 4). The digit of greatest value is 4, which is one less than the base. When counting, the sequence is 0 to 4. On the next count, we return to 0 in the units position, and a carry is generated. The resulting number is 10 and is read "one-zero." This number is equal to decimal 5 because the carry is equal to 5 (base 5). As this process continues, the sequence goes from 0 to 4 and back to 0 again. Each time a count greater than 4 is reached, a carry is generated.

In the quinary number system, digits in the units position possess the indicated value. Digits in the next position have *place value;* that is, digits in this position indicate the number of 5s present. This, then, is the 5s (5^1) position. The next position indicates the number of 25s (5^2) present, and so on, as shown in Figure 2-1.

Let's examine the number 324 in quinary. One may be tempted to read this as "three-hundred-twenty-four." However, such language implies base 10. This number should be read "three-two-four," which does not imply any base. It is very difficult for us to think in quinary, so let's convert the number to base 10, or:

$$324_5 = (75) + (10) + (4) = 89_{10}$$

or:

$$(3 \times 5^2) + (2 \times 5^1) + (4 \times 5^0) = 89_{10}$$

There is a 3 in the 25s position + a 2 in the 5s position + a 4 in the units position, which equals 89.

Example 2.1. Convert the following quinary numbers to decimal:

(a) 24_5 (b) 342_5 (c) 1442_5 (d) 20324_5

Solution: (a) $24_5 = (2 \times 5^1) + (4 \times 5^0) = 10 + 4 = 14_{10}$

(b) $342_5 = (3 \times 5^2) + (4 \times 5^1) + (2 \times 5^0) = 75 + 20 + 2 = 97_{10}$

(c) $1442_5 = (1 \times 5^3) + (4 \times 5^2) + (4 \times 5^1) + (2 \times 5^0) = 125 + 100 + 20 + 2 = 247_{10}$

(d) $20324_5 = (2 \times 5^4) + (0 \times 5^3) + (3 \times 5^2) + (2 \times 5^1) + (4 \times 5^0) = 1250 + 0 + 75 + 10 + 4 = 1339_{10}$

Let's reverse the process. In Example 2.2, the number 39 is converted from decimal to quinary.

Example 2.2. Convert 39_{10} to quinary.

Step 1. $\dfrac{39}{5} = 7 + 4$

Step 2. $\dfrac{7}{5} = 1 + 2$ LSD

Step 3. $\dfrac{1}{5} = 0 + 1$ —— MSD ——▶ 1 2 4 $39_{10} = 124_5$

Since we are converting to base 5, we first divide the decimal number by 5 (Step 1). The answer is 7 with a remainder of 4. This remainder is the LSD of our base 5 number. The whole number left after this initial

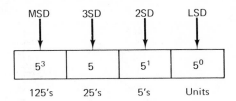

Figure 2-1.   Weight of position in quinary number system.

division (7) is again divided by 5 to determine the value of the digit in the next (5^1) position. This division leaves a remainder of 2, which becomes the digit in the 5^1 position. The whole number resulting from this division (1) is again divided by 5, and the remainder (1) is the digit in the 5^2 position. This digit is the MSD since no further division is possible.

Example 2.3. Convert the following decimal numbers to quinary numbers:

(a) 87   (b) 372   (c) 768

Solution:

(a) $\dfrac{87}{5} = 17 + 2$

$\dfrac{17}{5} = 3 + 2$

$\dfrac{3}{5} = 0 + 3 \longrightarrow 3 \quad 2 \quad 2 \qquad\qquad 87_{10} = 322_5$

(b) $\dfrac{372}{5} = 74 + 2$

$\dfrac{74}{5} = 14 + 4$

$\dfrac{14}{5} = 2 + 4$

$\dfrac{2}{5} = 0 + 2 \longrightarrow 2 \quad 4 \quad 4 \quad 2 \qquad 372_{10} = 2442_5$

(c) $\dfrac{768}{5} = 153 + 3$ ─────────────────────────┐

$\dfrac{153}{5} = 30 + 3$

$\dfrac{30}{5} = 6 + 0$

$\dfrac{6}{5} = 1 + 1$

$\dfrac{1}{5} = 0 + 1$ ──────▶ 1   1   0   3   3 $768_{10} = 11033_5$

2.3 OCTAL

Octal, or base 8, is an important numbering system in dealing with computers, as will be explained later. How many digits are used in octal? Since the base is 8, the number of digits is 8. What is the value of the largest digit in octal? The largest digit is one less than the base, or 7. Therefore, the 8 digits are 0 through 7. What is the value of a carry of 1 from the units position? A carry of 1 has a weight of 8 because the base is 8. Figure 2-2 shows positional value of the first four positions for the octal number system.

Let's convert the octal number 1025 to base 10. There are

One 512s which equals	512
+ zero 64s which equals	0
+ two 8s which equal	16
+ five 1s which equal	5
Sum	533

$1025_8 = 533_{10}$

Or:

$$(1 \times 8^3) + (0 \times 8^2) + (2 \times 8^1) + (5 \times 8^0) = 533_{10}$$

Example 2.4 Change the following octal numbers to decimal:

(a) 73_8 (b) 432_8 (c) 600_8 (d) 1234_8

Solution: (a) $73_8 = (7 \times 8^1) + (3 \times 8^0) = (56 + 3) = 59_{10}$

(b) $432_8 = (4 \times 8^2) + (3 \times 8^1) + (2 \times 8^0) = 256 + 24 + 2 = 282_{10}$

(c) $600_8 = (6 \times 8^2) + (0 \times 8^1) + (0 \times 8^0) = 384 + 0 + 0 = 384_{10}$

(d) $1234_8 = (1 \times 8^3) + (2 \times 8^2) + (3 \times 8^1) + (4 \times 8^0) = 512 + 128 + 24 + 4 = 668_{10}$

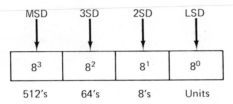

MSD	3SD	2SD	LSD
8^3	8^2	8^1	8^0
512's	64's	8's	Units

Figure 2-2.   Positional value of the first four positions for the octal number system.

Let's reverse the process. In Example 2.5, the number 189 is converted from decimal to octal.

Example 2.5. $189_{10} = (?)_8$

Step 1. $\dfrac{189}{8} = 23 + 5$

Step 2. $\dfrac{23}{8} = 2 + 7$ LSD

Step 3. $\dfrac{2}{8} = 0 + 2 \longrightarrow$ MSD $\rightarrow$ 2   7   5 $189_{10} = 275_8$

The method is identical to the one discussed in Example 2.2 except that division is by 8 since we are working in base 8.

Example 2.6. Convert the following decimal numbers to octal numbers.

(a) 78   (b) 376   (c) 1463

Solution: (a) $\dfrac{78}{8} = 9 + 6$

$\dfrac{9}{8} = 1 + 1$

$\dfrac{1}{8} = 0 + 1 \longrightarrow$ 1   1   6 $78_{10} = 116_8$

(b) $\dfrac{376}{8} = 47 + 0$ ─────────────┐

$\dfrac{47}{8} = 5 + 7$

$\dfrac{5}{8} = 0 + 5$ ────────▶ 5   7   0 $376_{10} = 570_8$

(c) $\dfrac{1463}{8} = 182 + 7$ ──────────┐

$\dfrac{182}{8} = 22 + 6$

$\dfrac{22}{8} = 2 + 6$

$\dfrac{2}{8} = 0 + 2$ ──▶ 2   6   6   7 $1463_{10} = 2667_8$

2.4 HEXADECIMAL

Another important number system used in computers is the hexadecimal system (base 16). How many digits exist in hexadecimal? Sixteen because the base is 16. What is the value of the largest digit in hexadecimal? Fifteen, one less than the base. However, 15 is a decimal number and requires two digits (15). In the hexadecimal number system, sixteen different symbols must be used. Zero through 9 are used for the first ten, and A through F are used for the remaining six, as shown in Figure 2-3. Figure 2-4 indicates positional value of the first four positions of the hexadecimal number system.

The hexadecimal number 3AC2 may be converted to base 10 as follows:

	Three 4096s which equal	12,288
+	ten 256s which equal	2,560
+	twelve 16s which equal	192
+	two 1s which equal	2
	Sum	$15,042_{10}$

Or:

$$(3 \times 16^3) + (10 \times 16^2) + (12 \times 16^1) + (2 \times 16^0) = 15,042_{10}$$

$$3AC2_{16} = 15,042_{10}$$

Decimal	Hexadecimal
0	0
1	1
2	2
3	3
4	4
5	5
6	6
7	7
8	8
9	9
10	A
11	B
12	C
13	D
14	E
15	F

Figure 2-3.   Hexadecimal digits.

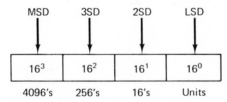

Figure 2-4.   Positional value of a four-digit hexadecimal number.

Example 2.7. Convert the following hexadecimal numbers to decimal numbers:

(a) 72   (b) C29   (c) 12AB   (d) 1A2A

Solution: (a) $72_{16} = (7 \times 16^1) + (2 \times 16^0) = 112 + 2 = 114_{10}$

(b) $C29_{16} = (12 \times 16^2) + (2 \times 16^1) + (9 \times 16^0) = 3113_{10}$

(c) $12AB_{16} = (1 \times 16^3) + (2 \times 16^2) + (10 \times 16^1) + (11 \times 16^0) = 4779_{10}$

(d) $1A2A = (1 \times 16^3) + (10 \times 16^2) + (2 \times 16^1) + (10 \times 16^0) = 6698_{10}$

The conversion of decimal numbers to hexadecimal is identical to conversion using other bases. In Example 2.8, the number 1324 is converted from hexadecimal to decimal. Since the hexadecimal number system has a base of 16, division is by 16. When working with remainders greater than 9, substitute the characters A, B, C, D, E, and F as required.

Example 2.8. $1324_{10} = (?)_{16}$

Step 1. $\dfrac{1324}{16} = 82 + 12$ ─────────────────────┐

Step 2. $\dfrac{82}{16} = 5 + 2$ ────────────┐   LSD

Step 3. $\dfrac{5}{16} = 0 + 5$ ──── MSD ──➤ 5   2   C $1324_{10} = 52C_{16}$

Example 2.9. Convert the following decimal numbers to hexadecimal numbers:

(a) 672   (b) 1763   (c) 12,760

Solution: (a) $\dfrac{672}{16} = 42 + 0$ ──────────────┐

$\dfrac{42}{16} = 2 + 10$

$\dfrac{2}{16} = 0 + 2$ ──────➤ 2   A   0 $672_{10} = 2A0_{16}$

(b) $\dfrac{1763}{16} = 110 + 3$ ──────────┐

$\dfrac{110}{16} = 6 + 14$

$\dfrac{6}{16} = 0 + 6$ ──────➤ 6   E   3 $1763_{10} = 6E3_{16}$

(c) $\dfrac{12760}{16} = 797 + 8$ ────────────┐

$\dfrac{797}{16} = 49 + 13$

$\dfrac{49}{16} = 3 + 1$

$\dfrac{3}{16} = 0 + 3 \longrightarrow 3$ 1   D   ↓8 $12760_{10} = 31D8_{16}$

2.5   BINARY NUMBER SYSTEM

In the binary number system, there are two digits, 0 and 1. The binary system is used for internal computer operations because only two signal levels are required—as opposed to decimal where 10 signal levels would be necessary. Since digits in the units position have a value of 0 or 1, numbers greater than 1 cause a carry to the next position. Each position represents the base raised to a power. In base 10, the units position had a power of 10^0, the next position 10^1, and so on. In binary or base 2, the same reasoning applies. The units position has a power of 2^0, the next position 2^1, and so on, as illustrated in Figure 2-5. Binary digits are called *bits* (a contraction of "binary digits"). Therefore, the digit in the units position is called the *least significant bit* (LSB), and so on until we reach the *most significant bit* (MSB).

MSB	4SB	3SB	2SB	LSB
2^4	2^3	2^2	2^1	2^0
16	8	4	2	1

Figure 2-5.   Positional value of the first five positions for the binary number system.

Converting from base 2 to base 10 is quite easy. If a 1 is present in a given position, the weight of that bit is added. If a 0 is present, the weight of that bit is not added. Consider the number 11010 in Example 2.10. The decimal equivalent is 26.

Example 2.10. $11010_2 = (?)_{10}$

Solution: $(1 \times 2^4) + (1 \times 2^3) + (0 \times 2^2) + (1 \times 2^1) + (0 \times 2^0) =$

 16 + 8 + 0 + 2 + 0 = 26

Example 2.11. Change the following binary numbers to decimal numbers:

(a) 1100   (b) 10001   (c) 101011   (d) 111101

Solution:

(a) $1100_2 = (1 \times 2^3) + (1 \times 2^2) + (0 \times 2^1) + (0 \times 2^0) =$

 8 + 4 + 0 + 0 = 12_{10}

(b) $10001_2 = (1 \times 2^4) + (0 \times 2^3) + (0 \times 2^2) + (0 \times 2^1) + (1 \times 2^0)$

 16 + 0 + 0 + 0 + 1 = 17_{10}

(c) $101011_2 = (1 \times 2^5) + (0 \times 2^4) + (1 \times 2^3) + (0 \times 2^2) + (1 \times 2^1) +$

(1×2^0)

 32 + 0 + 8 + 0 + 2

$= 43_{10}$

(d) $111101_2 = (1 \times 2^5) + (1 \times 2^4) + (1 \times 2^3) + (1 \times 2^2) + (0 \times 2^1) +$

(1×2^0)

 32 + 16 + 8 + 4 + 0

 + 1 = 61_{10}

Let's turn the process around and find the binary equivalent of a decimal number. The process of conversion from decimal to binary is identical to that of conversion from decimal to other bases.

Example 2.12. Convert 167_{10} to binary.

Solution: $\dfrac{167}{2} = 83 + 1$

$\dfrac{83}{2} = 41 + 1$

$\dfrac{41}{2} = 20 + 1$

$\dfrac{20}{2} = 10 + 0$

$\dfrac{10}{2} = 5 + 0$ $\qquad\qquad\qquad 167_{10} = 10100111_2$

$\dfrac{5}{2} = 2 + 1$ $\qquad\qquad$ LSB

$\dfrac{2}{2} = 1 + 0$

$\dfrac{1}{2} = 0 + 1 \rightarrow$ MSB $\rightarrow$ 1   0   1   0   0   1   1   1

Example 2.13. Convert the following decimal numbers to binary numbers:

(a) 7   (b) 34   (c) 89   (d) 203

Solution:

(a) $\qquad \dfrac{7}{2} = 3 + 1$

$\dfrac{3}{2} = 1 + 1$ $\qquad\qquad 7_{10} = 111_2$

$\dfrac{1}{2} = 0 + 1 \longrightarrow$ 1   1   1

(b)

$$\frac{34}{2} = 17 + 0$$

$$\frac{17}{2} = 8 + 1$$

$$\frac{8}{2} = 4 + 0$$

$$\frac{4}{2} = 2 + 0$$

$34_{10} = 100010_2$

$$\frac{2}{2} = 1 + 0$$

$$\frac{1}{2} = 0 + 1 \longrightarrow 1 \quad 0 \quad 0 \quad 0 \quad 1 \quad 0$$

(c)

$$\frac{89}{2} = 44 + 1$$

$$\frac{44}{2} = 22 + 0$$

$$\frac{22}{2} = 11 + 0$$

$89_{10} = 1011001_2$

$$\frac{11}{2} = 5 + 1$$

$$\frac{5}{2} = 2 + 1$$

$$\frac{2}{2} = 1 + 0$$

$$\frac{1}{2} = 0 + 1 \longrightarrow 1 \quad 0 \quad 1 \quad 1 \quad 0 \quad 0 \quad 1$$

(d) $\quad$ 203

$$\frac{203}{2} = 101 + 1$$

$$\frac{101}{2} = 50 + 1$$

$$\frac{50}{2} = 25 + 0$$

$$\frac{25}{2} = 12 + 1$$

$203_{10} = 11001011_2$

$$\frac{12}{2} = 6 + 0$$

$$\frac{6}{2} = 3 + 0$$

$$\frac{3}{2} = 1 + 1$$

$$\frac{1}{2} = 0 + 1 \longrightarrow 1 \quad 1 \quad 0 \quad 0 \quad 1 \quad 0 \quad 1 \quad 1$$

2.6 BINARY TO OCTAL CONVERSION

When using binary numbers to represent large quantities, you will require many ones and zeros. This is tedious and awkward. Other number systems are often used as a shorthand notation for binary numbers. You will note that if only three bit positions are considered, eight combinations are possible (000 through 111). Therefore, octal numbers may be substituted directly for 3-bit binary numbers. For example, the binary number 110111100001011001 may be converted to octal by the simple procedure of separating the bits into groups of three and substituting octal digits for each group of three bits, as follows:

110	111	100	001	011	001	(binary number divided into groups of three)
6	7	4	1	3	1	(octal substitution for each three-bit group)

Therefore, 110111100001011001_2 equals 674131_8.

Convert binary 1011000111100101 to octal.

001	011	000	111	100	101	(binary number)
1	3	0	7	4	5	(octal number)

To separate the binary number into 3-bit groups, the procedure begins from the LSB position. Therefore, two zeros were added to fill the most significant octal digit position (each group of three bits represents one octal digit).

The procedure may be reversed in order to convert from octal to binary. Let's convert octal 167432_8 to binary:

1	6	7	4	3	2	(octal)
001	110	111	100	011	010	(binary)

Three bits are substituted for each octal digit. The above procedure exemplifies the use of the octal numbering system in computers.

2.7   BINARY TO HEXADECIMAL CONVERSION

Hexadecimal is another base commonly used as a shorthand for binary. Sixteen possible combinations of four bits exist ($2^4 = 16$), and because there are 16 digits in hexadecimal, one hexadecimal digit is substituted for four binary bits. The procedure used to convert from binary to hexadecimal is similar to that used for converting from binary to octal except that the binary number is broken into groups of four bits instead of groups of three (as in conversion to octal). For example, let's convert 1101100110110011 to hexadecimal:

1101	1001	1011	0011
D	9	B	3

The number is first separated into groups of four bits, and then a hexadecimal digit is substituted for each group of four.

Convert 11010010110001 to hexadecimal:

0011	0100	1011	0001
3	4	B	1

Two zeros are added to the MSB positions in order to fill the most significant hexadecimal position.

In order to convert from hexadecimal to binary, the reverse procedure is used.

Convert $960B_{16}$ to binary:

9	6	0	B
1001	0110	0000	1011

$960B_{16} =$
1001011000001011_2

Convert $BEAF_{16}$ to binary:

B	E	A	F
1011	1110	1010	1111

$BEAF_{16} =$
1011111010101111_2

Figure 2-6 shows a printout from the main storage of a System/370 IBM computer. Notice that although the printout is in hexadecimal, the data in storage is actually in binary.

2.8 FRACTIONAL NUMBERS

How are numbers with values of less than 1 represented in binary? As in decimal, a point is used to separate the whole number from the fraction. The first bit position to the right of the binary point represents the decimal fraction $0.5(2^{-1})$; the second digit to the right represents 0.5 divided by 2, or $0.25(2^{-2})$; the third, 0.25 divided by 2, or $0.125(2^{-3})$; and so on. Figure 2-7 is a table of powers of 2 from 2^{16} to 2^{-16}.

The decimal number 0.1875 would be represented in binary as 0.0011. In the same manner, decimal 56.375 is represented in binary as 111000.011. Fractional binary numbers may be converted to hexadecimal just as they were before. However, the bits are separated, beginning with the point, into groups of four. For example, 0.11010011_2 is converted as follows:

. 1101	0011
. D	3

$0.11010011_2 = 0.D3_{16}$

Convert 0.0000011101_2 to hexadecimal:

.0000	0111	0100
. 0	7	4

$0.0000011101_2 = 0.074_{16}$

In this example, two 0s are added in the LSB positions in order to fill the four bit positions of the least significant hexadecimal digit position.

Let's convert 1010011.110011_2 to hexadecimal:

0101	0011.	1100	1100
5	3 .	C	C

$1010011.110011_2 = 53.CC_{16}$

```
          BFS2013         02/25/74

033EE0   00014177  000147F0  F3364170  004847FC   F3360203  F7A4F7AC  D201F6D0  F6D4D203
033F00   F7B0F7B8  5810F7B0  9101F7A1  4710F322   9601F7A1  90E2D144  4110F7D0  4500F314
033F20   000341F8  0A0298E2  D1444870  F6D047FC   F2A29104  80034780  F2845860  F7A44166
033F40   00004170  0048581D  015C47F8  0004411D   00784111  00045010  F5389209  F538581D
033F60   015C47F8  00080200  60016000  91021000   4710F3CE  95021001  4770F37C  1800D201
033F80   D0604000  480D0060  47F0F38E  95041001   4770F39C  D203D060  4005080D  00604E0D
033FA0   0060F395  D0680062  47F0F3FC  501D00FC   D207D060  40009801  D0605D00  F7C04E1D
033FC0   00684E0D  0070F384  47F0F3FC  96F0D068   F384D069  F420F384  00FC47F0  F3FC1B88
033FE0   43810001  D7090060  D060D06B  006A1B58   06804480  F384D069  D069D065  F154D060
034000   D060F384  D060D061  96F0D068  414D0071   91304000  4740F40C  96F04000  41440001
034020   1B554851  00021B45  9601D15A  47F0F0F0   D2005000  40009201  D1059400  F7C91233
034040   4780F43C  9400D106  47F0F440  9601D106   58A20000  41220004  41900020  1EAA0690
034060   4730F458  47F0F512  91101000  4780F47E   910F1000  4770F47E  9604D105  1EAA4730
034080   F4909101  D1064710  F4904710  F4BC1EAA   4730F4C4  9101D106  4710F4C4  47F0F4C8
0340A0   45E0F08A  4140F7C6  41500003  45E0F0F0   4111000A  45E0F08A  12774780  F51A9240
0340C0   60004166  00106670  47F0F51A  4111000A   47F0F51A  47F0F4A4  9602D105  45E0F08A
0340E0   19574FC0  F4EA1277  4770F4E4  45E0F126   47F0F4D0  188747F0  F4EC1885  06809240
034100   60004480  F50C45E0  F1B61255  4780F504   47F0F4D0  9401D105  47F0F4A8  D2006001
034120   600089A0  000147F0  F4A44111  000AD501   1000F7C4  4770F52C  47F0F1E8  469DF44C
034140   47F0F440  00000000  09034148  00000001   00000000  00000004  00034148  00000000
034160   00008400  08000000  00034188  00000000   00032A80  3200D3E2  E3C4E3C6  40400000
034180   C0000000  00000000  002020F3  24034337   80000000  80000000  00000000  0000FF00
```

Figure 2-6. Printout from the main storage (called a *core dump*) of a System/370 IBM computer. The six hexadecimal digits in the left-hand column are the addresses of the data. Two hexadecimal digits are stored at each address.

Power	Value	Power	Value
2^0	1	2	
2^1	2	2^{-1}	0.5
2^2	4	2^{-2}	0.25
2^3	8	2^{-3}	0.125
2^4	16	2^{-4}	0.062 5
2^5	32	2^{-5}	0.031 25
2^6	64	2^{-6}	0.015 625
2^7	128	2^{-7}	0.007 812 5
2^8	256	2^{-8}	0.003 906 25
2^9	512	2^{-9}	0.001 953 125
2^{10}	1024	2^{-10}	0.000 976 562 5
2^{11}	2048	2^{-11}	0.000 488 281 25
2^{12}	4096	2^{-12}	0.000 244 140 625
2^{13}	8192	2^{-13}	0.000 122 070 312 5
2^{14}	16,384	2^{-14}	0.000 061 035 156 25
2^{15}	32,768	2^{-15}	0.000 030 517 578 125
2^{16}	65,536	2^{-16}	0.000 015 258 789 062 5

Figure 2-7.   Table of powers of 2.

Convert $110111000110.1010111101_2$ to hexadecimal.

1101	1100	0110.	1010	1111	0100
D	C	6	A	F	4

$$110111000110.1010111101_2 = DC6.AF4_{16}$$

Quite often, two registers are used to store binary numbers. The first register stores the whole number; and the second stores the fraction. For example, if the binary number 1110101.1011001 were stored in two 16-bit registers, it would be stored like this:

1st Register	2nd Register
0000000001110101	1011001000000000

The hexadecimal notation for the above binary number would be

$$75.B2_{16}$$

The octal notation for this binary number is

$$165.544_8$$

A fixed number of bits, like those in the above example, is called a *word*. The length of the word depends on the computer and the way data is handled in the computer. For example, in the IBM System/360 computers, each word is 32 bits in length. The 7600 Control Data Corporation computers use 65-bit words.

2.9   BCD CODES

Substituting binary bits for characters in other number systems is called *encoding*. When three binary bits are substituted for an octal character, *Binary-Coded-Octal* (BCO) is produced. Substituting four bits for a hexadecimal character produces *Binary-Coded-Hexadecimal* (BCH). As you know, BCO and BCH are straight binary numbers.

What about encoding decimal? Substituting four binary bits for a decimal character produces a code called *Binary-Coded-Decimal* (BCD). Because ten decimal characters exist, at least four bits are required. Figure 2-8 shows the most common form of BCD. Because of the natural binary number system used, this code is sometimes referred to as NBCD.

DECIMAL	BCD
0	0000
1	0001
2	0010
3	0011
4	0100
5	0101
6	0110
7	0111
8	1000
9	1001

Figure 2-8.   Binary coded decimal (BCD).

There are 16 possible combinations of four bits (0000–1111). This means that six combinations must be deleted for BCD. For natural BCD, the last six combinations are not used (1010 through 1111). For example, 1001 is the code for decimal 9; but what does 1010 represent? In straight binary, it has a value of 10; but in decimal, there is no single character for 10. Two characters are required, a one and a zero (10). Ten in BCD would appear as follows:

$$0001 \quad 0000$$

The six combinations that have been deleted are invalid because they do not represent one of the ten decimal characters. The procedure for converting from decimal to BCD (encoding) is the same as that used for converting from hexadecimal or octal to binary (as discussed earlier). Let's convert 563_{10} to BCD:

5	6	3	
0101	0110	0011	$563_{10} = 010101100011_{(BCD)}$

One must remember that this BCD number is *not* a binary number, and 563_{10} does not equal 010101100011_2! The decimal number 563 equals 1000110011 in binary.

Convert 1769_{10} to BCD:

1	7	6	9	
0001	0111	0110	1001	$1769_{10} =$

$$0001011101101001_{(BCD)}$$

Convert 63.7608_{10} to BCD:

6	3	.	7	6	0	8
0110	0011	.	0111	0110	0000	1000

$$63.7608_{10} =$$
$$01100011.011101101000001000_{(BCD)}$$

You may ask, "Why not use BCD in the computer?" Some computers do use BCD. However, the trend is toward straight binary for two important reasons. First, more data can be stored in a given number of memory cells when binary is used. This is because for every four bits of memory, only ten decimal digits can be stored; whereas, four bits of memory can store 16 digits of hexadecimal. Remember, 6 of the 16 possible combinations of bits are invalid for BCD. Second, BCD complicates the arithmetic process.

The computer must first find the binary sum of the BCD digits and then decide if this sum is valid. If the sum is invalid, 0110 must be added.

You may have guessed that another advantage of using straight binary is that less time is required.

Where is BCD used? BCD is used as a binary code to represent decimal digits. Whenever decimal information is required, such as in digital voltmeters and electronic calculators, BCD is used. The process of encoding BCD from decimal is simple, whereas changing from decimal to binary (or vice versa) is complex and requires a great deal of circuitry. Thus, calculators use BCD because the input data (keyboard) as well as the output data (display) is decimal. As a result, calculators have rather complex arithmetic circuitry and are considerably slower than computers.

Since any ten combinations of four bits are required for BCD, other BCD codes are possible. The *excess-three code* is used when it is desirable to obtain the nine's complement of the decimal digit that is represented by the code. Figure 2-9 shows the excess-three code. The code is three greater than the natural binary code. This code is used when one desires to perform subtraction using the nine's complement method. Note that the nine's complement is obtained by taking the one's complement of the BCD code. For example, the nine's complement of 4 (0111) is 5 (1000); the nine's complement of 2 (0101) is 7 (1010); etc.

DECIMAL	EXCESS–THREE CODE
0	0011
1	0100
2	0101
3	0110
4	0111
5	1000
6	1001
7	1010
8	1011
9	1100

Nine's complement is obtained by taking one's complement.

Figure 2-9.   Excess-three code.

Another BCD code is called the 2'421 code (Figure 2-10) because of the unusual weighting of the bits. Instead of the MSB position having a weight of 8 as in the 8421 code, the MSB position in this code has a weight of 2. There are two bit positions (2 and 2') which have a weight of 2. For example, $5 = 2' + 2 + 1 = 1011$. This code is also used when the nine's complement is required.

Other codes are possible and come into play whenever certain patterns of bits can be useful in the solution of a particular problem. For example, the Gray code shown in Figure 2-11 is used for analog to digital conversion. You will notice that only one bit changes between counts. Using this code for certain analog to digital conversion techniques results in an error of not more than one count.

DECIMAL	GRAY CODE
0	0000
1	0001
2	0011
3	0010
4	0110
5	0111
6	0101
7	0100
8	1100
9	1101
10	1111
11	1110
12	1010
13	1011
14	1001
15	1000

DECIMAL	2'421
0	0000
1	0001
2	0010
3	0011
4	0100
5	1011
6	1100
7	1101
8	1110
9	1111

Nines complement is obtained by taking ones complement.

Figure 2-10.   2' 421 code.   **Figure 2-11.**   Gray code.

2.10 ALPHANUMERIC CODES

The computer also uses alphabetic data as well as special characters such as punctuation marks and mathematic symbols. If six bits are used in a code, a total of 64 characters can be represented (2^6). Figure 2-12 shows the *Binary Coded Decimal Interchange Code* (BCDIC) used by IBM. This code was used by certain IBM computers such as the 1401 and the 1410. However, it is still used as an interchange code for the transmission of data between computers, and between peripheral equipment such as terminals and the central processing unit.

The four LSBs of the code (8421) are referred to as the *numeric bits*. The next two significant bits are called the *zone bits* as shown in Figure 2-13. The *check bit* (C) will be discussed later in this chapter.

When a numeric character is represented, the zone bits are set to 00. For example, the number 5 would be represented as in Figure 2-14.

For the alphabetic characters A through I, the zone bits are set to 11. The letter D looks like Figure 2-15.

The letters J through R use 10 for the zone bits, and S through Z use 01. Notice that other special characters are included. For example, the lozenge (□) is represented by the code in Figure 2-16.

Example 2.14. Write the expression GO TO 36. using the BCDIC.

G	0	b	T	0	b
110111	100110	010000	010011	100110	010000

			3	6	
			000011	000110	111011

An alphanumeric code that can represent as many as 256 different characters is called the *Extended-Binary Coded Decimal Interchange Code* (EBCDIC). All characters in the EBCDIC code are represented by eight bits or two hexadecimal characters. The EBCDIC code is shown in Figure 2-17. EBCDIC permits the use of both upper-case and lower-case letters as well as many special characters including control characters such as NULL and PF. These control characters are interpreted by certain peripheral devices such as printers or typewriters. Many of the bit combinations do not have character assignments. The computer itself may make special assignments to some of these combinations; however, some are reserved for later assignment.

CHARACTER Report	Program	C	B	A	8	4	2	1
b (Low)		C						
.			B	A	8		2	1
□	)	C	B	A	8	4		
[			B	A	8	4		1
<			B	A	8	4	2	
‡		C	B	A	8	4	2	1
&	+	C	B	A				
$	.	C	B		8		2	1
*			B		8	4		
]		C	B		8	4		1
;		C	B		8	4	2	
Δ			B		8	4	2	1
—			B					
/		C		A				1
,		C		A	8		2	1
%	(			A	8	4		
m		C		A	8	4		1
\		C		A	8	4	2	
⧺				A	8	4	2	1
b				A				
≠	=				8		2	1
@		C			8	4		
:					8	4		1
>					8	4	2	
√		C			8	4	2	1
?		C	B	A	8		2	
A			B	A				1
B			B	A			2	
C		C	B	A			2	1
D			B	A		4		
E		C	B	A		4		1
F		C	B	A		4	2	
G			B	A		4	2	1
H			B	A	8			
I		C	B	A	8			1
!			B		8		2	
J		C	B					1
K		C	B				2	
L			B				2	1
M		C	B			4		
N			B			4		1
O			B			4	2	
P		C	B			4	2	1
Q		C	B		8			
R			B		8			1
‡				A	8		2	
S		C		A			2	
T				A			2	1
I		C		A		4		
V				A		4		1
W				A		4	2	
X		C		A		4	2	1
Y		C		A	8			
Z				A	8			1
Ø		C			8		2	
1								1
2							2	
3		C					2	1
4						4		
5		C				4		1
6		C				4	2	
7						4	2	1
8					8			
9 (High)		C			8			1

COLLATING SEQUENCE (Low → High)

Symbol	Name
‡	Group Mark
‡	Record Mark
⧺	Segment Mark
m	Word Separator
@	At Sign
#	Number Sign
&	Ampersand
+	Plus
*	Asterisk
%	Percent
/	Slash
\	Backslash
□	Lozenge
b	Blank
b	Substitute Blank
(	Left Parenthesis
)	Right Parenthesis
[	Left Bracket
]	Right bracket
√	Tape Mark
<	Less than
>	Greater than
=	Equal to
;	Semicolon
:	Colon
.	Period or Point
'	Prime or Apostrophe
—	Minus or Hyphen (Dash)
Δ	Delta

Figure 2-12. BCD Interchange Code (BCDIC).

Figure 2-13.   Seven-bit register showing position of numeric and zone bits.

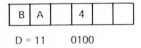

Figure 2-14.   Display of the number 5.

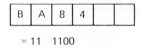

Figure 2-15.   Display of the letter D.

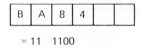

Figure 2-16.   Display of the lozenge.

The first four bits (first hexadecimal character) of the code are the zone bits, and the last four bits (second hexadecimal character) are the digit bits. The letter A, for example, is represented by the binary combination 11000001, or C1 in hexadecimal; the letter S is represented by 11100010, or E2; and the numeral 1 is represented by 11110001, or F1.

Example 2.15.   Write the expression GO TO 36. using the EBCDIC.

G	O	b	T	O	b	3	6	.
C7	D6	40	E3	D6	40	F3	F6	4B

In an attempt to standardize on an alphanumeric code, an association of manufacturers and users has published the American Standard Code for Information Interchange (ASCII). This code appears in graphical form in Figure 2-18. Notice that this is a seven-bit code. The three most significant bits are the zone bits and the remaining four are the digit bits. For example, zone 011 contains all of the numeric characters as well as six special characters. The numeral four is represented as 011 0100; numeral nine as 011 1001; a question mark (?) as 011 1111;

EBCDIC	Bit Configuration	Hex	EBCDIC	Bit Configuration	Hex	EBCDIC	Bit Configuration	Hex	EBCDIC	Bit Configuration	Hex
NULL	0000 0000		b (blank)	0100 0000	40		1000 0000			1100 0000	C0
	0000 0001			0100 0001		a	1000 0001	81	A	1100 0001	C1
	0000 0010			0100 0010		b	1000 0010	82	B	1100 0010	C2
	0000 0011			0100 0011		c	1000 0011	83	C	1100 0011	C3
PF	0000 0100	04		0100 0100		d	1000 0100	84	D	1100 0100	C4
HT	0000 0101	05		0100 0101		e	1000 0101	85	E	1100 0101	C5
LC	0000 0110	06		0100 0110		f	1000 0110	86	F	1100 0110	C6
DEL	0000 0111	07		0100 0111		g	1000 0111	87	G	1100 0111	C7
	0000 1000		C	0100 1000		h	1000 1000	88	H	1100 1000	C8
	0000 1001			0100 1001	49	i	1000 1001	89	I	1100 1001	C9
	0000 1010		¢	0100 1010	4A		1000 1010			1100 1010	
	0000 1011		.	0100 1011	4B		1000 1011			1100 1011	
	0000 1100		<	0100 1100	4C		1000 1100			1100 1100	
	0000 1101		(	0100 1101	4D		1000 1101			1100 1101	
	0000 1110		+	0100 1110	4E		1000 1110			1100 1110	
	0000 1111		!	0100 1111	4F		1000 1111			1100 1111	
	0001 0000		&	0101 0000			1001 0000			1101 0000	D0
	0001 0001			0101 0001		j	1001 0001	91	J	1101 0001	D1
	0001 0010			0101 0010		k	1001 0010	92	K	1101 0010	D2
	0001 0011			0101 0011		l	1001 0011	93	L	1101 0011	D3
RES	0001 0100	14		0101 0100		m	1001 0100	94	M	1101 0100	D4
NL	0001 0101	15		0101 0101		n	1001 0101	95	N	1101 0101	D5
BS	0001 0110	16		0101 0110		o	1001 0110	96	O	1101 0110	D6
IDL	0001 0111	17		0101 0111		p	1001 0111	97	P	1101 0111	D7
	0001 1000			0101 1000		q	1001 1000	98	Q	1101 1000	D8
	0001 1001			0101 1001		r	1001 1001	99	R	1101 1001	D9
	0001 1010		!	0101 1010	5A		1001 1010			1101 1010	
	0001 1011		$	0101 1011	5B		1001 1011			1101 1011	
	0001 1100		*	0101 1100	5C		1001 1100			1101 1100	
	0001 1101		)	0101 1101	5D		1001 1101			1101 1101	
	0001 1110		:	0101 1110	5E		1001 1110			1101 1110	
	0001 1111		¬	0101 1111	5F		1001 1111			1101 1111	
	0010 0000		−	0110 0000	60		1010 0000			1110 0000	E0
	0010 0001		/	0110 0001	61		1010 0001			1110 0001	
	0010 0010			0110 0010		s	1010 0010	A2	S	1110 0010	E2
	0010 0011			0110 0011		t	1010 0011	A3	T	1110 0011	E3
BYP	0010 0100	24		0110 0100		u	1010 0100	A4	U	1110 0100	E4
LF	0010 0101	25		0110 0101		v	1010 0101	A5	V	1110 0101	E5
EOB	0010 0110	26		0110 0110		w	1010 0110	A6	W	1110 0110	E6
PRE	0010 0111	27		0110 0111		x	1010 0111	A7	X	1110 0111	E7
	0010 1000			0110 1000		y	1010 1000	A8	Y	1110 1000	E8
	0010 1001			0110 1001		z	1010 1001	A9	Z	1110 1001	E9
	0010 1010			0110 1010			1010 1010			1110 1010	
	0010 1011		,	0110 1011	6B		1010 1011			1110 1011	
	0010 1100		%	0110 1100	6C		1010 1100			1110 1100	
	0010 1101		−	0110 1101	6D		1010 1101			1110 1101	
	0010 1110		>	0110 1110	6E		1010 1110			1110 1110	
	0010 1111		?	0110 1111	6F		1010 1111			1110 1111	
	0011 0000			0111 0000			1011 0000		0	1111 0000	F0
	0011 0001			0111 0001			1011 0001		1	1111 0001	F1
	0011 0010			0111 0010			1011 0010		2	1111 0010	F2
	0011 0011			0111 0011			1011 0011		3	1111 0011	F3
PN	0011 0100	34		0111 0100			1011 0100		4	1111 0100	F4
RS	0011 0101	35		0111 0101			1011 0101		5	1111 0101	F5
UC	0011 0110	36		0111 0110			1011 0110		6	1111 0110	F6
EOT	0011 0111	37		0111 0111			1011 0111		7	1111 0111	F7
	0011 1000			0111 1000			1011 1000		8	1111 1000	F8
	0011 1001			0111 1001	79		1011 1001		9	1111 1001	F9
	0011 1010			0111 1010	7A		1011 1010			1111 1010	
	0011 1011		#	0111 1011	7B		1011 1011			1111 1011	
	0011 1100		@	0111 1100	7C		1011 1100			1111 1100	
	0011 1101		'	0111 1101	7D		1011 1101			1111 1101	
	0011 1110		=	0111 1110	7E		1011 1110			1111 1110	
	0011 1111		"	0111 1111	7F		1011 1111			1111 1111	

Figure 2-17.   Extended Binary Coded Decimal Interchange Code (EBCDIC).

	000	001	010	011	100	101	110	111
0000	NULL	① DC$_0$	ƀ	0	@	P		
0001	SOM	DC$_1$	!	1	A	Q		
0010	EOA	DC$_2$	"	2	B	R		
0011	EOM	DC$_3$	#	3	C	S		
0100	EOT	DC$_4$ (stop)	$	4	D	T		
0101	WRU	ERR	%	5	E	U		
0110	RU	SYNC	&	6	F	V		
0111	BELL	LEM	'	7	G	W		
1000	FE$_0$	S$_0$	(	8	H	X	Unassigned	
1001	HT SK	S$_1$	)	9	I	Y		
1010	LF	S$_2$	*	:	J	Z		
1011	V$_{TAB}$	S$_3$	+	;	K	[		
1100	FF	S$_4$	(comma) ,	<	L	\		ACK
1101	CR	S$_5$	−	=	M	]		②
1110	SO	S$_6$	.	>	N	↑		ESC
1111	SI	S$_7$	/	?	O	←		DEL

Example: | 100 | 0001 | = A
b$_7$ ——————— b$_1$

The abbreviations used in the figure mean:

NULL	Null Idle	CR	Carriage return
SOM	Start of message	SO	Shift out
EOA	End of address	SI	Shift in
EOM	End of message	DC$_0$	Device control ①
			Reserved for data
			Link escape
EOT	End of transmission	DC$_1$ - DC$_3$	Device control
WRU	"Who are you?"	ERR	Error
RU	"Are you . . . . ?"	SYNC	Synchronous idle
BELL	Audible signal	LEM	Logical end of media
FE	Format effector	SO$_0$ - SO$_7$	Separator (information)
HT	Horizontal tabulation		Word separator (blank, normally non-printing)
SK	Skip (punched card)	ACK	Acknowledge
LF	Line feed	②	Unassigned control
V/TAB	Vertical tabulation	ESC	Escape
FF	Form feed	DEL	Delete Idle

Figure 2-18.   American Standard Code for Information Interchange (ASCII).

etc. The letter A is represented as 100 0001; J is 100 1010, etc. Control characters are also included in this code.

Example 2.16. Write the expression GO TO 36. using the ASCII code.

G	O	b	T	O	b	3	6	
47	4F	20	54	4F	20	33	36	2E

2.11 PARITY CHECKING

In most interchange codes, an additional bit called a *parity* or *check bit* is used with the code to determine if the data has been somehow altered during the process of transmission. This bit is sometimes referred to as a "redundant bit" because it is not part of the code itself but is used to check the code. The parity bit is used to make the total number of one bits either odd or even, depending on the type of parity used. Figure 2-19 shows how odd parity is used with the BCD code. If you count the number of one bits in each of the 5-bit combinations, you will notice that there is always an odd number. The object is that if the data has been accidentally altered, one of the bits will either be lost or gained. Therefore, when the data is received, it can be checked to determine if there is an odd number of one bits in every 5-bit combination. If not, data received

8	4	2	1	Parity Bit
0	0	0	0	1
0	0	0	1	0
0	0	1	0	0
0	0	1	1	1
0	1	0	0	0
0	1	0	1	1
0	1	1	0	1
0	1	1	1	0
1	0	0	0	0
1	0	0	1	1

Figure 2-19. Odd parity.

has been altered and is in error (such data is called "garbage"). Figure 2-20 shows a BCD code using *even* parity.

The BCD interchange code shown in Figure 2-16 used *odd* parity. The letter "C" indicates the parity or check bit. Notice that for all of the characters there is always an odd number of one bits. The EBCDIC shown in Figure 2-17 and the ASCII code in Figure 2-18 may be checked in the same manner.

Parity can also be checked after a large number of characters have been transmitted. Figure 2-21 shows how eight bits (10010101) are used to check parity in columnar form. The number of one bits in each column, including the parity bit, must equal an odd number (odd parity). This method of parity checking is often used when data is recorded on magnetic tape.

The process of parity checking goes on continuously whenever data is transferred between the central processing unit and peripherals or between the various sections of the central processing unit itself. This is done so that the user can be satisfied that his data is accurate and contains no garbage. The operator is made aware of garbage when the CPU stops and a red light goes on indicating parity error. The computer refuses to accept data that will not pass a parity check.

8	4	2	1	Parity Bit
0	0	0	0	0
0	0	0	1	1
0	0	1	0	1
0	0	1	1	0
0	1	0	0	1
0	1	0	1	0
0	1	1	1	1
1	0	0	0	1
1	0	0	1	0

Figure 2-20.   Even parity.

G	1100	0111
O	1101	0110
b	0100	0000
T	1110	0011
O	1101	0110
b	0100	0000
3	1111	0011
6	1111	0110
·	0100	1011
Parity bit	1001	0101

Figure 2-21.   Eight bits used to check parity.

2.12 COMPUTER WORDS

As we have just discussed, binary bits may be organized to form characters. Larger groups of bits, such as those used to represent entire numbers, are called *words*. Some computers use a fixed number of memory cells to store a word at a given location or *address* in its memory. Such computers are referred to as "word-addressable" and are called *fixed-word-length machines*. In other computers, single characters are "addressed" and are called *variable-word-length* machines because a word may have any number of characters.

Figure 2-22 is an example of how data would be stored in a fixed-word-length machine. Suppose Joe Blow receives $4.50 an hour, has 18 percent withheld for taxes, pays $10 a week for health insurance, and works 38 hours in a given week. Joe's name is stored at address 100, his hourly rate at address 101, his tax withholding rate at address 102, his weekly health insurance at 103, and the number of hours worked at address 104. Figure 2-23 shows how the same data is stored in a variable-word-length machine. It should be apparent that the variable-word-length

100	J	O	E	B	L	O	W	
101					4	5	0	Hourly Rate
102						1	8	Income Taxes
103				1	0	0	0	Health Insurance
104						3	8	≠ of Hours

Figure 2-22.   Data stored in fixed-word format.

format makes more efficient use of the computer's memory. However, speed of arithmetic calculation is the primary benefit of the fixed-word-length format.

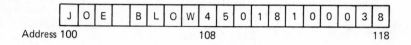

Address 100   108   118

Figure 2-23.   Data stored in variable-word-length format.

Another term commonly used is *byte*. Byte refers to a group of bits, usually eight in number. Data stored on punched tape or magnetic tape usually appears in groups or rows of eight bits. In ASCII, a byte can represent one character; the seven bits for the code and a parity bit. A byte can also represent one EBCDIC character.

QUESTIONS AND PROBLEMS—CHAPTER 2

In problems 1 through 14, convert the numbers to decimal.

1. 203_5
2. 1341_5
3. 44302_5
4. 14432_5
5. 272_8
6. 57_8
7. 717_8
8. 206_8
9. 1070_8
10. 1277_8
11. $DA37_{16}$
12. 35_{16}
13. $9A8B0_{16}$
14. $2B6_{16}$

In problems 15 through 26, change the decimal numbers to quinary, octal and hexadecimal numbers.

15. 33

16. 9

17. 276

18. 11

19. 999

20. 93

21. 4701

22. 765

23. 10,010

24. 1860

25. 6075

26. 22,462

27. (a) $111_2 = (?)_{10}$ (b) $1001_2 = (?)_{10}$ (c) $110011_2 = (?)_{10}$
    (d) $76_{10} = (?)_2$ (e) $273_{10} = (?)_2$ (f) $7620_{10} = (?)_2$

28. (a) $101_2 = (?)_{10}$ (b) $1101_2 = (?)_{10}$ (c) $10110_2 = (?)_{10}$
    (d) $101101_2 = (?)_{10}$ (e) $14_{10} = (?)_2$ (f) $83_{10} = (?)_2$
    (g) $351_{10} = (?)_2$ (h) $5648_{10} = (?)_2$

29. (a) $37_8 = (?)_2$ (b) $634_8 = (?)_2$ (c) $6732_8 = (?)_2$
    (d) $A6_{16} = (?)_2$ (e) $7BC_{16} = (?)_2$ (f) $A34D_{16} = (?)_2$
    (g) $110010111010_2 = (?)_8$ (h) $11101001110101_2 = (?)_2$
    (i) $110111000010_2 = (?)_{16}$ (j) $100111011001001_2 = (?)_{16}$

30. (a) $47_8 = (?)_2$ (b) $743_8 = (?)_2$ (c) $4073_8 = (?)_8$
    (d) $B9_{16} = (?)_2$ (e) $9A3_{16} = (?)_2$ (f) $AC4F_{16} = (?)_2$
    (g) $110010111010_2 = (?)_8$ (h) $1101001110101_2 = (?)_8$
    (i) $111001101100_2 = (?)_{16}$ (j) $10101100011011_2 = (?)_{16}$

31. Perform the following conversions.
    (a) 101.0101110111001_2 to octal and hexadecimal
    (b) 11001011.0100110011010_2 to octal and hexadecimal
    (c) $11111110.000000111001100_2$ to octal and hexadecimal
    (d) 7637.473_{10} to BCD
    (e) 17639.99073_{10} to BCD
    (f) $DAFE12.7CA63_{16}$ to binary

32. Perform the following conversions.
    (a) 110.1100101101101_2 to octal and hexadecimal
    (b) 10110111.0011101100101_2 to octal and hexadecimal
    (c) $101110110.10000000110110_2$ to octal and hexadecimal
    (d) 763.207_{10} to BCD

(e) 634.909_{10} to BCD

(f) $C3F.A6D_{16}$ to binary

33. Show how the number 1100101110.0101 would be stored in two 16-bit registers.

34. What is *encoding?*

35. How many combinations of bits are possible in BCD? How many are invalid? Why are these combinations invalid in BCD?

36. How would the number 740 be written in: (a) excess-three code, (b) $2'\ 421$ code, (c) Gray code?

37. How would the following alphanumeric characters be represented in EBCDIC?

(a) G	(c) $	(e) 6%
(b) Period	(d) 0	(f) #8

38. How would the following alphanumeric characters be represented in BCDIC?

(a) N	(c) 7
(b) Comma	(d) 59

39. Write the expression "FIND 13" in BCDIC.

40. Write the expression "ERASE 41." in: (a) EBCDIC and (b) ASCII.

41. What are the control characters in ASCII?

3

Binary Arithmetic

Data is stored in computers in the form of binary digits (BITS). In the computer, the arithmetic logic unit (ALU) reduces all arithmetic operations to a series of additions and/or subtractions. In this chapter some of the various methods by which the computer adds, subtracts, or converts data for the various operations are discussed.

The topics covered in this chapter are:

3.1 Addition Using Various Number Systems

3.2 Subtraction Using Various Number Systems

3.3 Addition in Binary

3.4 Subtraction in Binary

3.5 Signed Numbers

3.6 Adding and Subtracting Signed Numbers

3.7 Overflow

3.8 Sixteen's and Eight's Complement Methods

3.9 BCD Addition and Subtraction

3.1 ADDITION USING VARIOUS NUMBER SYSTEMS

3.1.1 Adding decimal numbers. Add the numbers 23 and 45. When numbers are added, they are added in columns with one number placed below the other. Either number may be placed first. In Example 3.1, the numbers 23 and 45 are placed in correct position for addition. The digits are separated to emphasize positional differences. Eight is in the units position $(3 + 5)$, and 6 is in the 10s position $(2 + 4)$. The answer is 68. (Six 10s and eight 1s.)

Example 3.1.   23 + 45 = ?

$$
\begin{array}{cc}
2 & 3 \\
4 & 5 \\
\hline
6 & 8
\end{array}
$$

Example 3.2.   Add the numbers 64 and 87 in base 10.

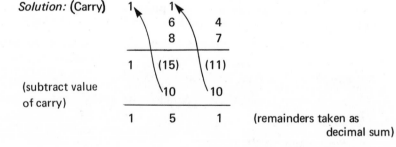

In Example 3.2, there is an 11 in the units position. This number is greater than 9, which tells us there must be a carry (10). The digit remaining in the units position will be the difference of 10 and 11, or 1. The 10s position now contains $6 + 8 +$ a carry of 1, or 15. Again, the number is greater than 9, so a carry is indicated. Subtracting 10 from 15, we get 5 with a carry of 1. The third position contains only the carry. Of course, we have been adding for years in decimal so our experience allows us to perform these operations automatically. We have taken time to discuss the arithmetic steps because this is the addition process in all bases.

Example 3.3. Add the decimal numbers below following the procedure in Example 3.2 :

(a) 738 + 417　　(b) 9706 + 463　　(c) 1774 + 7268

Solution:

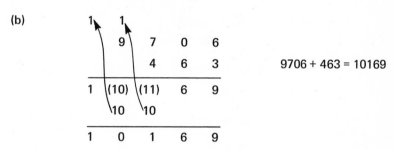

(a)

		7	3	8
		4	1	7
	1	(11)	5	(15)
		10		10
	1	1	5	5

738 + 417 = 1155

(b)

	9	7	0	6
		4	6	3
1	(10)	(11)	6	9
	10	10		
1	0	1	6	9

9706 + 463 = 10169

(c)

1	7	7	4
7	2	6	8
9	(10)	(14)	(12)
	10	10	10
9	0	4	2

1774 + 7268 = 9042

3.1.2 Adding quinary numbers. Let's add the quinary numbers 243 and 122 in Example 3.4.

Example 3.4. (Carry)

		2	4	3
		1	2	2
(decimal sum)		4	(7)	(5)
(subtract value of carry)			5	5
		4	2	0

$243_5 + 122_5 = 420_5$

As in base 10, the LSD positions are added first. The sum of 3 and 2 is 5. The number is greater than 4 (4 is the largest digit in base 5), which tells us there must be a carry. Subtracting 5 (the value of the carry) leaves 0 with a carry of 1. The 5^1 position now contains $4 + 2 +$ a carry of 1, or decimal 7. Again, the sum is greater than 4, indicating a carry. Subtracting 5 from 7 leaves a remainder of 2 with a carry of 1 into the next position (5^2). This position now contains $2 + 1 +$ a carry of 1, or 4. No carry exists (since the number is 4 or less), so 5 is not subtracted. The answer is 420_5. There are four 25s, two 5s, and no 1s.

Example 3.5. Add the following quinary numbers:

(a) 42   (b) 233   (c) 104   (d) 444
    14   142   44   444

Solution:

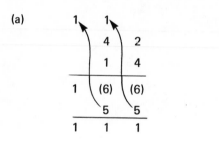

(a)

$42_5 + 14_5 = 111_5$

(b)

$233_5 + 142_5 = 430_5$

(c)

$104_5 + 44_5 = 203_5$

(d)

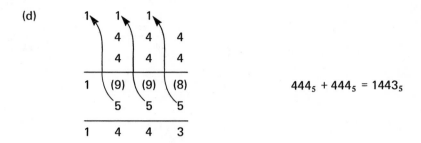

$$444_5 + 444_5 = 1443_5$$

3.1.3 Adding octal numbers.   The octal numbers 736 and 215 are added in Example 3.6.

Example 3.6. (Carry)

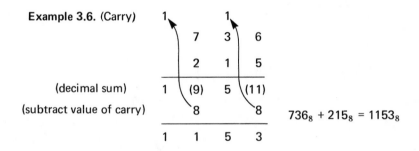

$$736_8 + 215_8 = 1153_8$$

Again, the LSD or units position is added first. The sum of 6 and 5 is 11. The number is greater than 7 (7 is the largest digit in base 8), which tells us there must be a carry. Subtracting 8 (the value of the carry) leaves 3 with a carry of 1. The 8^1 position now contains $3 + 1 +$ a carry of 1, or 5. Since 5 is less than 7, no carry exists. In the 8^2 position, $7 + 2$ is 9. The number is greater than 7, so a carry is generated. Subtracting the carry (8) yields a remainder of 1 with a carry of 1. The 8^3 position contains only the carry, so the answer is 1153_8.

Example 3.7.   Add the following octal numbers:

(a) 243   (b) 764   (c) 604
     172   414   777

Solution: (a)

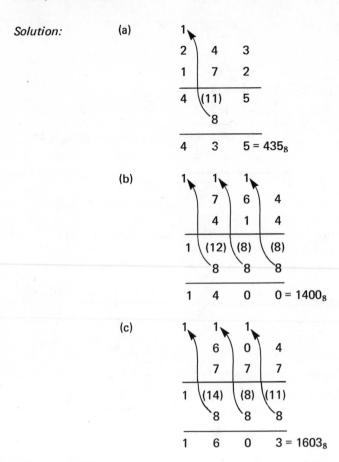

$$4 \quad 3 \quad 5 = 435_8$$

(b)

$$1 \quad 4 \quad 0 \quad 0 = 1400_8$$

(c)

$$1 \quad 6 \quad 0 \quad 3 = 1603_8$$

Because octal is often used in computers, tables of octal addition, such as the one in Figure 3-1, are available to simplify the addition process.

3.1.4 Adding hexadecimal numbers.   Add the hexadecimal numbers 7AF and 579 in Example 3.8.

Example 3.8. Add 7AF and 579.

(carry)

7   A   F
5   7   9

(decimal sum)   13   (18)   (24)

(subtract value of carry)   16   16

$$D \quad 2 \quad 8 = D28_{16}$$

Adding the units position yields 24. Twenty-four is greater than the largest digit (F), which is equal to decimal 15. This indicates that a carry should exist. Subtracting the value of a carry (16) leaves a remainder of 8 with a carry of 1. In the 16^1 position, the sum of A (decimal 10) + 7 + a carry of 1 is 18. Again, a carry is generated, leaving a remainder of 2 with a carry of 1. In the 16^2 position, the carry + 7 + 5 equals decimal 13. In hexadecimal, 13 is D, so the answer is $D28_{16}$.

Example 3.9. Add the following hexadecimal numbers:

(a) 4D3   (b) 789   (c) F347
    818   C47   E006

Solution:

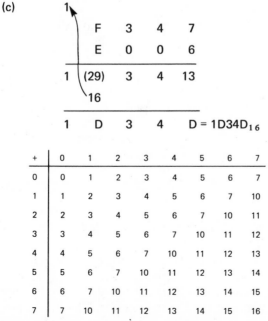

(a)

	4	D	3
	8	1	8
	12	14	11

C   E   B = CEB_{16}

(b)

	1↖		1↖	
		7	8	9
		C	4	7
	1	(19)	13	(16)
		16		16

1   3   D   0 = $13D0_{16}$

(c)

	1↖				
		F	3	4	7
		E	0	0	6
	1	(29)	3	4	13
		16			

1   D   3   4   D = $1D34D_{16}$

+	0	1	2	3	4	5	6	7
0	0	1	2	3	4	5	6	7
1	1	2	3	4	5	6	7	10
2	2	3	4	5	6	7	10	11
3	3	4	5	6	7	10	11	12
4	4	5	6	7	10	11	12	13
5	5	6	7	10	11	12	13	14
6	6	7	10	11	12	13	14	15
7	7	10	11	12	13	14	15	16

Figure 3-1. Table of octal addition.

	0	1	2	3	4	5	6	7	8	9	A	B	C	D	E	F
0	0	1	2	3	4	5	6	7	8	9	A	B	C	D	E	F
1	1	2	3	4	5	6	7	8	9	A	B	C	D	E	F	10
2	2	3	4	5	6	7	8	9	A	B	C	D	E	F	10	11
3	3	4	5	6	7	8	9	A	B	C	D	E	F	10	11	12
4	4	5	6	7	8	9	A	B	C	D	E	F	10	11	12	13
5	5	6	7	8	9	A	B	C	D	E	F	10	11	12	13	14
6	6	7	8	9	A	B	C	D	E	F	10	11	12	13	14	15
7	7	8	9	A	B	C	D	E	F	10	11	12	13	14	15	16
8	8	9	A	B	C	D	E	F	10	11	12	13	14	15	16	17
9	9	A	B	C	D	E	F	10	11	12	13	14	15	16	17	18
A	A	B	C	D	E	F	10	11	12	13	14	15	16	17	18	19
B	B	C	D	E	F	10	11	12	13	14	15	16	17	18	19	1A
C	C	D	E	F	10	11	12	13	14	15	16	17	18	19	1A	1B
D	D	E	F	10	11	12	13	14	15	16	17	18	19	1A	1B	1C
E	E	F	10	11	12	13	14	15	16	17	18	19	1A	1B	1C	1D
F	F	10	11	12	13	14	15	16	17	18	19	1A	1B	1C	1D	1E

Figure 3-2. Table of hexadecimal addition.

The table in Figure 3-2 may be used to simplify hexadecimal addition.

3.2 SUBTRACTION USING VARIOUS NUMBER SYSTEMS

3.2.1 Subtracting decimal numbers. Subtraction in base 10 is so automatic for us that we may do it without thinking of the rules involved. Since these rules in base 10 are identical to those for other bases, and since we are more familiar with base 10 than with other bases, let's review the process of subtraction in base 10.

Example 3.10. Subtract 25 from 43.

Solution:

```
                   3              ⟋ — Borrow 1.
(minuend)          4  → (1)3
(subtrahend)       2       5
                  _____
(difference)       1       8 = 18
```

As you know, we start with the least significant position and subtract. We cannot subtract 5 from 3. Therefore we "borrow" from the next significant position of the minuend. We may now interpret the 3 as 13 and find the difference, which is 8. Moving to the next position, we must reduce the minuend by 1 because of the "borrow." We now subtract the number in the subtrahend from the new minuend, and the difference is 1. Remember, the "borrow" carries with it a weight equal to that of the base.

3.2.2 Subtracting octal numbers. In Example 3.11, the number 267_8 is subtracted from 512_8. Notice that a borrow is necessary in the first position since 7 is greater than 2. The borrow creates a new number in the minuend, 12. (This number is read "one-two"—*not* "twelve" since 12 is a decimal number. One-two in base 8 means $8 + 2$ in base 10.)

Example 3.11. Subtract 267_8 from 512_8.

```
                   4    (1)0
(minuend)          5 ⟋    1 → (1)2
(subtrahend)       2    6    7
                  _____
(difference)       2    2    3 = 223₈
```

One-two minus 7 equals 3. If you think in base 10, then $(8 + 2) - 7$ equals 3. In the next position, 1 was borrowed from the number in the minuend, making its value 0. A borrow is again required; it makes the value of the minuend 10 (or $8 + 0$ in base 10). Subtracting 6 from 10 yields a difference of 2. In the next position, 1 was borrowed from the minuend, making its value 4. The difference in this position is 2.

Figure 3-1, which was previously used for octal addition, may also be used for octal subtraction as illustrated in Figure 3-3. The numbers across the top now represent numbers in the subtrahend. The numbers in the left-hand column are interpreted as the difference, and the numbers in the remaining columns, which are the sums in the addition table, are used as the minuends.

	0	1	2	3	4	5	6	7
0	0	1	2	3	4	5	6	7
1	1	2	3	4	5	6	7	10
2	2	3	4	5	6	7	10	11
3	3	4	5	6	7	10	11	12
4	4	5	6	7	10	11	12	13
5	5	6	7	10	11	12	13	14
6	6	7	10	11	12	13	14	15
7	7	10	11	12	13	14	15	16

Figure 3-3. Table of octal subtraction.

Example 3.12. Solve the following subtraction problems in base 8:

(a)   43   (b)   161   (c) 2172
     −16   − 72   −1717

Solution:

(a)   3
           4 ⟶ (1)3
          −1   6
          _____
           2   5 = 25_8

(b)   0   (1)5
           1   6 ⟶ (1)1
           −   7   2
          _____
                    6   7 = 67_8

(c)   1   6
        2 ⟶ (1)1   7 ⟶ (1)2
       −1   7   1   7
       _____
               2   5   3 = 253_8

3.2.3 *Subtracting hexadecimal numbers.*   Applying the same rules as in base 8, let's subtract the hexadecimal number 85E from the number C37.

Example 3.13.   C37 − 85E = ?

Solution:

	B	(1)2	
(minuend)	Ȼ ↗	ꝫ ⟶	(1)7
(subtrahend)	8	5	E
(difference)	3	D	9 = $3D9_{16}$

Notice that a borrow is necessary in the first position since E is greater than 7. One-seven (*not* seventeen) minus E equals 9; or, in base 10, (16 + 7) − 14 equals 9. (Remember, the borrow has a weight of 16.) In the next position, a borrow is required so that 5 is subtracted from 12 (one-two) which equals D. In base 10, it would be (16 +2) − 5 = 13. Finally, in the next position, subtracting 8 from B (remember that we borrowed 1, making C a B) yields 3; or, in base 10, 11 − 8 = 3. Figure 3-2, which was previously used for hexadecimal addition, may also be used for hexadecimal subtraction as shown in Figure 3-4.

Example 3.14.   Subtract the following hexadecimal numbers:

(a) A29	(b) 3A12	(c) F1B2
−7BF	−29A0	−ABC7

Solution:

(a)

	9	(1)1	
	Ⱥ ↗	2 ⟶	(1)9
	7	B	F
	2	6	A = $26A_{16}$

(b)

	9		
3	Ⱥ ⟶	(1)1	2
2	9	A	0
1	0	7	2 = 1072_{16}

(c)

	E	(1)0	(1)A	
	Ⅎ ↗	1 ↗	ᵬ ⟶	(1)2
	A	B	C	7
	4	5	E	B = $45EB_{16}$

	0	1	2	3	4	5	6	7	8	9	A	B	C	D	E	F
0	0	1	2	3	4	5	6	7	8	9	A	B	C	D	E	F
1	1	2	3	4	5	6	7	8	9	A	B	C	D	E	F	10
2	2	3	4	5	6	7	8	9	A	B	C	D	E	F	10	11
3	3	4	5	6	7	8	9	A	B	C	D	E	F	10	11	12
4	4	5	6	7	8	9	A	B	C	D	E	F	10	11	12	13
5	5	6	7	8	9	A	B	C	D	E	F	10	11	12	13	14
6	6	7	8	9	A	B	C	D	E	F	10	11	12	13	14	15
7	7	8	9	A	B	C	D	E	F	10	11	12	13	14	15	16
8	8	9	A	B	C	D	E	F	10	11	12	13	14	15	16	17
9	9	A	B	C	D	E	F	10	11	12	13	14	15	16	17	18
A	A	B	C	D	E	F	10	11	12	13	14	15	16	17	18	19
B	B	C	D	E	F	10	11	12	13	14	15	16	17	18	19	1A
C	C	D	E	F	10	11	12	13	14	15	16	17	18	19	1A	1B
D	D	E	F	10	11	12	13	14	15	16	17	18	19	1A	1B	1C
E	E	F	10	11	12	13	14	15	16	17	18	19	1A	1B	1C	1D
F	F	10	11	12	13	14	15	16	17	18	19	1A	1B	1C	1D	1E

Figure 3-4.   Table of hexadecimal subtraction.

3.3   *ADDITION IN BINARY*

Add the binary numbers in Example 3.15.

Example 3.15.   (a) $1101_2 + 1001_2 = (?)$

	(carry)	1			1	
		1	1	0	1
		1	0	0	1
(decimal sum)		(2)	1	1	(2)
(subtract value of carry)		2			2

$$1 \quad 0 \quad 1 \quad 1 \quad 0 = 10110_2$$

(b) $1101_2 + 1011_2 = (?)$

$$
\begin{array}{ccccc}
1 & 1 & 1 & 1 & \\
  & 1 & 1 & 0 & 1 \\
  & 1 & 0 & 1 & 1 \\
\hline
1 & (3) & (2) & (2) & (2) \\
  & 2 & 2 & 2 & 2 \\
\hline
1 & 1 & 0 & 0 & 0 = 11000_2
\end{array}
$$

The method of addition is the same as that used with other bases. Notice that $1 + 1$ equals decimal 2, but 2 is invalid in base 2, so a carry is generated into the next position. We see from the example that a carry is generated whenever a sum greater than 1 results from the addition of numbers in a column. From working these and other examples of binary addition, we should note that: (1) $0 + 0$ always equals 0; (2) $0 + 1$ always equals 1; (3) $1 + 1$ always equals 0 and a carry; and (4) $1 + 1 +$ a carry always equals a 1 and a carry.

Example 3.16. Add the following binary numbers:

(a) 10   (b) 111   (c) 1001   (d) 11100
    11   101   1010   01010

Solution:

(a)

$$
\begin{array}{ccc}
1 & & \\
 & 1 & 0 \\
 & 1 & 1 \\
\hline
1 & (2) & 1 \\
 & 2 & \\
\hline
1 & 0 & 1 = 101_2
\end{array}
$$

(b)

$$
\begin{array}{cccc}
1 & 1 & 1 & \\
 & 1 & 1 & 1 \\
 & 1 & 0 & 1 \\
\hline
1 & (3) & (2) & (2) \\
 & 2 & 2 & 2 \\
\hline
1 & 1 & 0 & 0 = 1100_2
\end{array}
$$

(c)

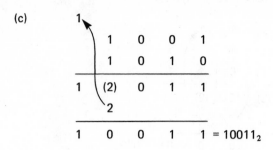

$$1 \quad 0 \quad 0 \quad 1 \quad 1 = 10011_2$$

(d)

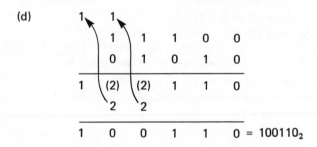

$$1 \quad 0 \quad 0 \quad 1 \quad 1 \quad 0 = 100110_2$$

3.4   SUBTRACTION IN BINARY

The process of subtraction in base 2 is similar to that in the other bases, as was discussed earlier in this chapter.

Example 3.17. Subtract 0101 from 1110.

Solution:	Borrow ——————		0	
(minuend)	1	1	✗→ (1)0	
(subtrahend)	0	1	0	1
(difference)	1	0	0	1 = 1001₂

$$= 1001_2$$

Example 3.18. Subtract 01101 from 10011.

Solution: Borrow ╲ 0   1

(minuend)   ✗→(1)0̸→(1)0   1   1

(subtrahend)   0   1   1   0   1

0   0   1   1   0 = 00110₂

$$= 00110_2$$

In Example 3.17, a borrow is necessary in the least significant position since the minuend is 0. The borrow has a weight of 2 because we are working in base 2. Note that in Example 3.18 a borrow is required in the 2^2 position. However, the 2^3 position contains a 0, so that the borrow must come from the 2^4 position to the 2^3 position. A borrow is now possible from the 2^3 position, and the minuend in the 2^2 position is 10 (one-zero), resulting in a difference of 1. The new minuend in the next position (2^3) is now 1, and the difference is 0. Our original borrow came from the next position (2^4), so the minuend is 0 and the difference is 0.

3.4.1 Ten's complement.   In the modern computer, subtraction is usually performed by addition of the complement of the subtrahend. This is done in order to simplify the design of the arithmetic section of the computer.

Example 3.19.   Subtract 676 from 835.

$$
\begin{array}{r}
835 \\
-676 \\
\hline
159
\end{array}
$$

This same operation can be performed by adding the ten's complement of 676 to 835. The ten's complement of a decimal number is the difference between that number and the next higher power of ten. Since 676 occupies the units, tens, and hundreds position, 10^3 would be the next higher power of ten. Therefore, the ten's complement of 676 equals 10^3 or 1000 minus 676.

$$
\begin{array}{r}
1000 \\
676 \\
\hline
324
\end{array}
$$
    (ten's complement of 676)

The next step is to add the ten's complement to the minuend and subtract 1000 from the result, as follows:

$$
\begin{array}{r}
1000 \\
-\ 676 \\
\hline
324
\end{array}
\qquad
\begin{array}{r}
835 \ \text{(minuend)} \\
+\ 324 \ \text{(ten's complement)} \\
\hline
1159 \\
-1000 \\
\hline
159 \ \text{(difference)}
\end{array}
$$

Notice that the last operation (subtracting 1000) is just a matter of dropping the 1 in the most significant position. Let's try another problem.

Example 3.20. Subtract 4678 from 6392.

Solution:

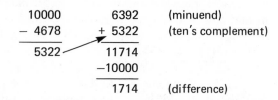

10000	6392	(minuend)
− 4678	+ 5322	(ten's complement)
5322	11714	
	−10000	
	1714	(difference)

3.4.2 Nine's complement. Finding the ten's complement is complicated by the fact that borrowing is required. The nine's complement of a digit is the difference between 9 and that digit and can be found by a simple inspection procedure. Let's find the nine's complement of 235.

```
999
235
764     (nine's complement of 235)
```

The nine's complement may be converted to the ten's complement by simply adding 1. Find the ten's complement of 5789.

```
9999
5789
4210     (nine's complement )
+   1
4211     (ten's complement)
```

3.4.3 One's complement. The one's complement of a binary number is similar to the nine's complement of a decimal number. The binary number to be complemented is subtracted from a binary number made up of all ones. In the following example, the one's complement of 10110010 is found.

```
       1  1  1  1  1  1  1  1
(−)    1  0  1  1  0  0  1  0
       0  1  0  0  1  1  0  1    (one's complement
                                  of 10110010)
```

Careful inspection reveals that the one's complement of a binary number may be produced by changing all of the ones in the number to zeros and changing all of the zeros to ones.

								(invert this number
1	0	1	1	0	0	1	0	to get one's
0	1	0	0	1	1	0	1	complement)

The computer is able to perform this operation very simply with a circuit called an *inverter*. Notice that carries or borrows are not involved in this operation.

In order to subtract using the one's complement method, the one's complement of the subtrahend is added to the minuend. The carry out of the highest bit position is then added to the *lowest bit position* (LSB). This is called *end-around carry*. Find the difference between 10110011 and 01101101.

```
(minuend)     1 0 1 1 0 0 1 1 ──► 1 0 1 1 0 0 1 1
(subtrahend)  0 1 1 0 1 1 0 1 ──► 1 0 0 1 0 0 1 0    (one's
                                                      complement)
                              1 0 1 0 0 0 1 0 1
                              └──────────────► 1      (end-around
                                                       carry)
                                0 1 0 0 0 1 1 0       (difference)
```

Find the difference between 11011000 and 10110011 using the one's complement method.

```
(minuend      1 1 0 1 1 0 0 0 ──► 1 1 0 1 1 0 0 0
(subtrahend)  1 0 1 1 0 0 1 1 ──► 0 1 0 0 1 1 0 0    (one's
                                                      complement)
                              1 0 0 1 0 0 1 0 0
                              └──────────────► 1      (end-around
                                                       carry)
                                0 0 1 0 0 1 0 1       (difference)
```

3.4.4 Two's complement. Perhaps the most popular method of subtraction used by computers is the two's complement method. The two's complement of a number may be found by adding 1 to the one's complement. Let's find the two's complement of 10011101. First, change all 0s to 1s and all 1s to 0s (one's complement). Then add 1 as follows:

```
1 0 0 1 1 1 0 1 ──► 0 1 1 0 0 0 1 0    (one's complement)
            (+)                    1
                    0 1 1 0 0 0 1 1    (two's complement)
```

In order to subtract, the two's complement of the subtrahend is obtained and then added to the minuend. Let's subtract 01001010 from 01100111.

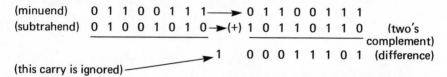

The carry resulting from the most significant bit position is ignored. There is no end-around carry as with the one's complement method. The computer obtains the one's complement simply by using an inverter. The two's complement is then obtained by starting the addition process with a carry as shown in Example 3.21.

Example 3.21.　(a) Subtract 00110110 from 01011011

Solution:

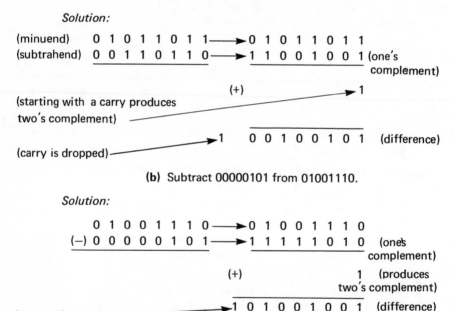

(b) Subtract 00000101 from 01001110.

Solution:

The number of bit positions in the subtrahend must agree with the number of bit positions in the minuend. This is usually accomplished by providing a fixed number of storage cells in the computer. In the preceding examples, eight bit positions were used; therefore, eight storage cells would be used to store the subtrahend. Eight cells are also used to store the difference, causing the carry out of the most significant bit position to be dropped because there is no place to store it. Each of the above eight-cell combinations is called a *register*. Example 3.22 is intended to convey the concept of registers with fixed numbers of bit positions.

Example 3.22. Subtract 00001100 from 01001010 (Figure 3-5).

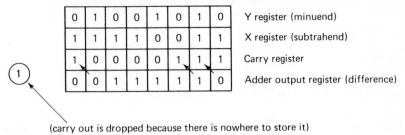

0	1	0	0	1	0	1	0	Y register (minuend)
1	1	1	1	0	0	1	1	X register (subtrahend)
1	0	0	0	0	1	1	1	Carry register
0	0	1	1	1	1	1	0	Adder output register (difference)

(carry out is dropped because there is nowhere to store it)

Figure 3-5. Concept of registers with fixed numbers of bit positions.

The minuend is placed in the Y register, and the one's complement of the subtrahend is placed in the X register. A 1 is placed in the LSB position of the carry register. All the other positions of the carry register store a zero at the start. The addition proceeds from the LSB position to the MSB position. If a carry is produced with the addition of a column, a 1 is stored in the *next* bit position of the carry register. Since the output register is limited to eight bit positions, the carry out of the MSB position is dropped.

Example 3.23. Subtract the following binary numbers using the two's complement method:

(a)
11011011
−10001101

(b)
11110000
−01010101

(c)
10000011
−01110110

Solution:

(a) (minuend   11011011 ⟶ 11011011
    (subtrahend) 10001101 (complement)⟶ 01110010
                (add)   1

                      1 01001110 = 01001110

(b)   11110000 ⟶ 11110000
                        01010101 ⟶ 10101010
                                      1

                      1 10011011 = 10011011

(c)
```
10000011 ——————→ 10000011
01110110 ——————→ 10001001
                          1
```

$\overline{1}$ 00001101 = 00001101

3.5 SIGNED NUMBERS

Within any numbering system, positive and negative values can exist. Numbers that are preceded by a plus ($+$) or a minus ($-$) sign are called *signed numbers*. When binary numbers are stored in a computer, the MSB position of a register is reserved for the *sign bit*, which is used to indicate the sign of the number stored in that register. If the MSB is a 0, the number stored has a positive value. If the MSB is a 1, the number has a negative value. Figure 3-6 demonstrates how the number 23 would be stored in an 8-bit position register. Negative 23 could be stored in binary as in Figure 3-7.

The negative number in the above example is said to be stored in *true form*. However, most computers store negative numbers in either the one's or two's complement form. Negative 23 in the one's complement form would be stored as shown in Figure 3-8. The inversion process used to produce the one's complement automatically produces the 1 indicating a minus number.

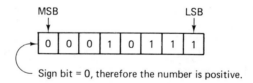

Figure 3-6.   The number 23 stored in an 8-bit position register.

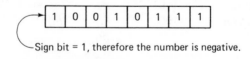

Figure 3-7.   The number -23 stored in an 8-bit position register.

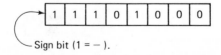

Figure 3-8.   Storing -23 in one's complement form.

Figure 3-9 demonstrates how negative 23 is stored in the two's complement form:

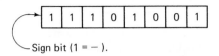

Sign bit (1 = −).

Figure 3-9. Storing −23 in two's complement form.

In the above examples, one bit position is used to store the sign bit, leaving only seven bit positions for the absolute value of the number. Eight-bit registers are used in this chapter to provide ease of reading. Most computers use registers longer than eight bits. However, the MSB position is used for the sign bit regardless of the length of the register.

3.6 ADDING AND SUBTRACTING SIGNED NUMBERS

The rules for adding or subtracting signed binary numbers are the same as those employed with signed decimal numbers:

1. To add numbers having like signs, give the sum the same sign; that is, two positive numbers produce a positive sum, and two negative numbers produce a negative sum.
2. To add numbers having different signs, find the difference and give it the sign of the larger number.
3. To subtract signed numbers, change the sign of the subtrahend and add the subtrahend to the minuend according to the preceding rules.

Most large computers store negative numbers in the two's complement form. The following examples demonstrate the advantage of the two's complement form. Let's find the sum of positive 6 and positive 8.

```
     +6     00000110    (positive numbers are stored in true form)
(+) +8     00001000
     ───    ────────
    +14     00001110    (sum is positive and in true form)
```

Now let's find the sum of negative 6 and negative 8.

```
     −6     11111010    (negative numbers are stored in the two's
(+) −8     11111000                               complement form)
     ───    ────────
    −14     11110010    (sum is negative and in two's complement form)
```

Remember, the MSB is the sign bit! This sum is negative and the computer stores it in the two's complement form. There is no need to recomplement this number unless it is to be printed out.

Now let's add negative 6 and positive 8.

```
         -6      11111010     (two's complement of 6)
   (+)   +8      00001000     (true form of 8)
         ──      ────────
         +2      00000010     (true form of positive 2)
```

OK! Let's add positive 6 and negative 8.

```
         +6      00000110     (true form of 6)
   (+)   -8      11111000     (two's complement of 8)
         ──      ────────
         -2      11111110     (two's complement of 2)
```

Again, the sign bit reveals a negative number (−2) in the two's complement form.

Now let's try subtracting positive 6 from positive 8.

```
         +8      00001000     00001000     (true form of 8)
   (−)   +6      00000110     11111010     (two's complement of 6)
         ──      ────────     ────────
         +2                   00000010     (true form of 2)
```

To subtract, the subtrahend is complemented and added.

Subtract positive 8 from positive 6.

```
         +6      00000110     00000110     (true form of 6)
   (−)   +8      00001000     11111000     (two's complement of 8)
         ──      ────────     ────────
         -2                   11111110     (two's complement of 2)
```

In this case, the difference is negative and appears in the two's complement form. Remember, the computer stores negative numbers in the two's complement form. Now let's subtract negative 8 from positive 6.

```
         +6      00000110     00000110     (true form of 6)
   (−)   -8      11111000     00001000     (true form of 8)
         ───     ────────     ────────
        +14                   00001110     (true form of 14)
```

To subtract, the subtrahend is complemented and added.

In the last example, because 8 was a negative number, it was stored in the two's complement form. Then, in order to subtract, it was recomplemented before adding.

The final example demonstrates subtracting a positive number from a negative number.

−8	11111000	11111000	(two's complement of 8)
(−) +6	00000110	11111010	(two's complement of 6)
−14		11110010	(two's complement of 14)

When performing addition and subtraction using the two's complement method, the computer needs *only* circuitry to perform the add operation and the two's complementation. Multiplication and division are performed by routines of repeated addition and subtraction.

3.7 OVERFLOW

The largest positive number that can be stored in an 8-bit register is 01111111 (+127). The largest negative number is 10000000 (−128). If the result of an arithmetic operation exceeds these numbers, the effect will be an *overflow*. All such overflows are erroneous and can be detected if:

1. there is a carry into the sign bit position with no carry out of it, or

2. there is no carry into the sign bit position with a carry out of it.

3.8 SIXTEEN'S AND EIGHT'S COMPLEMENT METHODS

When large signed binary numbers are added or subtracted, the student may find it convenient to use the sixteen's complement method. The sixteen's complement is obtained by taking the fifteen's complement and adding one. The fifteen's complement is obtained by substituting each digit with the difference between its value and 15. For example the sixteen's complement of 3B8 is:

C47 (fifteen's complement)
  1 (plus 1)
C48 (sixteen's complement)

Assuming that this number represents a binary number and the MSB is the sign bit, then an 8 or greater in the most significant digit position will represent a negative number. Therefore, 3B8 is a positive number in the true form and C48 is a negative number in the sixteen's complement form.

In many systems, octal is used to represent large binary numbers. Therefore, the student may find it convenient to use the eight's complement method for performing signed binary number additions and subtractions. The seven's complement is first obtained by substituting each octal digit with the difference between its value and 7. Then 1 is added to produce the eight's complement. For example, the eight's complement of 136_8 is:

$$\begin{array}{l} 641 \text{ (seven's complement)} \\ \underline{1 \text{ (plus 1)}} \\ 642 \end{array}$$

If the above octal number represents a binary number, and the MSB is the sign bit, then a 4 or greater in the most significant digit (MSD) position will represent a negative number. Therefore, 136 is a positive number in the true form, and 642 is a negative number in the eight's complement form.

The following subtraction is performed using the two's complement method.

0011	1100	1011 ⟶	0011	1100	1011	(minuend in true form)
(-)0010	0110	0111 ⟶	1101	1001	1001	(two's complement)
		⟶	⚡0001	0110	0100	(difference in true form)

Carry is dropped ⟶

The same problem may be solved by using the sixteen's complement method.

$$\begin{array}{ll} 3CB \longrightarrow 3CB & \text{(minuend is true form)} \\ (-) \ 267 \longrightarrow D99 & \text{(subtrahend in sixteen's complement form)} \\ \hspace{1.6cm} \cancel{1}\ 164 & \text{(difference in true form)} \end{array}$$

Carry is dropped

If the above binary numbers are expressed in octal, the eight's complement method may be used:

$$\begin{array}{ll} 1713_8 \longrightarrow 1713 & \text{(minuend in true form)} \\ (-) \ 1147_8 \longrightarrow 6631 & \text{(subtrahend in eight's complement form)} \\ \hspace{1.6cm} \cancel{1}\ 0544 & \text{(difference in true form)} \end{array}$$

Carry is dropped

The solution of the following subtraction produces a negative number which appears in the two's complement form.

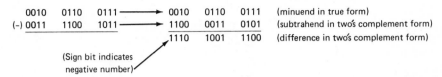

```
0010    0110    0111 ────────►  0010    0110    0111    (minuend in true form)
(-) 0011  1100    1011 ────────►  1100    0011    0101    (subtrahend in two's complement form)
                                 ─────────────────────
                                  1110    1001    1100    (difference in two's complement form)

        (Sign bit indicates
        negative number)
```

Substituting the sixteen's complement method, this problem would be solved as follows:

```
      267 ────────►  267    (minuend in true form)
(-)   3CB ────────►  C35    (subtrahend in sixteen's complement form)
                     ────
                     E9C    (difference in sixteen's complement form)
```

The MSD (E) is 8 or greater; therefore the difference is taken as a negative number in the sixteen's complement form.

Assuming that octal is used to represent the above binary numbers, the eight's complement method is used as follows:

```
      1147₈ ────────► 1147    (minuend in true form)
(-)   1713₈ ────────► 6065    (subtrahend in eight's complement form)
                      ────
                      7234    (difference in eight's complement form)
```

Because the MSD is 4 or greater, the difference is taken as a negative number in the eight's complement form.

Although the above procedures may seem difficult at first, they become timesavers whenever many calculations are to be performed.

3.9 BCD ADDITION AND SUBTRACTION

Although most computers use straight binary for their arithmetic operations, some earlier computers as well as present-day processors and calculators use BCD. BCD addition and/or subtraction is complicated by the fact that some sums or differences are invalid. When two BCD digits are added, it is possible to produce 16 different sums, 6 of which are invalid. Furthermore, when a carry-in is considered, an additional 4 more conditions are invalid.

When two BCD digits are added as binary numbers, sums greater than 9 can result as indicated in the table of BCD sums in Figure 3-10. Each invalid sum is corrected by adding BCD 6 (0110). BCD 6 (0110)

Decimal	Uncorrected BCD sum $C_4\Sigma_4\Sigma_3\Sigma_2\Sigma_1$	Corrected BCD sum $C_4\Sigma_4\Sigma_3\Sigma_2\Sigma_1$	
0	0000	0000	
1	0001	0001	
2	0010	0010	
3	0011	0011	
4	0100	0100	No correction required
5	0101	0101	
6	0110	0110	
7	0111	0111	
8	1000	1000	
9	1001	1001	
10	1010	10000	
11	1011	10001	
12	1100	10010	
13	1101	10011	
14	1110	10100	
15	1111	10101	Correction required
16	10000	10110	
17	10001	10111	
18	10010	11000	
19	10011	11001	

Figure 3-10.   Table of BCD sums.

is required because there are 6 invalid combinations of the 4-bits used in BCD. For example, when we add BCD 0101 to 0011:

$$
\begin{array}{rr}
0101 & 5 \\
(+)\ 0011 & (+)\ 3 \\
\hline
1000 & 8 \\
\end{array}
$$

a valid sum will result. However, when we add 0101 to 0111, we get:

$$
\begin{array}{rr}
0101 & 5 \\
(+)\ 0111 & (+)\ 7 \\
\hline
1000 & 12 \\
\end{array}
$$

The sum 1100 does not exist in BCD since this number is invalid for BCD. A second operation is required to correct this situation. Six (the number of invalid combinations) must be added to every invalid sum.

```
    5              0101
  + 7          (+) 0111
   1̄2             1100      (invalid sum)
                  0110      (6, the number of invalid combinations)
         0001     0010      (valid sum)
           1        2
```

Let's now apply these principles in adding multidigit numbers. Remember, although we are using binary numbers to add, we are in fact adding decimal numbers. Let's add 676 to 743. The procedure is shown in Example 3-24.

Example 3-24. Add 676 and 743.

Thousands	Hundreds	Tens	Ones	
0000	0111	0100	0011	Addend
0000	0110	0111	0110	Augend
				Add ones column
		0	1001	Valid BCD sum
				Add tens column plus the carry
		1011		Invalid binary sum
		0110		Correct by adding 6
	1	0001		Valid BCD sum
				Add hundreds column plus the carry
	1110			Invalid binary sum
	0110			Correct by adding 6
1	0100			Valid BCD sum
				Add thousands column plus the carry
0001	0100	0001	1001	Valid BCD sum representing
(1)	(4)	(1)	(9)	decimal 743 + 676 = 1419.

Now let's subtract 676 from 743, using the nine's complement method.

Example 3-25. Subtract 676 from 743.

Thousands	Hundreds	Tens	Ones	
0000	0111	0100	0011	Minuend
1001	0011	0010	0011	Subtrahend in nine's complement
				Add ones column
		0	0110	Valid BCD sum
				Add tens column plus the carry
		0110		Valid BCD sum
				Add the hundreds column
	1010			Invalid binary sum
	0110			Correct by adding 6
1	0000			Valid BCD sum
				Add thousands column plus the carry
1010				Invalid binary sum
0110				Correct by adding 6
1 0000				Valid BCD sum
			1	End-around carry
0000	0000	0110	0111	Valid BCD sum representing
		(6)	(7)	decimal 743 – 676 = 67.

Finally, let's subtract 974 from 835 using the nine's complement method.

Example 3-26. Subtract 974 from 835.

Thousands	Hundreds	Tens	Ones	
0000	1000	0011	0101	Minuend
1001	0000	0010	0101	Subtrahend in nine's complement
				Add ones column
			1010	Invalid binary sum
			0110	Correct by adding 6
		1	0000	Valid BCD sum
				Add tens column plus the carry
		0110		Valid BCD sum
				Add hundreds column plus the carry
	1000			Valid BCD sum
				Add thousands column plus the carry
0 1001	1000	0110	0000	Valid BCD sum
				Zero carry from MSD column indicates the sum is negative and in the nine's complement form.
− 0000	0001	0011	1001	Nine's complement of BCD sum
−	(1)	(3)	(9)	equals valid negative difference representing 835 − 974 = −139.

3.10 BCD-TO-BINARY CONVERSION

Most large computers perform arithmetic operations using binary numbers. Data entering the computer is usually in the form of an interchange code, such as EBCDIC. You will recall that the numerical portion of these codes is BCD. Therefore, before the numerical data can be processed, it must be converted to binary.

Figure 3-11 illustrates the process by which BCD (0101 1001 0110) is converted to binary (1001010100). Each time the data is shifted to the right, a division by 2 is accomplished. For example, if a bit is shifted from the 8-bit position to the 4-bit position, its weight is divided by 2. Suppose that a bit is shifted from the 1-bit position of a BCD digit to the 8-bit position of the next BCD digit. Its weight reduces from 10 to 8 instead of from 10 to 5. This means that the resulting BCD number will have a value of 3 greater than it should be (8 instead of 5). It is possible to subtract 3 (0011) after the shift to make the appropriate correction. You will notice in the example that if an 8-bit is detected in any decade, 3 (0011) is subtracted from that decade.

The process continues with as many shifts as there are bit positions

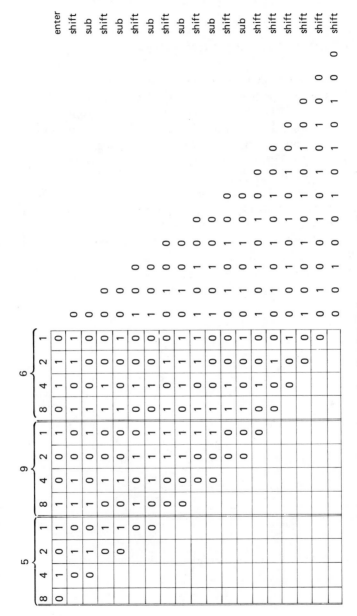

Figure 3-11. Example of BCD-to-binary conversion.

in the number. In practice, the subtraction is usually accomplished by adding the two's complement of 3 (1101).

3.11   BINARY-TO-BCD CONVERSION

The process of binary-to-BCD conversion is similar to BCD-to-binary conversion except that the shifting process is from right to left and the correction is somewhat different. A left shift results in a multiplication by 2. If a logical 1 is shifted from the 8-bit position to the 1-bit position of the next higher digit, its value should increase from 8 to 16 instead of from 8 to 10, as it would for BCD. Therefore, to correct for binary, a 6 must be added after the shift. This correction can be made before the shift by adding 3 whenever an 8 bit exists. However, after the shift, if a number greater than 9 exists in any decade, 10 must be subtracted and a value of 1 must be added to the next decade.

The entire process of correction can be simplified by adding 3 (0011) to any digit having a value greater than 4 *prior* to the shift. Figure 3-12 illustrates the binary-to-BCD conversion process. The process is complete when the last bit has been shifted in.

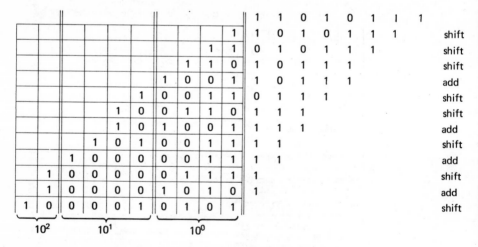

Figure 3-12.   Example of binary-to-BCD conversion.

QUESTIONS AND PROBLEMS—CHAPTER 3

1. Add the following octal numbers; convert the answers to decimal numbers.

   (a) 763
        214
        ___

   (b) 277
        641
        ___

   (c) 763
        367
        ___

2. Add the following octal numbers; convert the answers to decimal numbers.

   (a) 273
        171
        ___

   (b) 444
        367
        ___

   (c) 637
        214
        ___

3. Add the following hexadecimal numbers; convert the answers to decimal numbers.

   (a) 4A3
        D13
        ___

   (b) ABCD
        EF10
        ____

   (c) 109A
        2D1C
        ____

4. Add the following hexadecimal numbers; convert the answers to decimal numbers.

   (a) A14
        3D7
        ___

   (b) B29C
        ED4F
        ____

   (c) FACE
        BADE
        ____

5. Subtract the following octal numbers.

   (a)   73
       (−) 56
           ___

   (b)   376
       (−) 277
           ___

   (c)   2133
       (−) 1574
           ____

6. Subtract the following octal numbers.

   (a)   72
       (−) 15
           ___

   (b)   673
       (−) 176
           ___

   (c)   6014
       (−) 3247
           ____

7. Subtract the following hexadecimal numbers.

   (a)   A17
       (−) 26B
           ___

   (b)   7A29
       (−) 100F
           ____

   (c)   FADE
       (−) 2C3F
           ____

8. Subtract the following hexadecimal numbers.

   (a)   B27
       (−) 6A4
           ___

   (b)   4A9B
       (−) 2BC9
           ____

   (c)   FACE
       (−) AB3F
           ____

9. Add the following binary numbers; convert the answers to decimal numbers.

   (a) 101
        110
        ___

   (b) 1011
        1101
        ____

   (c) 110011
        100011
        _____

10. Add the following binary numbers; convert the answers to decimal numbers.

    (a) 1111   (b) 10011   (c) 101111
        1111   11011   110011

11. Subtract the following binary numbers using the one's complement method.

    (a)   1010   (b)   110010   (c)   1100110
        (−) 0101   (−) 100111   (−) 1011001

12. Subtract the following binary numbers using the one's complement method.

    (a)   1100   (b)   10110   (c)   100111
        (−) 0110   (−) 01101   (−) 001101

13. Repeat Problem 11 using the two's complement method.

14. Repeat Problem 12 using the two's complement method.

15. Subtract the following decimal numbers using the ten's complement method.

    (a)   547   (b)   7613
        (−) 279   (−) 2724

16. Repeat Problem 15 using the nine's complement method.

17. Perform the indicated operations in binary; give answer in both binary and decimal.

    (a)   −9   (b)   13   (c)   9   (d)   −13
        (+) 6   (−) −5   (+) −4   (+) 8

    (e)   23   (f)   −6   (g)   15   (h)   −7
        (+) 14   (−) −9   (+) −12   (+) −4

18. Perform the indicated operations in binary; give answer in both binary and decimal.

    (a)   −8   (b)   16   (c)   7   (d)   −17
        (+) 3   (−) −7   (+) −3   (−) 6

    (e)   36   (f)   −8   (g)   27   (h)   −12
        (+) 27   (−) −8   (+) −9   (+) −15

19. Perform the following subtractions using the eight's complement method. Assume that the octal numbers represent binary numbers. Assume the two's complement method is used and 12 bit positions are used where the MSB is the sign bit. (Twelve bit positions require 4 octal digits.)

(a) 361_8 (b) 3273_8 (c) 3106_8
    $(-)\ 254_8$ $(-)\ 2426_8$ $(-)\ 1727_8$

(d) 2267_8 (e) 1776_8 (f) 4063_8
    $(-)\ 4365_8$ $(-)\ 7177_8$ $(-)\ 7274_8$

20. Perform the following subtractions using the eight's complement method. Assume that the octal numbers represent binary numbers. Assume the two's complement method is used and 12 bit positions are used where the MSB is the sign bit. (Twelve bit positions require 4 octal digits.)

(a) 723_8 (b) 5023_8 (c) 6063_8
    $(-)\ 273_8$ $(-)\ 3127_8$ $(-)\ 2333_8$

(d) 3477_8 (e) 1630_8 (f) 6004_8
    $(-)\ 6652_8$ $(-)\ 6742_8$ $(-)\ 6175_8$

21. Perform the following subtractions using the sixteen's complement method. Assume that the hexadecimal numbers represent binary numbers. Assume the two's complement method is used and 16 bit positions are used where the MSB is the sign bit. (Sixteen bit positions require 4 hexadecimal digits.)

(a) $4C_{16}$ (b) $A27_{16}$ (c) $CD27_{16}$
    $(-)\ 2D_{16}$ $(-)\ 8AB_{16}$ $(-)\ AA4D_{16}$

(d) $4A_{16}$ (e) $27B_{16}$ (f) $72A6_{16}$
    $(-)\ A9_{16}$ $(-)\ ABC_{16}$ $(-)\ A4BC_{16}$

22. Perform the following subtractions using the sixteen's complement method. Assume that the hexadecimal numbers represent binary

numbers. Assume the two's complement method is used and 16 bit positions are used where the MSB is the sign bit. (Sixteen bit positions require 4 hexadecimal digits.)

(a) $A7_{16}$
     $(-)\ 89_{16}$

(b) $C74_{16}$
     $(-)\ ABF_{16}$

(c) $D0A6_{16}$
     $(-)\ A1A9_{16}$

(d) $6B_{16}$
     $(-)\ A6_{16}$

(e) $7A60_{16}$
     $(-)\ 9A6A_{16}$

(f) $A07E_{16}$
     $(-)\ EACF_{16}$

23. Perform the following additions and subtractions using BCD.

(a)   463
     $(+)\ 642$

(b)   207
     $(+)\ 463$

(c)   843
     $(-)\ 656$

(d)   906
     $(-)\ 627$

(e)   206
     $(-)\ 673$

(f)   463
     $(-)\ 674$

24. Perform the following additions and subtractions using BCD.

(a)   562
     $(+)\ 273$

(b)   673
     $(+)\ 345$

(c)   723
     $(-)\ 542$

(d)   642
     $(-)\ 456$

(e)   317
     $(-)\ 426$

(f)   106
     $(-)\ 769$

4

Logic Functions and Diagrams— Boolean Algebra

This chapter deals with logic, logic functions, logic symbology, and Boolean algebra. Logic diagrams that consider only the concepts are called *first-level logic diagrams*. Logic diagrams that use real circuits are called *second-level* or *implemented logic diagrams*. The student is asked to consider the symbology in this chapter at the concept level only (first level). If he understands the circuits (gates—Chapter 5 and Chapter 6) used to perform these functions, he is encouraged to study Chapter 8 and this chapter together. Chapter 8 considers the concepts of this chapter but uses real circuits.

Boolean algebra was developed by the Englishman George Boole around 1847. It is sometimes referred to as the "algebra of logic." In 1938, Claude E. Shannon, a research assistant at M.I.T., published a paper describing how Boolean algebra could be used to represent switching, or two-state, circuits. The techniques in Shannon's work have continuously been improved, and they are used in the design of all types of logic circuits today.

The topics covered in this chapter are:

4.1   Logic Variables

4.2   Logic Operations

4.1   LOGIC VARIABLES

Boolean algebra is used in manipulating logic variables. A variable is either completely true or completely false; partly true or partly false values are not allowed. When a variable is not true, by implication it must be false. Conversely, if the variable is not false, it must be true. Because of this characteristic, Boolean algebra is ideally suited to variables that have two states or values, such as YES and NO, or for a number system that has two values, 1 and 0 (i.e., binary number system).

A *variable* is a quantity represented by a symbol. For example, B (the variable) could represent the presence of Bob. B has two values: if Bob is present, B equals "true"; if Bob is absent, B equals "false." Note that Bob is not the variable; B is the variable that represents *the presence of Bob.*

A switch is ideally suited to represent the value of any two-state variable because it can only be "off" or "on." Consider the SPST switch in Figure 4-1. When the switch is in the closed position, it indicates that Bob is present (B = true). When it is in the open position, it then represents that Bob is absent (B = false).

It should be obvious that a closed switch could also represent values such as true, yes, one (1), HIGH (H), go, etc.; and in the open position: false, no, zero (0), LOW (L), no go, etc.

Figure 4-1.   SPST switch: (a) closed, (b) open.

4.2 LOGIC OPERATIONS

There are only three basic logic operations:

1. The *conjunction* (logical product) commonly called AND, symbolized by ($\cdot$).
2. The *disjunction* (logical sum) commonly called OR, symbolized by ($+$).
3. The *negation,* commonly called NOT symbolized by ($'$) or ($^-$).

These operations are performed by logic circuits. All functions within a computer can be performed by combinations of these three basic logic operations.

4.3 THE AND FUNCTION

The AND function can be illustrated by the following analogy. The members of group A are Bob, Charley, and Dick. Note that the names in this group are combined by the conjunction "and." That means group A equals the presence of Bob, *and* Charley *and* Dick. This may be symbolized as

$$A = B \cdot C \cdot D$$

A is true (group A is present) when B is true (Bob is present); AND C is true (Charley is present); AND D is true (Dick is present). A is not true if any one or more of the members are absent.

The circuit in Figure 4-2 can be used to produce the AND function of the above example. The light indicates that group A is present only when *all* members of the group are present. A logic circuit producing the AND function is called an AND gate and is symbolized in Figure 4-3.

Figure 4-4(a) shows the logic symbol for a 2-input AND gate. The function (A•B) is produced when A is true AND B is true. In order to show all of the conditions that may exist at the input and output of a logic gate, *truth tables* are used. Since there are two inputs and each input

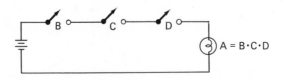

Figure 4-2.   Switches connected in series to produce the AND function.

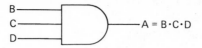

Figure 4-3.   Logic symbol for a three-input AND gate.

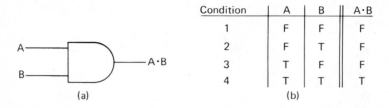

Condition	A	B	A·B
1	F	F	F
2	F	T	F
3	T	F	F
4	T	T	T

(a)   (b)

Figure 4-4.   Two-input AND gate: (a) logic symbol; (b) truth table.

has two possible states (true and false), the number of possible conditions at the input would be 2 raised to the second power (2^2), or 4. The truth table for a 2-input AND gate is shown in Figure 4-4(b). Note that condition 4 is the only one in which all of the inputs are true so that the AND function is produced and a true output appears.

AND gates may have two or more inputs. Figure 4-5 shows a three-input AND gate and its truth table. Ones (1) and zeros (0) are used for the values of the variables. Because three variables are used, eight possible conditions exist ($2^3 = 8$). Condition 8 is the only one that will produce a true output (1) because all of the input variables are true (equal to 1).

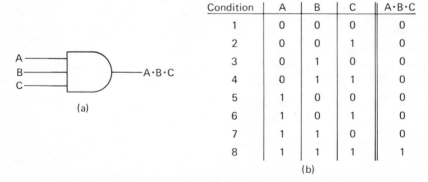

Condition	A	B	C	A·B·C
1	0	0	0	0
2	0	0	1	0
3	0	1	0	0
4	0	1	1	0
5	1	0	0	0
6	1	0	1	0
7	1	1	0	0
8	1	1	1	1

(a)   (b)

Figure 4-5.   Three-input AND gate: (a) logic symbol; (b) truth table.

At this point, it should be stated that the (•) symbol used in the expression A•B•C is the AND operator and indicates logical multiplication. For example, condition 8 in Figure 4-5 can be interpreted as $1 \times 1 \times 1 = 1$; whereas condition 7 can be interpreted as $1 \times 1 \times 0 = 0$.

Using the same approach, we find that all other conditions (1 through 6) also produce a zero. It is important to remember that in the binary system, a 2 cannot exist—only 1s and 0s. As with ordinary algebra, the operator symbol (•) can be omitted; thus, A•B•C = ABC and is read "A AND B AND C."

4.4 THE OR FUNCTION

The OR function is exemplified by the following analogy. The members of group A are Bob, Charley, and Dick. A representative of this group could be Bob or Charley or Dick, or any combination of them. This expression may be symbolized as

$$R = B + C + D$$

Where R is a representative of group A, R is true (a representative of group A is present) when B is true (that is, when Bob is present); OR C is true (Charley is present); OR D is true (Dick is present). Only one of the members needs to be present in order that group A be represented. However, group A is also represented when more than one member is present. Group A will not be represented (false) when *all* members are absent.

The circuit in Figure 4-6 can be used to produce the OR function of the above example. The light (R) indicates that a representative of group A is present when one OR more members of the group are present. A circuit producing the OR function is called an *OR gate;* it is symbolized in Figure 4-7.

Figure 4-8(a) shows the logic symbol for a two-input OR gate. The function (A + B) is produced when A is true OR B is true. The OR function mentioned above is *inclusive;* that is, it allows for the possibility that a true output is produced when both inputs are true.

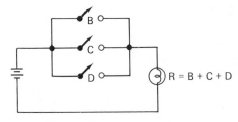

Figure 4-6. Switches connected in parallel to produce the OR function.

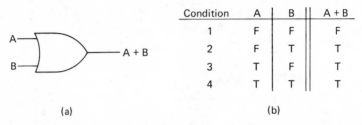

Figure 4-7.   Logic symbol for a three-input OR gate.

Condition	A	B	A + B
1	F	F	F
2	F	T	T
3	T	F	T
4	T	T	T

(a)   (b)

Figure 4-8.   Two-input OR gate: (a) logic symbol; (b) truth table.

A truth table for the two-input OR gate appears in Figure 4-8(b). Because there are two input variables, a total of four conditions are possible ($2^2 = 4$). Note that conditions 2, 3, and 4 produce a true output because at least one of the input variables is true. Condition 1 produces a false output because neither one of the input variables is true.

OR gates may have two or more inputs. Figure 4-9 shows a three-input OR gate and its truth table. Ones (1) and zeros (0) are used for the values of the variables. Because three variables are used, eight possible conditions occur ($2^3 = 8$). Conditions 2 through 8 produce a true (1) output because at least one of the input variables is true (1). Condition 1 is the only one that does not produce a true output.

It should be pointed out at this time that the "+" symbol used in the expression A + B + C is the OR operator and indicates logical addition. In logical addition, $1 + 0 = 1$ and $1 + 1 = 1$. *One OR one does not*

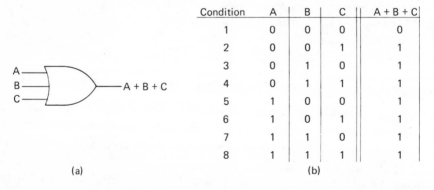

Condition	A	B	C	A + B + C
1	0	0	0	0
2	0	0	1	1
3	0	1	0	1
4	0	1	1	1
5	1	0	0	1
6	1	0	1	1
7	1	1	0	1
8	1	1	1	1

(a)   (b)

Figure 4-9.   Three-input OR gate: (a) logic symbol; (b) truth table.

equal two! Condition 8 in Figure 4-9 may be interpreted as $1 + 1 + 1 = 1$; whereas condition 5 can be interpreted as $1 + 0 + 0 = 1$. Condition 1 produces a zero (0) output because $0 + 0 + 0 = 0$.

4.5 THE NOT FUNCTION

The NOT function is produced by the inverting operation. The concept can be illustrated by the circuit in Figure 4-10. The switch is normally closed; therefore, the indicator lamp will light (true). Activating the switch (turning it off), however, will break the circuit, and the lamp will go out (not true). We may then say that the lamp is on (true) only when the switch is *not* activated. This condition can be expressed as

$$L = \overline{A}$$

This condition is verbalized as "L equals not A."

A logic circuit producing the NOT function is called an *inverter;* its symbol appears in Figure 4-11. An inverter converts the state or value of a variable to its complement. Thus, if variable A appears at the input, $\overline{A}$ (not A) is produced at the output; and conversely, when $\overline{A}$ appears at the input, A is produced at the output. When performing the NOT function, a 1 will be changed to a 0 and vice versa, as shown in the truth table in Figure 4-11(b).

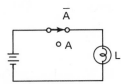

Figure 4-10. Representation of the NOT function.

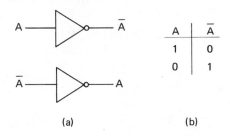

Figure 4-11. Inverter: (a) logic symbol; (b) truth table.

4.6 BOOLEAN EXPRESSIONS

The three basic logic functions discussed—AND, OR, and NOT—either individually or in various combinations, are the basic building blocks for all computer logic circuits. A few of these combinations will be illustrated by our group "A" analogy consisting of the presence of Bob (B), Charley (C), and Dick (D). Suppose we wish to describe a situation where the group (A) is represented by the presence of Bob (B) and Charley (C), but not Dick (D). This situation could be expressed as $X = BC\overline{D}$, which states that Bob and Charley are present, but Dick is NOT present; it is illustrated in Figure 4-12. Note the use of the inverter.

How would you describe a condition whereby the group (A) is represented by at least two members—in other words, a majority? This situation could be expressed in Boolean algebra as

$$X = BC + BD + CD$$

and is symbolized in Figure 4-13.

How could the group be represented by the presence of Charley (C) and the absence of both Bob (B) and Dick (D)? One possible method would be

$$A = \overline{B}C\overline{D}$$

It is shown in Figure 4-14.

Suppose we wish to express the situation whereby the entire group is NOT present—in other words, one or more members are absent. The Boolean expression would be

$$A = \overline{BCD}$$

It is illustrated in Figure 4-15. Note that the entire group is affected by the inverter.

Finally, how would you express a situation where the group is represented by either Bob and Charley or by Dick alone? One possible method is shown in Figure 4-16.

Logic diagrams are drawn in order to symbolize logic expressions.

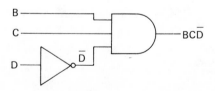

Figure 4-12. Logic circuit used to represent the expression $A = BC\overline{D}$.

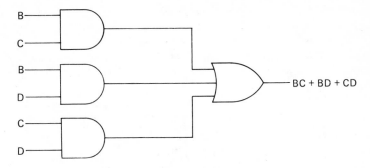

Figure 4-13. Logic circuit used to represent the expression BC + BD + CD.

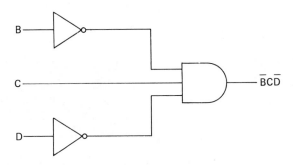

Figure 4-14. Logic circuit used to represent the expression $\overline{B}C\overline{D}$.

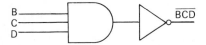

Figure 4-15. Logic circuit used to represent the expression $\overline{BCD}$.

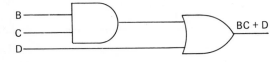

Figure 4-16. Logic circuit used to represent the expression BC + D.

When the student interprets the diagrams, he should be able to verbalize the expression. For example, let's interpret the diagram in Figure 4-17. In order to get a true output, Charley must be present (C), AND the other input to the AND gate must also be true. This input may be interpreted as true when B OR D are not present ($\overline{B + D}$). Therefore, a true output is indicated when Charley (but not Bob and Dick) is present C ($\overline{B + D}$). The student must understand that both Bob and Dick must be absent for there to be a true output.

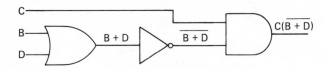

Figure 4-17.   Logic circuit used to represent the expression C($\overline{B + D}$).

4.7   THE EXCLUSIVE-OR FUNCTION

The EXCLUSIVE-OR function is an extension of the AND, OR, and NOT functions. The function may be illustrated by the following analogy. Suppose we wanted group A to be represented by the presence of either Bob or Charley but not both. The expression may be symbolized as

$$R = B \oplus C$$

where R is a representative of group A and is true only when B is true or C is true and is false when B and C (B•C) are true. The EXCLUSIVE-OR operator symbol is $\oplus$, an encircled plus sign.

The logic symbol for the EXCLUSIVE-OR gate appears in Figure 4-18(a), and its truth table is shown in Figure 4-18(b). Notice that this truth table is identical to the truth table for the inclusive-OR gate in Figure 4-8(b) with the exception of the last condition where an F appears instead of a T. The only difference between an EXCLUSIVE-OR and an inclusive OR is that for the EXCLUSIVE-OR, when both inputs are true, the output is false.

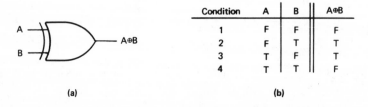

Condition	A	B	A⊕B
1	F	F	F
2	F	T	T
3	T	F	T
4	T	T	F

(a)   (b)

Figure 4-18.   Two-input EXCLUSIVE-OR gate: (a) logic symbol; (b) truth table.

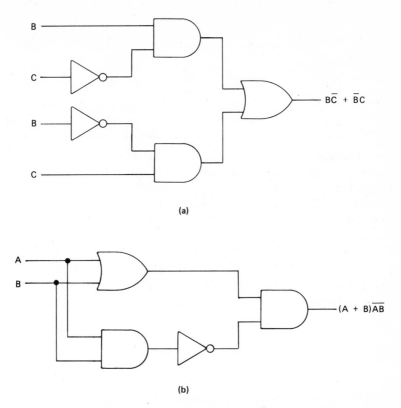

Figure 4-19. Logic diagrams for the EXCLUSIVE-OR function.

As previously mentioned, the AND, OR, and NOT functions are basic logic building blocks. These may be used to produce the EXCLUSIVE-OR function. One Boolean expression for the EXCLUSIVE-OR function is $B\overline{C} + \overline{B}C$. The logic diagram for this expression is shown in Figure 4-19(a).

Another expression is $(B + C)\overline{BC}$. The logic diagram for this expression appears in Figure 4-19(b).

4.8 *BOOLEAN POSTULATES AND THEOREMS*

The following postulates, laws, and theorems are important in the simplification and manipulation of logic expressions. The student is encouraged to become familiar with these as they will be used throughout this chapter. Note that each postulate, law, or theorem is described in two parts. These are *duals* of each other; that is, the dual of the OR operator is the AND, while the dual of a given variable is its complement.

Postulates	
1a. $A = 1$ (if $A \neq 0$)	1b. $A = 0$ (if $A \neq 1$)
2a. $0 \cdot 0 = 0$	2b. $0 + 0 = 0$
3a. $1 \cdot 1 = 1$	3b. $1 + 1 = 1$
4a. $1 \cdot 0 = 0$	4b. $1 + 0 = 1$
5a. $\overline{1} = 0$	5b. $\overline{0} = 1$

Postulates are self-evident truths. Consider postulates 1a and 1b. A variable is either true (1) or false (0). Postulates 2a, 3a, and 4a represent the conjunctive (AND) form and define the function of the AND operator, as shown in Figure 4-20. Postulates 2b, 3b, and 4b represent the disjunctive (OR) form and define the function of the OR operator as shown in Figure 4-21. Postulates 5a and 5b define the function of the NOT operator as illustrated in Figure 4-22.

The following are properties of ordinary algebra that also apply to Boolean algebra. Remember, Boolean expressions contain variables having only two possible values.

Algebraic Properties	
Commutative	
6a. $AB = BA$	6b. $A + B = B + A$
Associative	
7a. $A(BC) = AB(C)$	7b. $A + (B + C) = (A + B) + C$
Distributive	
8a. $A(B + C) = AB + AC$	8b. $A + BC = (A + B)(A + C)$

The *commutative* property simply means that the circuit is not affected by the order or sequence of the variables. This is illustrated in Figure 4-23.

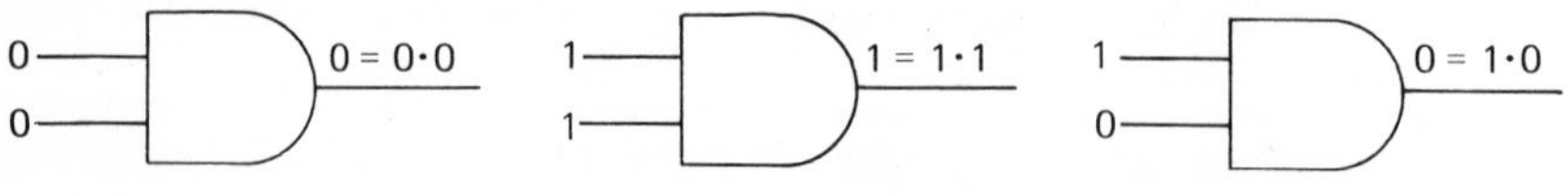

Figure 4-20.   Logic symbolization of postulates 2a, 3a, and 4a.

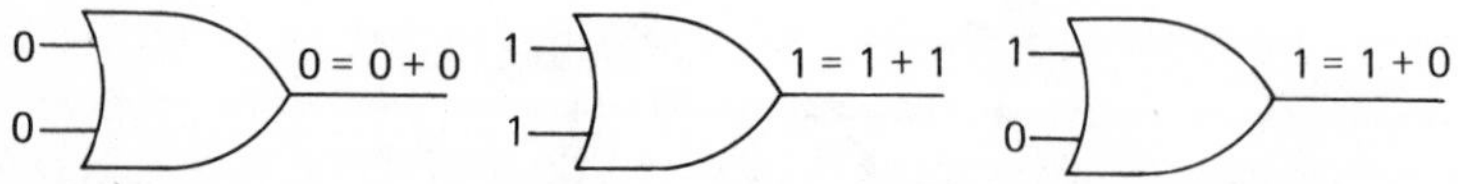

Figure 4-21.   Logic symbolization of postulates 2b, 3b, and 4b.

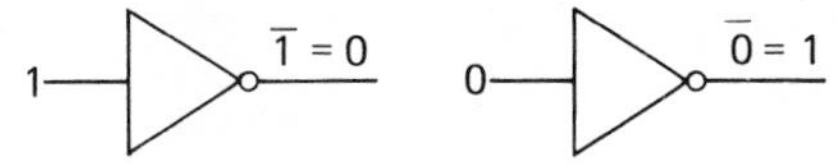

Figure 4-22.   Logic symbolization of postulates 5a and 5b.

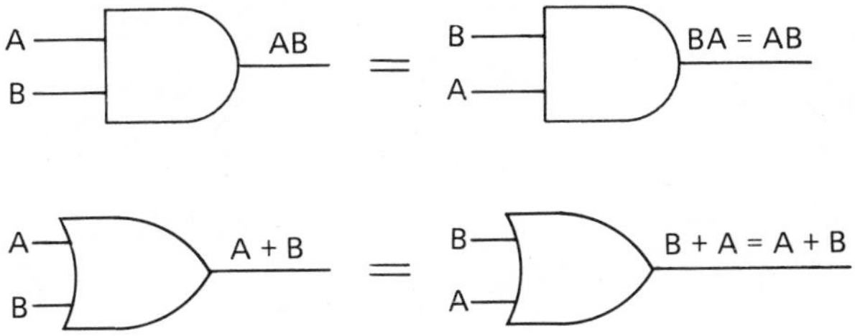

Figure 4-23.   Logic symbols used to demonstrate commutative property.

The *associative* property pertains to the parentheses. It shows that a sequence exclusively of AND functions (7a) or a sequence exclusively of OR functions (7b) is not affected by the placement of the parentheses, as indicated in Figure 4-24.

The *distributive* property for 8a and 8b may be proven by performing the algebraic multiplication or by factoring. Figure 4-25 shows that the application of the distributive property produces two forms of an expression, and thus, two circuits may be realized for each.

The following theorems define the application of the operators to variables:

Theorems	
9a. $A \cdot 0 = 0$	9b. $A + 0 = A$
10a. $A \cdot 1 = A$	10b. $A + 1 = 1$
11a. $A \cdot A = A$	11b. $A + A = A$
12a. $A \cdot \overline{A} = 0$	12b. $A + \overline{A} = 1$
13a. $\overline{\overline{A}} = A$	13b. $A = \overline{\overline{A}}$

Theorems 9a, 10a, 11a, and 12a pertain to the AND function. These are symbolized in Figure 4-26. For theorem 9a, one of the variables is always a zero. Therefore, the output will be a zero regardless of the value of A. Theorem 10a tells us the output will be determined by the input variable A. If $A = 1$, then $1 \cdot 1 = 1$. If $A = 0$, then $1 \cdot 0 = 0$. For theorem 11a, if the input variable $A = 1$, the output will be $1 \cdot 1 = 1$. If the input variable $A = 0$, the output will be $0 \cdot 0 = 0$.

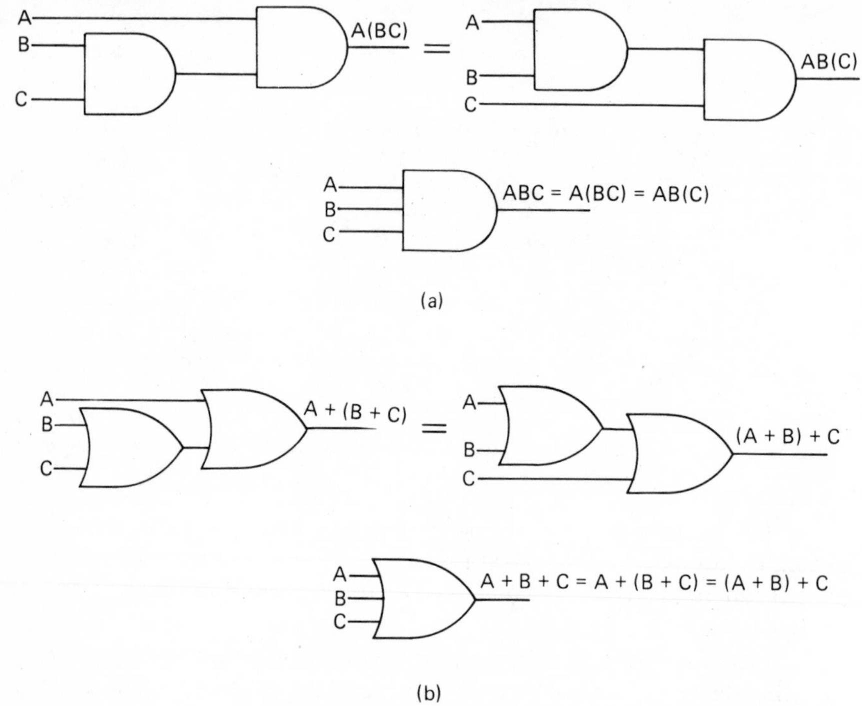

(a)

(b)

Figure 4-24.   Logic symbols and circuits used to demonstrate the associative property: (a) logic circuit to demonstrate AND; (b) logic circuit to demonstrate OR.

For theorem 12a, the output will always be 0 because when the input variable $A = 1$, the other input variable will be $0(\overline{A})$, causing the output to be $1 \cdot 0 = 0$. The same output will be produced when the value of the input variables are reversed. Theorem 12a is called a *self-contradiction*.

Theorem 13a expresses double negation. An expression that has been inverted twice equals its original value, as shown in Figure 4-27.

Theorems 9b, 10b, 11b, and 12b pertain to the OR function and are illustrated in Figure 4-28. For the OR function, one or more of the input variables must be true in order that a true output be present. For theorems 9b and 11b, the output is determined by the input variable A. For 10b and 12b the output will be a 1 regardless of the value of A. For theorem 12b, either A or $\overline{A}$ must always equal 1.

DeMorgan's Theorem	
14a. $\overline{ABC} = \overline{A} + \overline{B} + \overline{C}$	14b. $\overline{A} + \overline{B} + \overline{C} = \overline{ABC}$

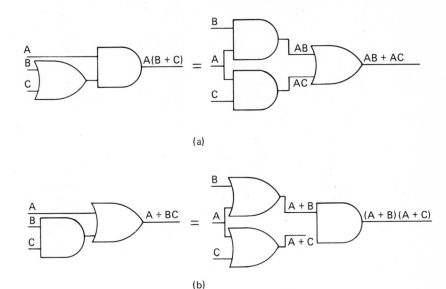

(a)

(b)

Figure 4-25. Logic symbols and circuits used to demonstrate distributive property: (a) logic circuit to demonstrate A(B +C) = AB + AC; (b) logic circuit to demonstrate A + BC = (A + B) (A + C).

The equivalent of the Boolean expression can be found by using *DeMorgan's theorem*. Three steps are required: First, replace all of the OR operator symbols (+) with AND operator symbols (•) and all of the AND operator symbols with OR operator symbols; next, replace all variables with their complements; finally, complement the entire expression. For example, determine DeMorgan's equivalent of the expression A + B.

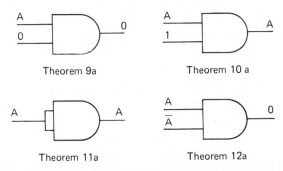

Theorem 9a

Theorem 10 a

Theorem 11a

Theorem 12a

Figure 4-26. Logic symbolization of the AND function.

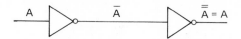

Figure 4-27.   Logic symbolization of double negation.

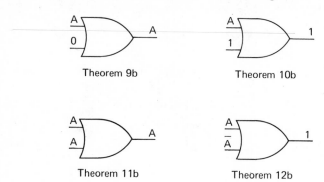

Figure 4-28.   Logic symbolization of the OR function.

Step 1. Replace the OR operator with the AND operator $A \cdot B$
Step 2. Complement each variable $\overline{A} \cdot \overline{B}$
Step 3. Complement the entire expression $\overline{\overline{A} \cdot \overline{B}}$

Therefore, $A + B = \overline{\overline{A} \cdot \overline{B}}$ and is symbolized in Figure 4-29.
Determine DeMorgan's equivalent of the expression AB.

Step 1. Replace the AND operator with the OR operator $A + B$
Step 2. Complement each variable $\overline{A} + \overline{B}$
Step 3. Complement the entire expression $\overline{\overline{A} + \overline{B}}$

Therefore, $AB = \overline{\overline{A} + \overline{B}}$ and is symbolized in Figure 4-30.
Let's find the DeMorgan's equivalent of the expression in theorem 14a, $\overline{ABC}$.

Step 1. $\overline{A} + \overline{B} + \overline{C}$
Step 2. $\overline{\overline{A}} + \overline{\overline{B}} + \overline{\overline{C}} = A + B + C$
Step 3. $\overline{A + B + C}$

Therefore

$$\overline{ABC} = \overline{A + B + C}$$

The symbols for the equivalent expressions above are given in Figure 4-31.

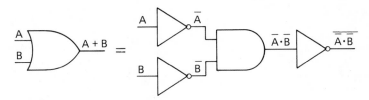

Figure 4-29.   Logic symbolization demonstrating that A + B = $\overline{\overline{A}\cdot\overline{B}}$.

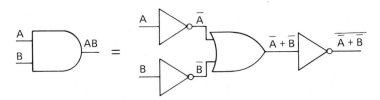

Figure 4-30.   Logic symbolization demonstrating that AB = $\overline{\overline{A} + \overline{B}}$.

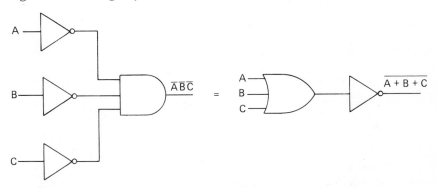

Figure 4-31.   Logic symbolization of the DeMorgan's equivalency $\overline{ABC} = \overline{A} + \overline{B} + \overline{C}$.

Now let's try finding DeMorgan's equivalent of the expression in theorem 14b, $\overline{A} + \overline{B} + \overline{C}$.

Step 1. $\overline{\overline{A}\cdot\overline{B}\cdot\overline{C}}$
Step 2. $\overline{\overline{\overline{A}}\cdot\overline{\overline{B}}\cdot\overline{\overline{C}}} = A\cdot B\cdot C$
Step 3. $\overline{A\cdot B\cdot C}$

Therefore

$$\overline{A} + \overline{B} + \overline{C} = \overline{A\cdot B\cdot C}$$

The logic symbols for these expressions are given in Figure 4-32.

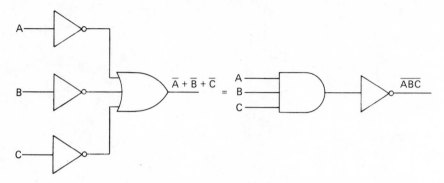

Figure 4-32.   Logic symbolization of the DeMorgan's equivalency
$\overline{A} + \overline{B} + \overline{C} = \overline{ABC}$.

Sometimes variables appear in a certain pattern that can obviously be simplified. The *absorption* theorem shows how substitution may be used on some commonly derived patterns.

Absorption Theorem	
15a. $A(A + B) = A$	15b. $A + AB = A$
16a. $A(\overline{A} + B) = AB$	16b. $A + \overline{A}B = A + B$

Theorems 15a and 15b show that the value of B is redundant. A true output occurs only when A is true, as shown in Figure 4-33.

Theorem 16a states that only when A AND B are true, the statement is true, as shown in Figure 4-34. Recall that A and $\overline{A}$ cannot be true simultaneously.

Theorem 16b is the dual of theorem 16a and is shown in Figure 4-35. For the expression to be true, either A must be true or B must be true. ($\overline{A}$ is redundant.)

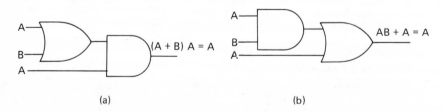

(a)   (b)

Figure 4-33.   Logic symbolization of: (a) theorem 15a; (b) theorem 15b.

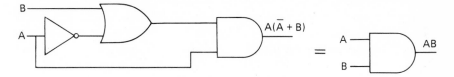

Figure 4-34. Logic symbolization of theorem 16a.

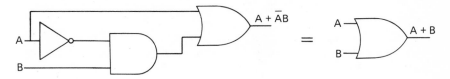

Figure 4-35. Logic symbolization of theorem 16b.

4.9 TRUTH TABLES

Logical expressions can be studied very conveniently by using truth tables. Truth tables present in columnar form all the possible combinations that can exist. The number of possible combinations is a function of the number of variables in question. Since each variable can attain a value of either 1 or 0, the number of possible combinations is equal to 2 raised to the number of variables. For example, if a gate has two variables, A and B, then there exist 2^2, or 4, possible combinations. A three-input gate would have 2^3, or 8, possible combinations, etc.

When constructing a truth table, it is important to keep the correct order of the original statement and not to skip any steps along the way. For example, the truth table that verifies the statement $AB = \overline{A} + \overline{B}$ is given in Figure 4-36.

A	B	AB	$\overline{A}$	$\overline{B}$	$\overline{A} + \overline{B}$	$\overline{\overline{A} + \overline{B}}$
0	0	0	1	1	1	0
0	1	0	1	0	1	0
1	0	0	0	1	1	0
1	1	1	0	0	0	1

————EQUAL————

Figure 4-36. Truth table to verify the statement $AB = \overline{\overline{A} + \overline{B}}$.

Note that the two input variables, A and B, are arranged such that all four possible combinations are represented. The AB column is a result of "ANDing" columns A and B according to postulates previously described. In the next two columns, A and then B are complemented. [The $\overline{A} + \overline{B}$ column is a result of ORing the $\overline{A}$ and $\overline{B}$ columns.] Finally $(\overline{A} + \overline{B})$ is complemented to produce the $\overline{\overline{A} + \overline{B}}$ column. Note that the contents of column AB and column $\overline{\overline{A} + \overline{B}}$ are identical. Therefore

$$AB = \overline{\overline{A} + \overline{B}} \qquad \text{(DeMorgan's theorem)}$$

Consider the expression $A + \overline{A}B = A + B$ (absorption theorem 16b). Since there are two variables, a total of four combinations exist, as shown in Figure 4-37.

The term $A + \overline{A}B$ is broken into three columns, $\overline{A}$, $\overline{A}B$, and $A + \overline{A}B$. Finally, A and B are ORed to form the $A + B$ column. Note that the order of column $A + \overline{A}B$ is the same as that of column $A + B$. Therefore, $A + \overline{A}B = A + B$.

Let us now construct a truth table for DeMorgan's theorem (14a), which states that $\overline{A + B + C} = \overline{A}\overline{B}\overline{C}$. Because there are three variables, 2^3, or 8, combinations exist, as shown in Figure 4-38. Note that column $\overline{A + B + C}$ is identical to column $\overline{A}\overline{B}\overline{C}$. Therefore

$$\overline{A + B + C} = \overline{A}\overline{B}\overline{C}$$

The truth table proving the associative property $A + B + C = A + (B + C) = (A + B) + C$ is shown in Figure 4-39.

The process of using truth tables to prove the equality of two expressions is called *proof by perfect induction*. The student is encouraged to use truth tables to evaluate logic expressions or to prove the results of logic operations. He can also use tables to define the function of various logic circuits.

| Input variables | | $\overline{A}$ | $\overline{A}B$ | $A + \overline{A}B$ | $A + B$ |
A	B				
0	0	1	0	0	0
0	1	1	1	1	1
1	0	0	0	1	1
1	1	0	0	1	1
				└─ EQUAL ─┘	

Figure 4-37.   Truth table to verify the statement $A + \overline{A}B = A + B$.

Input variables								
A	B	C	A + B + C	$\overline{A + B + C}$	$\overline{A}$	$\overline{B}$	$\overline{C}$	$\overline{A}\,\overline{B}\,\overline{C}$
0	0	0	0	1	1	1	1	1
0	0	1	1	0	1	1	0	0
0	1	0	1	0	1	0	1	0
0	1	1	1	0	1	0	0	0
1	0	0	1	0	0	1	1	0
1	0	1	1	0	0	1	0	0
1	1	0	1	0	0	0	1	0
1	1	1	1	0	0	0	0	0

EQUAL

Figure 4-38. Truth table to verify the statement $\overline{A + B + C} = \overline{A}\,\overline{B}\,\overline{C}$.

Input variables							
A	B	C	B + C	A + (B + C)	A + B	(A + B) + C	A + B + C
0	0	0	0	0	0	0	0
0	0	1	1	1	0	1	1
0	1	0	1	1	1	1	1
0	1	1	1	1	1	1	1
1	0	0	1	1	1	1	1
1	0	1	1	1	1	1	1
1	1	0	1	1	1	1	1
1	1	1	1	1	1	1	1

EQUAL

Figure 4-39. Truth table to verify the statement A + (B + C) = (A + B) + C.

4.10   SIMPLIFICATION OF LOGIC EXPRESSIONS

In order to demonstrate how logic expressions are produced and simplified, the following example will be given:

Goat-Wolf-Corn Problem

(From: *Introduction to the Basic Computer* by Donald Eadie, Prentice-Hall, Englewood Cliffs, N.J., 1968.)

A farmer employs as a hired hand a man who is not too bright. The hired hand is to keep the farmer's goat out of the barn while the door is open because the barn contains a pile of corn to be used as cattle feed. Also, a wolf, who considers the goat a tasty dish, lurks nearby.

The farmer decides to design a box which has three switches: one each for "door open," "goat in sight," and "wolf in sight." The hired man will merely flip the switches according to what he sees at a given time. If the situation is "dangerous," i.e., the ingredients are such as to cause harm to one of the items to be protected, the combination of switches flipped will ring an alarm and the farmer will take appropriate action.

It is assumed that:

1. the situation is safe if the goat and wolf are not visible; and

2. the wolf does not eat corn.

First, let's assign symbols to each of the variables. Since, when the barn door is open, the corn is available (door = corn), three variables are required:

$$D = \text{door open}$$
$$G = \text{goat visible}$$
$$W = \text{wolf visible}$$

Next, a truth table is constructed. Because there are three variables, a total of eight conditions are possible (2^3), as shown in Figure 4-40.

The next step is to define which conditions are dangerous. Condition 4 represents a dangerous situation because the wolf may devour the goat.

CONDITION	D	G	W	
1	0	0	0	
2	0	0	1	
3	0	1	0	
4	0	1	1	→ Dangerous Situation ($\overline{D}GW$)
5	1	0	0	
6	1	0	1	
7	1	1	0	→ Dangerous Situation ($DG\overline{W}$)
8	1	1	1	→ Dangerous Situation (DGW)

Figure 4-40.   Truth table for Door-Goat-Wolf Problem.

Condition 7 is dangerous because the corn is available to the goat. And in condition 8, anything may happen: the goat may eat the corn, or the wolf may get the goat. Therefore, conditions 4, 7, and 8 are dangerous. An expression for these conditions can now be written. Whenever a 0 appears in the truth table, the complement of the variable is used. Whenever a 1 appears, the true form of the variable is used. The expression therefore is

$$\textbf{Danger} = \overline{D}GW + DG\overline{W} + DGW$$

It should be apparent that the expression can be simplified because variables appear more than once. It is possible to simplify the expression by making substitutions using the appropriate theorems or postulates as shown in Figure 4-41. Therefore, the dangerous situations $= \mathrm{G}\,(\mathrm{D} + \mathrm{W})$ or $\mathrm{GD} + \mathrm{GW}$, and are symbolized in Figure 4-42.

1.	$\overline{D}GW + DG\overline{W} + DGW$	Original Expression
2.	$G\overline{D}W + GD\overline{W} + GDW$	Theorem 6b
3.	$G\,(\overline{D}W + D\overline{W} + DW)$	Theorem 8a
4.	$G[\overline{D}W + (D\overline{W} + DW)]$	Theorem 7b
5.	$G[\overline{D}W + D(\overline{W} + W)]$	Theorem 8a
6.	$G[\overline{D}W + D\,(1)]$	Theorem 12b
7.	$G(\overline{D}W + D)$	Theorem 10a
8.	$G\,(D + \overline{D}W)$	Theorem 6b
9.	$G\,(D + W)$	Theorem 16b
	or	
	$GD + GW$	Theorem 8a

Figure 4-41.   Simplification of Door-Goat-Wolf Problem.

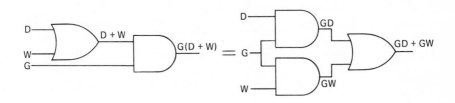

Figure 4-42.

4.11   KARNAUGH MAPS

When digital circuits are first designed they may be more complex than necessary. By using Boolean algebra, theorems, and laws previously discussed, it is often possible to simplify logic expressions without changing the original logic functions. Simplified expressions are very desirable because they will result in less complex circuits with increased reliability. In addition, the smaller circuits result in increased speeds and greater economy.

Karnaugh maps provide a method by which a Boolean expression can be visualized, and more important, they represent a technique for simplification. A Karnaugh map is made up of a group of squares. The number of squares required depends on the number of variables. A two-variable expression requires four squares; three variables, eight squares; four variables, 16 squares; etc. The number of squares required is equal to 2 raised to the number of variables. Karnaugh maps are seldom useful for more than five variables.

To begin, consider the truth table in Figure 4-43. The standard truth table for a two-input variable AND gate is shown in Figure 4-43(a). Note that a true output (1) exists only when A is true AND B is true. The other three outputs are indicated as not true (0).

A	B	AB
0	0	0
0	1	0
1	0	0
1	1	1

(a)

A	B	AB
$\overline{A}$	$\overline{B}$	$\overline{A}\,\overline{B}$
$\overline{A}$	B	$\overline{A}B$
A	$\overline{B}$	$A\overline{B}$
A	B	AB

(b)

Figure 4-43.

In Figure 4-43(b), a similar table is produced. In this table, 1 is replaced by the variable, and 0 is replaced by the complement of the variable. The variables are then ANDed. Again, there exist four possible combinations. These are $\overline{A}\overline{B} + \overline{A}B + A\overline{B} + AB$. When both tables of Figure 4-43 are combined, the result is the Karnaugh map illustrated in Figure 4-44.

Note that the squares represent each of the four possible conditions for the two variables. The top left square corresponds to the condition where both A and B are logical 0 and so is expressed as $\overline{A}\overline{B}$. The bottom right square corresponds to the condition where both A and B equal a logical 1 and therefore can be expressed as AB. In the top right square,

	A	
B	0	1
0	(0, 0) $\overline{A}\overline{B}$	(1, 0) $A\overline{B}$
1	(0, 1) $\overline{A}B$	(1, 1) AB

Figure 4-44.

A is equal to a logical 1, but B is equal to a logical 0 and is expressed as $A\overline{B}$. In the bottom left square, A is equal to 0, and B is equal to 1 and is expressed as $\overline{A}B$.

Two Karnaugh maps, each having two variables, appear in Figure 4-45. Note that a 1 appears in the top right and bottom right squares of Figure 4-45(a). The 1's indicate the presence of the term which corresponds to that square. The Karnaugh map expression for Figure 4-45(a) is

$$Map = A\overline{B} + AB \qquad \text{(refer to Figure 4-44)}$$

The logical expression for Figure 4-45(b) is

$$Map = \overline{A}\overline{B} + \overline{A}B + AB \qquad \text{(refer to Figure 4-44)}$$

The empty squares represent the absence of a term and are omitted to make the map easier to read.

Rules for a two-variable Karnaugh map:

1. Two adjacent squares are combined to represent a single variable.

2. The resulting variable is common to both terms.

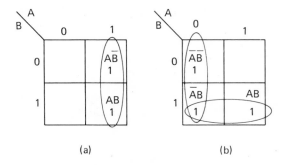

(a)   (b)

Figure 4-45.

Returning to Figure 4-45(a), note that there are two adjacent 1s which may be combined to represent one variable. This is shown in Figure 4-46.

In the expression $A\overline{B} + AB$, A is the common variable for both terms. Therefore, $A\overline{B} + AB = A$, which is much simpler than the original expression. That this is true may be easily verified through the use of some of the laws and theorems previously discussed.

$A\overline{B} + AB$	original expression
$A(\overline{B} + B)$	theorem 8a
$A(1)$	theorem 12b
A	theorem 10a
Therefore:	$A\overline{B} + AB = A$

For Figure 4-45(b), you will note that there are two sets of adjacent 1s, as shown in Figure 4-47. The two *adjacent vertical* 1s give $\overline{A}\overline{B} + \overline{A}B$, which simplifies to $\overline{A}$ (because $\overline{A}$ is common to both terms). The two *adjacent horizontal* 1s give $\overline{A}B + AB$, which simplifies to B (because B is common to both terms). Therefore

$$\overline{A}\overline{B} + \overline{A}B + AB = \overline{A} + B$$

You may verify this expression by using previous laws and theorems, as follows:

$\overline{A}\overline{B} + \overline{A}B + AB$	original expression
$\overline{A}(\overline{B} + B) + AB$	theorem 8a
$\overline{A}(1) + AB$	theorem 12b
$\overline{A} + AB$	theorem 10a
$\overline{A} + B$	theorem 16b

Therefore

$$\overline{A}\overline{B} + \overline{A}B + AB = \overline{A} + B$$

Let us now consider the truth table for a three-input AND gate, as shown in Figure 4-48. In the table for 4-48(b), 1s and 0s have been replaced by their respective variables. Note that for three variables, there

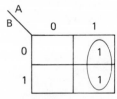

Figure 4-46.   Karnaugh map showing 1s in two adjacent squares.
MAP $= A\overline{B} + AB = A$.

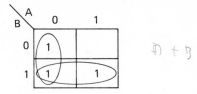

Figure 4-47. Karnaugh map showing two sets of adjacent squares. MAP $= \overline{A}\overline{B} + \overline{A}B + AB$.

A	B	C	ABC
0	0	0	0
0	0	1	0
0	1	0	0
0	1	1	0
1	0	0	0
1	0	1	0
1	1	0	0
1	1	1	1

(a)

A	B	C	ABC
$\overline{A}$	$\overline{B}$	$\overline{C}$	$\overline{A}\overline{B}\overline{C}$
$\overline{A}$	$\overline{B}$	C	$\overline{A}\overline{B}C$
$\overline{A}$	B	$\overline{C}$	$\overline{A}B\overline{C}$
$\overline{A}$	B	C	$\overline{A}BC$
A	$\overline{B}$	$\overline{C}$	$A\overline{B}\overline{C}$
A	$\overline{B}$	C	$A\overline{B}C$
A	B	$\overline{C}$	$AB\overline{C}$
A	B	C	ABC

(b)

Figure 4-48. Truth table for 3-input AND gate: (a) using 1s and 0s; (b) using variables.

are eight possible combinations. The Karnaugh map for the three variables is shown in Figure 4-49. Each square represents the logical product of the three variables.

Rules for a three-variable Karnaugh map:

1. A single square represents a three-variable term.
2. Two adjacent squares are combined to represent a two-variable term.
3. Four adjacent squares are combined to represent one variable.

Rule 1 for the three-variable Karnaugh map is exemplified in Figure 4-50. The terms in the squares cannot be combined. Therefore

$$\text{Map} = \overline{A}B\overline{C} + A\overline{B}\overline{C} + \overline{A}\overline{B}C$$

Figure 4-51 illustrates some possible combinations of the two adjacent squares (Rule 2). Figure 4-51(a) shows how two end squares may

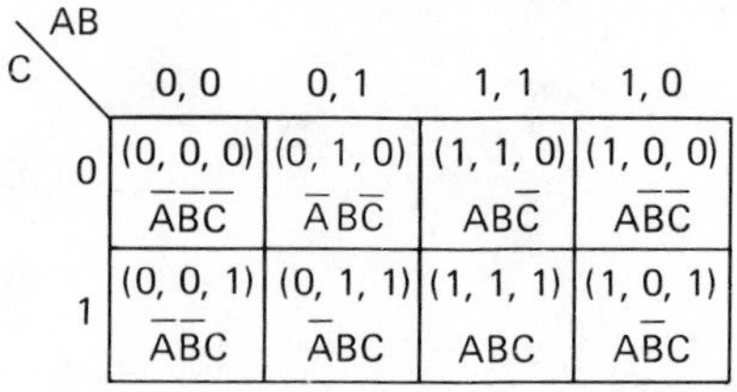

AB
C
0, 0
0, 1
1, 1
1, 0
0
(0, 0, 0)
$\bar{A}\bar{B}\bar{C}$
(0, 1, 0)
$\bar{A}B\bar{C}$
(1, 1, 0)
$AB\bar{C}$
(1, 0, 0)
$A\bar{B}\bar{C}$
1
(0, 0, 1)
$\bar{A}\bar{B}C$
(0, 1, 1)
$\bar{A}BC$
(1, 1, 1)
ABC
(1, 0, 1)
$A\bar{B}C$

Figure 4-49. General form of 3-variable Karnaugh map.

AB
C
0, 0
0, 1
1, 1
1, 0
0
1
1
1
1

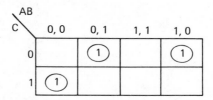

Figure 4-50. Single square represents a 3-variable term.

be considered as adjacent to each other if we think of the map as a wraparound cylinder.

The map reading for Figure 4-51(a) is $\bar{A}\bar{B}\bar{C} + A\bar{B}\bar{C} = \bar{B}\bar{C}$ because $\bar{B}\bar{C}$ is common to both squares. Figure 4-51(b) shows two sets of adjacent squares. The two top left squares read $\bar{A}\bar{B}\bar{C} + \bar{A}B\bar{C} = \bar{A}\bar{C}$; the two bottom right squares read $ABC + A\bar{B}C = AC$. Therefore, the map reading for Figure 4-51(b) is

$$\text{Map} = \bar{A}\bar{C} + AC$$

Rule 3 is illustrated by Figure 4-52.

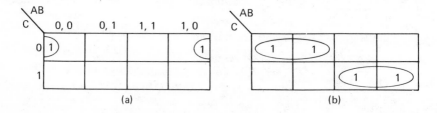

AB
C
0, 0
0, 1
1, 1
1, 0
0
1
1
1
1
(a)
AB
C
1
1
1
1
(b)

Figure 4-51. Two adjacent square combinations: (a) two adjacent squares rule (wraparound cylinder); (b) two adjacent squares rule.

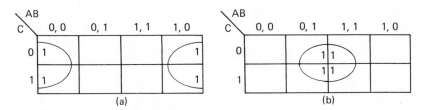

Figure 4-52. Four adjacent square combinations: (a) four adjacent squares (wraparound cylinder); (b) four adjacent squares.

In Figure 4-52(a), the four outer boxes may be combined because they are considered to be adjacent to each other. This is sometimes referred to as the "wraparound cylinder concept." The map reading for Figure 4-52(a) is $\overline{A}\overline{B}\overline{C} + A\overline{B}\overline{C} + \overline{A}\overline{B}C + A\overline{B}C = \overline{B}$. This is because $\overline{B}$ is the common variable for all of the terms and may be verified by the laws and theorems previously discussed:

$\overline{A}\overline{B}\overline{C} + A\overline{B}\overline{C} + \overline{A}\overline{B}C + A\overline{B}C$	original expression
$(\overline{A}\overline{B}\overline{C} + \overline{A}\overline{B}C) + (A\overline{B}\overline{C} + A\overline{B}C)$	theorem 7b
$\overline{A}\overline{B}\,(\overline{C} + C) + A\overline{B}\,(\overline{C} + C)$	theorem 8a
$\overline{A}\overline{B}\,(1) + A\overline{B}\,(1)$	theorem 10a
$\overline{A}\overline{B} + A\overline{B}$	theorem 8a
$\overline{B}\,(\overline{A} + A)$	theorem 8a
$\overline{B}(1)$	theorem 12b
$\overline{B}$	theorem 10a

Figure 4-52(b) shows another possible combination involving four adjacent squares. The map reading for Figure 4-52(b) is

$$\overline{A}B\overline{C} + AB\overline{C} + \overline{A}BC + ABC = B.$$

Note that B is the common variable in all of the terms and may be verified as follows:

$\overline{A}B\overline{C} + AB\overline{C} + \overline{A}BC + ABC$	original expression
$\overline{A}B\overline{C} + \overline{A}BC + AB\overline{C} + ABC$	theorem 7b
$\overline{A}B\,(\overline{C} + C) + AB\,(\overline{C} + C)$	theorem 8a
$\overline{A}B\,(1) + AB\,(1)$	theorem 10a
$\overline{A}B + AB$	theorem 8a
$B\,(\overline{A} + A)$	theorem 8a
$B\,(1)$	theorem 12b
B	theorem 10a

As another example of a three-variable Karnaugh map, consider Figure 4-53. Note that the four top squares (top row) may be combined to read

$$\overline{A}\overline{B}\overline{C} + \overline{A}B\overline{C} + AB\overline{C} + A\overline{B}\overline{C} = \overline{C}$$

because $\overline{C}$ is common to all the terms. The middle four squares may be combined to read

$$\overline{A}B\overline{C} + AB\overline{C} + \overline{A}BC + ABC = B$$

Therefore, the resulting expression for the Karnaugh map in Figure 4-53 is

$$Map = B + \overline{C}$$

Complex Boolean expressions may often be simplified by construction of a Karnaugh map. Suppose we wish to simplify the following logic statement:

$$X = A\overline{B}\overline{C} + ABC + \overline{A}\overline{B}C + A\overline{B}C + \overline{A}\overline{B}\overline{C}$$

Since there are three variables, a Karnaugh map having eight squares is constructed, and 1s are placed in the square corresponding to each term of the expression, as shown in Figure 4-54.

Note that the four outer squares may be combined because of the wraparound concept, resulting in

$$\overline{A}\overline{B}\overline{C} + A\overline{B}\overline{C} + \overline{A}\overline{B}C + A\overline{B}C = \overline{B}$$

The two lower right squares may be combined to give

$$ABC + A\overline{B}C = AC$$

Therefore, the simplified expression is

$$X = \overline{B} + AC$$

Recall the Door-Goat-Wolf problem previously discussed in this chapter. Its truth table is reproduced in Figure 4-55(a). A table using the

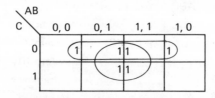

Figure 4-53.　Karnaugh map solution. Map $= B + \overline{C}$.

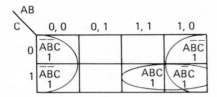

Figure 4-54. Karnaugh map solution. Map $= B + AC$.

D	G	W
0	0	0
0	0	1
0	1	0
0	1	1
1	0	0
1	0	1
1	1	0
1	1	1

(a)

D	G	W
$\bar{D}$	$\bar{G}$	$\bar{W}$
$\bar{D}$	$\bar{G}$	W
$\bar{D}$	G	$\bar{W}$
$\bar{D}$	G	W
D	$\bar{G}$	$\bar{W}$
D	$\bar{G}$	W
D	G	$\bar{W}$
D	G	W

(b)

The rows "0 1 1", "1 1 0", and "1 1 1" are marked **Dangerous Situation**.

Figure 4-55. Truth table for Door-Goat-Wolf problem: (a) using 1s and 0s; (b) using variables.

appropriate variables is shown in Figure 4-55(b). The expression for the dangerous situation was given as

$$\text{Danger} = \bar{D}GW + DG\bar{W} + DGW$$

It is possible to simplify this expression by using a three-variable Karnaugh map. This is shown in Figure 4-56. When the two adjacent vertical squares are combined, we read

$$DGW + DG\bar{W} = DG$$

When the two adjacent horizontal squares are combined, we have

$$\bar{D}GW + DGW = GW$$

Therefore, the simplified expression for the Door-Goat-Wolf Problem is

$$\text{Danger} = DG + GW \text{ or } G(D + W)$$

A four-variable Karnaugh map is shown in Figure 4-57. Because there are four variables, 16 combinations are possible.

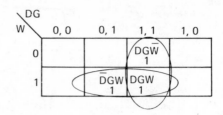

Figure 4-56. Karnaugh map representing Door-Goat-Wolf Problem.

	AB 0, 0	0, 1	1, 1	1, 0
0, 0	$\overline{A}\overline{B}\overline{C}\overline{D}$	$\overline{A}B\overline{C}\overline{D}$	$AB\overline{C}\overline{D}$	$A\overline{B}\overline{C}\overline{D}$
0, 1	$\overline{A}\overline{B}\overline{C}D$	$\overline{A}B\overline{C}D$	$AB\overline{C}D$	$A\overline{B}\overline{C}D$
1, 1	$\overline{A}\overline{B}CD$	$\overline{A}BCD$	$ABCD$	$A\overline{B}CD$
1, 0	$\overline{A}\overline{B}C\overline{D}$	$\overline{A}BC\overline{D}$	$ABC\overline{D}$	$A\overline{B}C\overline{D}$

Figure 4-57. General form of four-variable Karnaugh map.

Rules for a four-variable Karnaugh map:

1. A single square represents a four-variable term.
2. Two adjacent squares represent a three-variable term.
3. Four adjacent squares represent a two-variable term.
4. Eight adjacent squares represent one variable.

For example, consider the Karnaugh map in Figure 4-58. The eight adjacent squares, according to Rule 4, may be combined to represent a single variable. Thus

$$AB\overline{C}\overline{D} + A\overline{B}\overline{C}\overline{D} + AB\overline{C}D + A\overline{B}\overline{C}D + ABCD + A\overline{B}CD + ABC\overline{D} + A\overline{B}C\overline{D} = A$$

The four adjacent squares are combined to represent a two-variable term:

$$\overline{A}\overline{B}CD + \overline{A}BCD + ABCD + A\overline{B}CD = CD$$

The two adjacent terms (wraparound) combine to give a three-variable term:

$$\overline{A}\overline{B}C\overline{D} + A\overline{B}C\overline{D} = \overline{B}C\overline{D}$$

Therefore, the map expression for Figure 4-58 is

$$A + CD + \overline{B}C\overline{D}$$

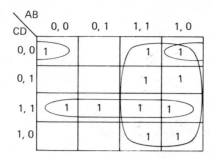

Figure 4-58. Karnaugh map solution. Map = A + CD + $\overline{BC}\overline{D}$.

Let's simplify the logic circuit in Figure 4-59. We will simplify first by using Boolean rules, then by the Karnaugh map method.

Simplification by Boolean laws, theorems, and postulates:

1. $ABC\overline{D} + A\overline{B}C\overline{D} + AB\overline{C}\overline{D} + A\overline{B}\overline{C}\overline{D}$	original expression
2. $\overline{D}(ABC + A\overline{B}C + AB\overline{C} + A\overline{B}\overline{C})$	theorem 8a
3. $\overline{D}(ABC + AB\overline{C} + A\overline{B}C + A\overline{B}\overline{C})$	theorem 6b
4. $\overline{D}[AB(C + \overline{C}) + A\overline{B}(C + \overline{C})]$	theorem 8a
5. $\overline{D}[AB(1) + A\overline{B}(1)]$	theorem 12b
6. $\overline{D}[AB + A\overline{B}]$	theorem 10a
7. $\overline{D}[A(B + \overline{B})]$	theorem 8a
8. $\overline{D}[A(1)]$	theorem 10a
9. $\overline{D}[A]$	theorem 6a
10. $A\overline{D}$	simplified expression

The terms are placed in the appropriate square, as shown in Figure 4-60(a). Figure 4-60(b) shows how the two top and two bottom squares are combined into the four adjacent square rule. Note that $A\overline{D}$ is common to all terms; therefore, Map = $A\overline{D}$. The simplified logic circuit is shown in Figure 4-61. Another example of a simplification process is shown in Figure 4-62.

Simplification by Boolean laws, theorems, and postulates:

1. $ABD + ACD + BCD + AB + \overline{A}CD +$ $\overline{A}\overline{B}CD$	original expression
2. $ACD + BCD + AB + \overline{A}CD + \overline{A}\overline{B}CD$	($AB = AB + ABD$, theorem 15b)

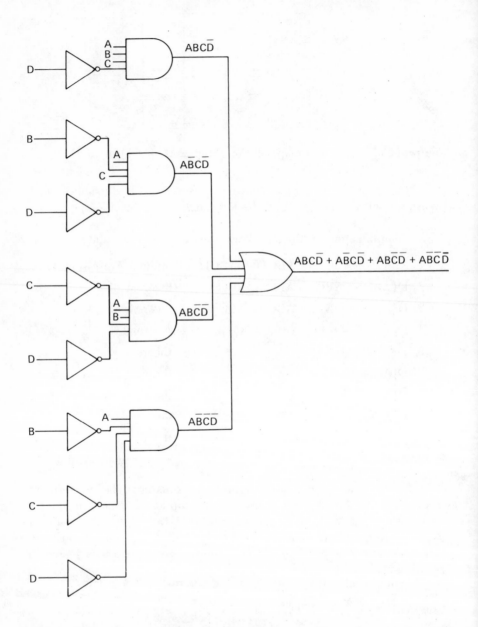

Figure 4-59. Logic circuit to be simplified.

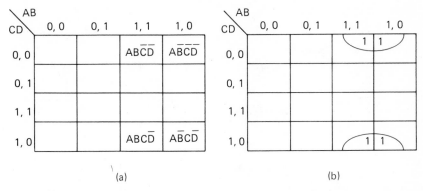

Figure 4-60. Karnaugh map solution for the logic circuit in Figure 4-59: (a) variables placed in appropriate squares; (b) four adjacent square rule applied.

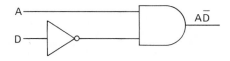

Figure 4-61. Simplified circuit of Figure 4-59.

3. $ACD + BCD + AB + \bar{A}CD$ $\qquad$ ($\bar{A}CD = \overline{ACD} + \overline{AB}CD$, theorem 15b)

4. $CD(A + \bar{A} + B) + AB$ $\qquad$ theorem 8a

5. $CD(1 + B) + AB$ $\qquad$ theorem 12b

6. $CD(1) + AB$ $\qquad$ theorem 10b

7. $AB + CD$ $\qquad$ simplified expression

The Karnaugh map for the expression is given in Figure 4-63(a). Note that each term must be appropriately represented. For example, ACD occurs in two squares, AB occurs in four squares, and so on. When the vertical four adjacent squares are combined [Figure 4-63(b)] note that AB is common to all terms. The common term for the horizontal four adjacent squares is CD; therefore, Map = AB + CD. The simplified logic circuit is shown in Figure 4-64.

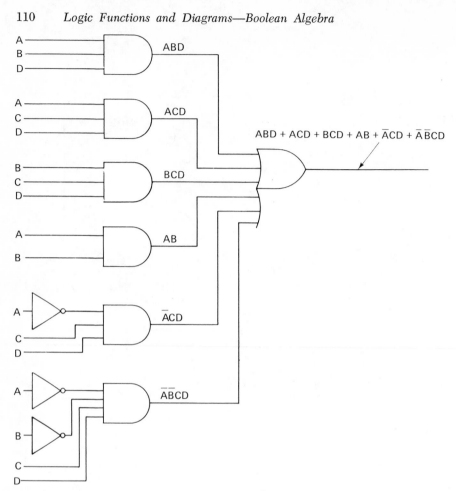

Figure 4-62.   Circuit simplification problem.

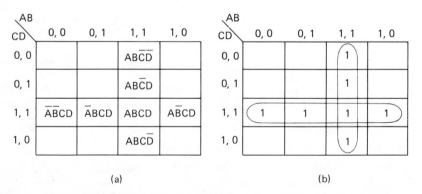

(a)   (b)

Figure 4-63.   Karnaugh map solution for the logic circuit in Figure 4-62: (a) variables placed in appropriate squares; (b) four adjacent squares rule applied.

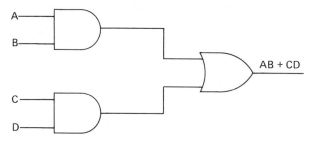

Figure 4-64.   Simplified circuit of Figure 4-62

QUESTIONS AND PROBLEMS—CHAPTER 4

1. What are the three basic logic operations?
2. How many inputs may an AND gate have?
3. How many inputs may an OR gate have?
4. How many inputs may an EXCLUSIVE-OR gate have?
5. Draw the logical symbol for a 4-input AND gate.
6. Draw the logical symbol for a 5-input OR gate.

Construct a truth table for the Boolean statements in problems 7 through 12. (Use 0s and 1s.)

7. $X = BC\overline{D}$
8. $X = BC + BD + CD$
9. $X = \overline{ABC}$
10. $X = C(\overline{B + D})$
11. $X = (A + \overline{A}B) + (A + B)$
12. $X = (A + BC) + (A + B)(A + C)$

Draw first-level logic diagrams to represent the expressions in problems 13 through 18.

13. $W = A\overline{B}$
14. $Z = \overline{A}B + C$
15. $Y = \overline{A} + BC$
16. $R = AC + BC$
17. $X = \overline{B} + C + ABC$
18. $T = \overline{\overline{A} + \overline{B}} + C$

19. Write the Boolean expression for the logic diagrams in Figure 4-65.

20. Write the Boolean expression for the logic diagrams in Figure 4-66.

21. In your own words, define each of the following.
    (a) commutative property
    (b) distributive property
    (c) DeMorgan's theorem

22. By using Boolean algebra, prove that $A\overline{B} + \overline{A}B = (A + B)\overline{AB}$.

    Use Boolean algebra to simplify the statements in problems 23 through 31.

23. $A + AB$

24. $AC + BC$

25. $\overline{\overline{\overline{AB}}}$

26. $\overline{ABC} + D$

27. $A + C + AB + BC$

28. $AB + AC + \overline{A}C + \overline{AC}$

29. $XYZ(\overline{A}BD + ABD) + X(\overline{A}X + AB + A\overline{B})$

30. $(A + B\overline{C})(\overline{A} + B + \overline{C})$

31. $Y(\overline{AB\overline{C} + \overline{A}B\overline{C} + ABC})$

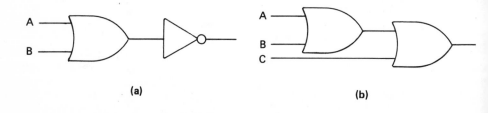

(a)   (b)

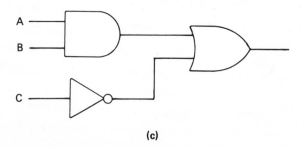

(c)

Figure 4-65.

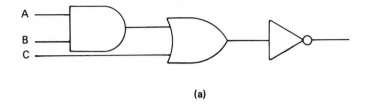

(a)

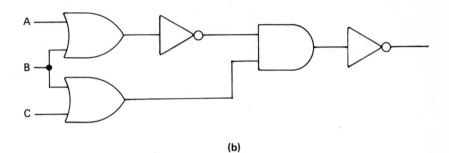

(b)

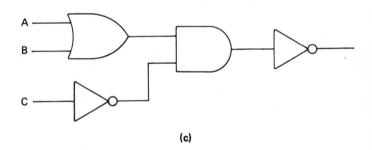

(c)

Figure 4-66.

Use Karnaugh maps to simplify the expressions in problems 32 through 35.

32. $AB + \overline{A}B + \overline{B}A + \overline{A}\overline{B}$

33. $A\overline{B}C + AB\overline{C} + \overline{A}B\overline{C}$

34. $AB\overline{C} + \overline{A}\overline{B}C + \overline{A}B\overline{C} + ABC$

35. $\overline{A}\overline{B}\overline{C}\overline{D} + A\overline{B}\overline{C}D + \overline{A}B\overline{C}D + A\overline{B}\overline{C}\overline{D} + A\overline{B}CD + A\overline{B}C\overline{D}$

5

Bipolar Gates

A *gate* is a device that has one output and one or more inputs. In digital circuits, the gate output is either HIGH or LOW, depending on the conditions present at the input. All digital circuits are made up of various combinations of logic gates. This chapter considers various kinds of bipolar gates. In later chapters, gates and inverters will be used as "building blocks" to form flip-flops, registers, counters, arithmetic units, etc.

The topics covered in this chapter are:

5.1 SWITCHING MODES

Bipolar digital ICs can be categorized into three technological group-
ings: (1) current sourcing, (2) current sinking, and (3) current mode
switching. These terms relate to the input conditions necessary to activate
the gate.

5.1.1 Current Sourcing.   Consider the circuit in Figure 5-1. Like in any
common emitter configuration, Q_2 is an inverter. In order to activate the
function (turn on the transistor), a current must be supplied to the base
of the transistor. This may be accomplished by an identical circuit (Q_1)
in the off state. A source of current is required to turn *on* the transistor.
When Q_2 is off, its output will supply (source) current to the next circuit.
This mode of switching is called *current sourcing.*

5.1.2 Current sinking.   Some circuits use internal components to turn
on the transistor, as shown in Figure 5-2. In circuit A, R_1 supplies the
required current to turn Q_1 on. Without an input connection, Q_1 is on,
and its output will be LOW level. This condition causes the input voltage
to circuit B to approach 0 volts, virtually shorting out its base current.
This shorting path can be described as a "sink" for the base current of
circuit B. This method of switching is called *current sinking.*

5.1.3 Current mode.   In current mode switching, the direction of the
current is switched, as shown in Figure 5-3. If V_{in} is greater than V_{BB},
Q_1 is *on* and Q_2 is *off*. This condition provides a HIGH level output voltage. If V_{in} is less than V_{BB}, Q_1 is *off* and Q_2 is *on*. This state provides a

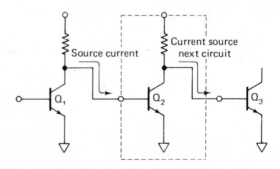

Figure 5-1.   Current sourcing.

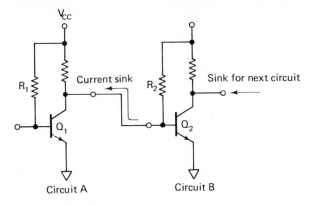

Figure 5-2.   Current sinking.

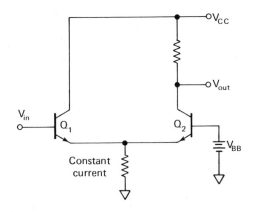

Figure 5-3.   Current mode switching.

LOW level output. Note that in this circuit, the total current is constant and only the direction of the current is switched; that is, the current path is through either Q_1 or Q_2, depending on the relationship between V_{in} and V_{BB}.

5.2   THE RTL FAMILY OF LOGIC GATES

The resistor-transistor logic (RTL) family was one of the first widely used integrated circuits. It consists only of resistors and transistors, thus its name—RTL. Although it is being displaced by other families, it is still available and currently in use.

The RTL gate performs the NOR function. Its symbol and truth table appear in Figure 5-4. The symbol is interpreted—"If A is HIGH

A	B	X
H	H	L
L	H	L
H	L	L
L	L	H

(a)   (b)

Figure 5-4.   RTL gate: (a) symbol; (b) truth table.

level, OR B is HIGH level, OR both are HIGH level, the output will be LOW level."

The basic circuit for the RTL gate appears in Figure 5-5. The resistor values have been standardized throughout the industry. The specified supply voltage (V_{CC}) is 3.6 volts. This circuit is an example of "current sourcing" because it requires an input current to turn on either transistor. If either transistor is *on* OR both transistors are *on*, the output voltage will be LOW level. LOW level is defined by the specifications as less than 0.4 volts. However, this voltage is usually about 0.2 volts, which is the saturation voltage of the on transistor. If both transistors are *off*, the output voltage will be HIGH level and will vary between V_{CC} of 3.6 volts and approximately 1 volt, depending on the load current. A *logical 1* is defined as a minimum HIGH level voltage of approximately 1 volt. A *logical 0* is defined as LOW level; it is specified as being less than 0.4 volts.

The load current is dependent upon the number of gate inputs that are driven by a gate output; that number is called the *fan-out*. As the number of driven gates increases, the output current increases, causing the output voltage level to decrease. Figure 5-6 shows the voltage transfer characteristics when one gate input is driven (fan-out = 1).

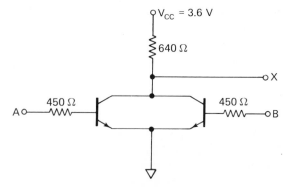

Figure 5-5.   Basic circuit for an RTL gate.

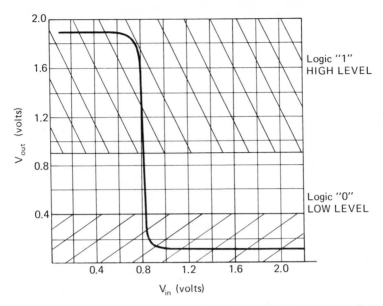

Figure 5-6. RTL transfer characteristics when the fan-out equals 1.

Figure 5-7 shows the voltage transfer characteristics when five gate inputs are driven (fan-out = 5). Note that the voltage level drops to nearly 1 volt with a fan-out of 5. Manufacturers specify that the maximum

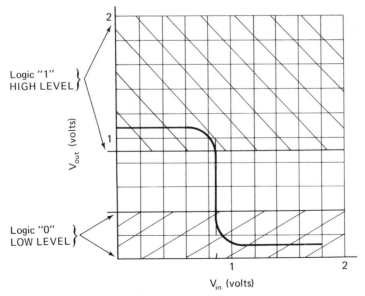

Figure 5-7. RTL transfer characteristics when the fan-out equals 5.

fan-out for RTL gates is 5. The number of inputs may be increased by adding more transistors, including base resistors, in parallel with existing transistors.

Figure 5-8 shows the RTL 9915 dual 3-input NOR gate. Component values were selected, based on the following criteria: (1) power dissipation, (2) speed, and (3) factors involving interconnection of circuits. In order that a fast switching speed be maintained, the resistance values selected should be quite low. This maintains short RC time constants. The low value of the base resistor tends to provide a high conductance path for the removal of stored charges in the base region. In order to keep currents and power dissipation at a minimum, low voltage power supplies ($V_{CC} = 3.6$ V) are used. Only NOR gates exist in the RTL family. However, if the DeMorgan's equivalent is used, as shown in Figure 5-9, all (AND, OR, and NOT) functions are available.

Figure 5-10 is a 9927 quad inverter. Note that this circuit is identical to the NOR gate, the only exception being that there is only one input.

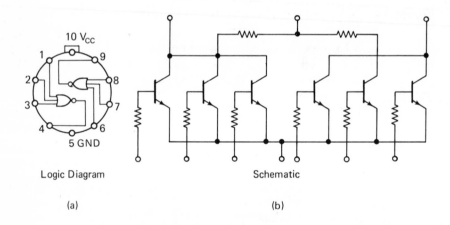

Logic Diagram (a)   Schematic (b)

Figure 5-8.   RTL 9915 dual 3-input NOR gate: (a) logic symbol; (b) logic diagram. (Courtesy of Fairchild Semiconductor.)

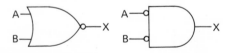

Figure 5-9.   NOR gate and its DeMorgan's equivalent symbols.

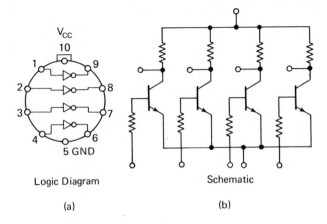

Logic Diagram   Schematic

(a)   (b)

Figure 5-10. RTL 9927 quad inverter: (a) logic symbol; (b) logic diagram. (Courtesy of Fairchild Semiconductor.)

5.3 THE DTL FAMILY OF LOGIC GATES

Another logic circuit used in ICs employs diodes and transistors; it is thus called *diode-transistor logic* (DTL). The DTL gate performs the NAND logic function. The NAND symbol, along with its DeMorgan's equivalent and truth table, are illustrated in Figure 5-11.

A basic circuit for the DTL NAND gate appears in Figure 5-12. This circuit belongs in the "current sinking" category because R_1 provides the necessary current for the base to turn *on* the transistor. If inputs A and B are connected to a HIGH level, D_1 and D_2 will be reverse-biased while D_3 and D_4 will be forward-biased. With D_3 and D_4 forward-biased, Q_1 is *on*, resulting in a LOW level output. This condition describes the

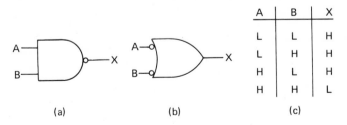

A	B	X
L	L	H
L	H	H
H	L	H
H	H	L

(a)   (b)   (c)

Figure 5-11. NAND and its equivalent symbols: (a) NAND; (b) equivalent; (c) truth table.

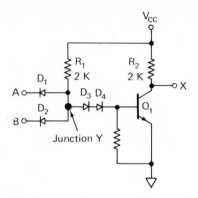

Figure 5-12.   Basic circuit for the DTL NAND gate.

NAND function of the gate—"If A AND B are HIGH level, then the output will be LOW level."

DeMorgan's equivalent function occurs when A is LOW level, OR B is LOW level, OR BOTH are LOW level. Referring to Figure 5-12, note that when either A OR B are LOW (0.4 volts or less), the potential at junction Y will be approximately 1 volt. This potential is the sum of the input voltage and the junction drop (0.6 volts) across D_1 and D_2. However, a potential at junction Y of at least 1.8 volts is needed to keep Q_1 *on* (0.6 volts for each diode D_3 and D_4 and 0.6 volts for the emitter-base junction of Q_1). Therefore, Q_1 will be turned *off* when A OR B OR BOTH are LOW level and the DeMorgan's equivalent function is realized—"If A OR B are LOW level, then the output will be HIGH level."

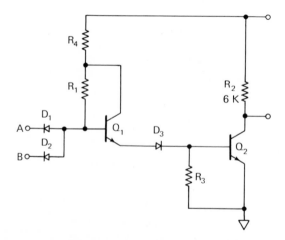

Figure 5-13.   Standard DTL circuit.

The circuit in Figure 5-13 is the diagram of a DTL NAND gate, which has been standardized by many manufacturers. The outstanding feature of this circuit is that one of the diodes (D_3) has been replaced with a transistor whose collector is returned to the junction of R_1 and R_4. This increases the current gain of the circuit and results in an increased input impedance and a corresponding reduction in sinking current requirements. The off-set transistor Q_1 provides the drive current required to turn Q_2 *on*. The biasing arrangement of Q_1 makes it always active; thus, turn-on and turn-off times are reduced. Although Q_2 is completely saturated, R_3 is provided to remove any possible stored charge. The voltage transfer characteristics of the basic DTL gate are shown in Figure 5-14(a) and (b).

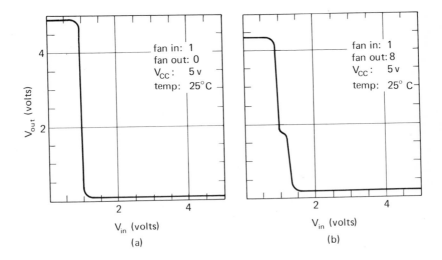

Figure 5-14. Voltage transistor characteristic of basic DTL gates: (a) with fan-out of 0; (b) with fan-out of 8.

5.4   THE TTL FAMILY OF LOGIC GATES

The most widely used family of logic is the transistor-transistor family (TTL). A unique characteristic of this gate is the multi-emitter input.

The basic TTL gate performs the NAND function, like the DTL gate shown in Figure 5-11. The TTL gate also belongs in the current sinking category. Let's examine the TTL gate in Figure 5-15. Note that Q_2 obtains its base current from the multi-emitter transistor Q_1. If both emitters of Q_1 are HIGH, both emitter-base junctions are reverse-biased.

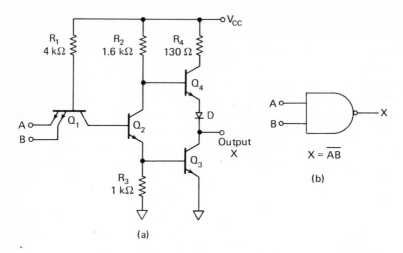

(a)

(b)

$X = \overline{AB}$

Figure 5-15. Standard TTL 2-input NAND gate: (a) schematic diagram; (b) logic symbol.

However, the collector-base junction is forward-biased, and the base of Q_2 is "current-sourced" through R_1, turning Q_2 *on*. Q_2 turns *on* Q_3 and turns *off* Q_4, producing a LOW level output. The NAND function is realized under these conditions: "If ALL inputs are HIGH, the output will be LOW."

The DeMorgan's equivalent function occurs when any one or more of the inputs are LOW level. Any LOW level input provides emitter current for Q_1 and Q_1 saturates, turning *off* Q_2. If Q_2 is *off*, Q_4 turns *on* and Q_3 turns *off*. This results in a HIGH level output—"If one or more inputs are LOW, the output will be HIGH."

The switching speed is improved in three ways. (1) When Q_1 is turned *on*, it provides a very low impedance to the base of Q_2. This low impedance allows fast removal of the stored base charge of Q_2, resulting in a much faster turn-*off* time. (2) The multi-emitter transistor replaces the combinations of diodes and resistors used in other logic circuits. Therefore, the geometrical size of the TTL gate is much smaller. This reduces parasitic capacitance, and greater switching speeds result. Smaller size means lower costs, more functions, or both per given IC. (3) The output characteristics are improved through the use of an active pull-up circuit. Q_3 and Q_4 form a circuit referred to as a *totem pole*. The advantage of this circuit is that it provides a low-driving source impedance. This permits high-capacitance loads to be driven without serious degradation of switching time.

5.5   *THE ECL FAMILY OF LOGIC GATES*

The mode of switching previously discussed relies on saturating a common-emitter transistor. A major problem with this mode of switching is that the turn-off time is relatively long due to the stored charge in the base region. When a transistor is driven into saturation, the base region is saturated with minority carriers. The transistor will not turn *off* until these charges have been removed.

In emitter-coupled logic (ECL), the problem of stored charge does not exist because none of the transistors are saturated. A basic ECL circuit is shown in Figure 5-16.

ECL employs current mode switching, whereby the direction of the current, rather than its magnitude, is switched. V_{BB} and R_1 establish the amount of current to be switched:

$$I_1 = \frac{V_{BB} - V_{BE3}}{R_1}$$

The OR function is produced when input A OR input B is HIGH. The result is ECL gates perform the OR function as described in Figure 5-17. The OR function is produced when input A OR input B is HIGH. The result is that all of I_1 is switched away from Q_3 causing the collector voltage of Q_3 to go HIGH. Q_4 is provided to increase the current gain and improve the output driving capabilities. Since the common collector configuration does not invert, the output is also HIGH. According to the OR symbol—"If A is HIGH OR if B is HIGH, the output will be HIGH."

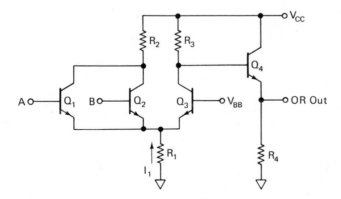

Figure 5-16.   Basic ECL circuit.

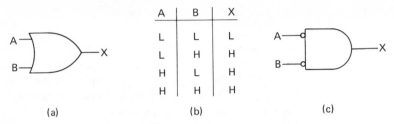

(a)   (b)   (c)

Figure 5-17.   Symbols and truth table for basic ECL gate: (a) OR symbol; (b) truth table; (c) equivalent symbol.

When A AND B are LOW, Q_1 and Q_2 are turned *off*. This causes Q_3 to turn *on* and conduct all of I_1, resulting in a LOW output. This describes the DeMorgan's equivalent function—"If all inputs are LOW, the output will be LOW."

Most ECL gates perform both the OR function and the NOR function, as shown in Figure 5-18. The NOR function is accomplished with a transistor (Q_5) connected in the common collector configuration.

There are two primary advantages of ECL: (1) faster switching time because none of the transistors are allowed to saturate; and (2) switching transients on the supply line are minimized due to the current mode switching.

An example of a recently introduced ECL circuit is the Fairchild 9500, shown in Figure 5-19. It uses the same configuration as the basic ECL circuit—with the addition of temperature-compensated internal reference and current source networks. The voltage transfer characteristics for the Fairchild 9500 are shown in Figure 5-20. The outstanding characteristic of the 9500 ECL family is a propagation delay of approximately 2 nsec.

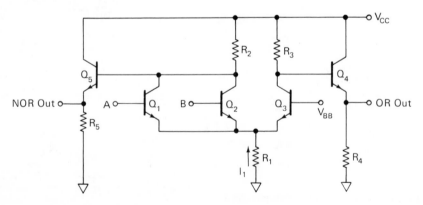

Figure 5-18.   Basic ECL circuit with NOR output.

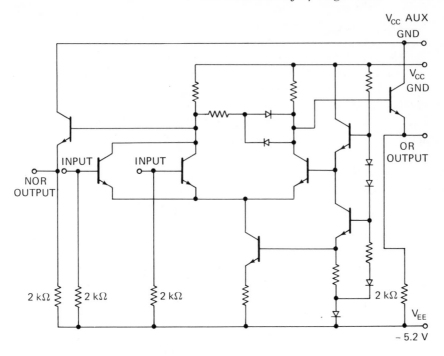

Figure 5-19. Fairchild 9500 series ECL standard gate. (Courtesy of Fairchild Semiconductor.)

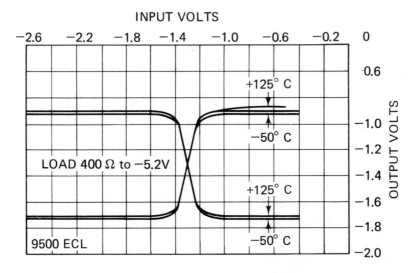

Figure 5-20. Voltage transfer characteristics for the Fairchild 9500 ECL gate. (Courtesy of Fairchild Semiconductor.)

5.6 SPECIFICATIONS

In their specification sheets, manufacturers guarantee that a device will operate within specifications over its entire temperature range and supply voltage range. These specs are usually expressed in "worst-case" terms: the specifications indicate how the worst-case tests were made. Worst-case testing provides a built-in margin of safety. In the authors' opinion, the most important specifications are:

1. A definition of logic voltage levels
2. Loading rules
3. Noise immunity
4. Propagation delay

Because transistor-transistor logic is currently the most popular logic family, TTL specifications have been selected for discussion. The student should realize that specifications will differ for other logic families.

5.7 LOGIC VOLTAGE LEVELS

Assuming that HIGH level is equal to logical 1 and LOW level is equal to logical 0, note that the following definitions exist:

V_{IL} is the voltage level required for a logical 0 at an input. For the TTL 7400 series, it is a guaranteed maximum of 0.8 volts.

V_{IH} is the voltage level required for a logical 1 at an input. For the TTL 7400 series, it is a guaranteed minimum of 2 volts.

V_{OL} is the voltage level output from an output in the logical 0 state. For the TTL 7400 series, it is a guaranteed maximum of 0.4 volts.

PARAMETER	DEFINITION	9300 9N/54.74 9H/54H.74H	93L00 9L00	93S00 9S/74S	9S/94S
V_{OH}	Minimum Output Voltage in the HIGH State	2.4 V	2.4 V	2.5 V	2.2 V
V_{OL}	Maximum Output Voltage in the LOW State	0.4 V	0.3 V	0.6 V	0.5 V
V_{IH}	Minimum Voltage Level Which is Guaranteed to be interpreted as a HIGH at the input	2.0 V	2.0 V	2.0 V	2.0 V
V_{IL}	Maximum Voltage Level Which is Guaranteed to be interpreted as a LOW at the input	0.8 V	0.7 V	0.8 V	0.8 V

Figure 5-21.   Worst-case DC logic levels for Fairchild TTL devices. (Courtesy of Fairchild Semiconductor.)

V_{OH} is the voltage level output from an output in the logical 1 state. For the TTL 7400 series, it is a guaranteed minimum of 2.4 volts.

Figure 5-21 lists the voltage level definitions for all the TTL families manufactured by Fairchild Semiconductor.

5.8 LOADING RULES

In order to simplify designing with TTL devices, certain rules have been devised. These rules are based on the normalization of input loading and output drive factors. For the TTL 7400 series:

1 unit TTL Load (U.L.) = 40 μA in the HIGH state

1.6 mA in the LOW state

The driving capabilities of a gate may then be expressed in terms of the number of loads it can drive. Figure 5-22 lists the input and output loading factors for all of the TTL families manufactured by Fairchild Semiconductor.

The fan-out drive capability of a TTL device reflects its ability to sink current in the output LOW state and to source or drive current in the output HIGH state. Referring to the diagram in Figure 5-23, note that the sinking current (I_{OL}) is defined as the maximum current into the driving gate in the LOW state. This is the sum of the currents from the inputs of the driven gates.

The sinking current required to drive one standard gate to a LOW state is 1.6 mA and is called I_{IL}. The standard TTL gate has a sinking current capability (I_{OL}) of 16 mA and requires 1.6 mA to drive its input to a LOW state. Therefore, one standard gate has the ability to drive ten standard gate inputs to the LOW state.

FAMILY	INPUT LOAD		OUTPUT DRIVE	
	HIGH	LOW	HIGH	LOW
9000	1 U.L.	1 U.L.	20 U.L.	10 U.L.
9H00/74H00	1.25 U.L.	1.25 U.L.	25 U.L.	12.5 U.L.
9L00	0.5 U.L.	0.25 U.L.	10 U.L.	2.5 U.L.
9N00/7400	1 U.L.	1 U.L.	20 U.L.	10 U.L.
9S00/74S00	1.25 U.L.	1.25 U.L.	25 U.L.	12.5 U.L.

Figure 5-22. Loading factors for Fairchild TTL devices. (Courtesy of Fairchild Semiconductor.)

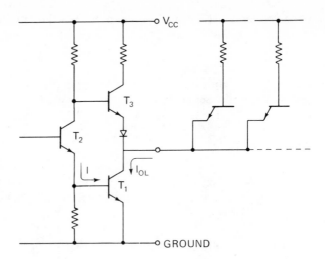

Figure 5-23.   TTL output circuit in LOW state. (Courtesy of Fairchild Semiconductor.)

In order to overcome leakage current between the emitter and base of the input device and stray capacitance, the input must be *driven* to the HIGH state. This is accomplished by the active pull-up transistor at the output of the driving gate (Q_3 in Figure 5-24); and it assures a fast

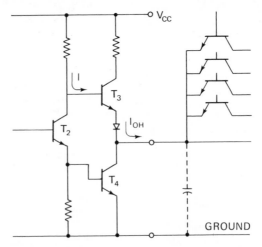

Figure 5-24.   TTL output circuit in HIGH state. (Courtesy of Fairchild Semiconductor.)

transition from the LOW to the HIGH state. Each input of a standard gate requires a maximum of 40 μA (I_{IH}) to achieve this transition. The circuit in Figure 5-24 shows a standard gate in the HIGH state.

The active HIGH drive current (I_{OH}) for the standard gate is 800 μA. Therefore, the output of a standard gate has the ability to drive 20 standard inputs. Although the static I_{IH} requirements of most circuits is about 40 μA, about 35 mA is made available at the instant of LOW to HIGH output transition to charge up the stray capacitances that appear between gates.

Example—Input Load:

1. A 9N00/7400 gate, which has a maximum I_{IL} of 1.6 mA and an I_{IH} of 40 μA, is specified as having an input load factor of 1 U.L. (also called a fan-in of 1).

2. The 93H72, which has a value of $I_{IL} = 3.2$ mA and an I_{IH} of 80 μA on the CP terminal, is specified as having an input load factor of $\dfrac{3.2\text{ mA}}{1.6\text{ mA}}$ or 2 U.L.

Example—Output Drive:

The output of the 9N00/7400 will sink 16 mA in the LOW (logic 0) state and source 800 μA in the HIGH (logic 1) state. The normalized output LOW drive factor is (16 mA/1.6 mA) = 10 U.L., and the output HIGH drive factor is (800 μA/40 μA) = 20 U.L.

Limits imposed by the loading rules are based on worst-case conditions. Values for MSI devices vary significantly. Consult the appropriate data sheets for actual characteristics. Figure 5-25 shows a TTL/MSI 9304 and its loading rules.

5.9 WIRED-OR APPLICATIONS

Certain TTL devices, such as the 7403 shown in Figure 5-26, are provided with an "open" collector output to permit the wired-OR (actually, the wired-AND) function. This is achieved by connecting open collector outputs together and adding an external pull-up resistor. The value of the pull-up resistor is determined by considering the fan-out of the OR tie and the number of devices in the OR tie.

LOGIC SYMBOL

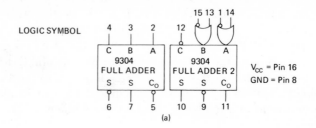

V_{CC} = Pin 16
GND = Pin 8

(a)

LOGIC DIAGRAM

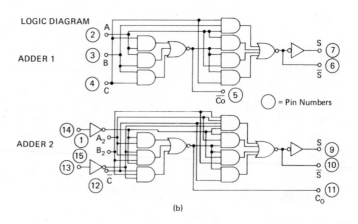

◯ = Pin Numbers

(b)

PIN NAMES		LOADING (Note a)
FULL ADDER 1		
A, B	Operand Inputs	4 U.L.
C	Carry Input	4 U.L.
S	Sum Output (Note c)	10 U.L.
$\overline{S}$	Complementary Sum Output (Note c)	10 U.L.
$\overline{C}_O$	Carry (Active LOW) Output (Note b)	7 U.L.
FULL ADDER 2		
A_1, B_1	OR Operand (Active HIGH) Input	1 U.L.
$\overline{A}_2, \overline{B}_2$	OR Operand (Active LOW) Input	4 U.L.
$\overline{C}$	Carry (Active LOW) Input	4 U.L.
S	Sum Output (Note c)	10 U.L.
$\overline{S}$	Complementary Sum Output (Note c)	10 U.L.
C_O	Carry (Active HIGH) Output (Note b)	7 U.L.

NOTES:
  a. 1 Unit Load (U.L.) = 40 μA HIGH/1.6 mA LOW.

  b. 7 U.L. is the output LOW drive factor and 14 U.L. is the output HIGH drive factor.

  c. 10 U.L. is the output LOW drive factor and 20 U.L. is the output HIGH drive factor.

(c)

Figure 5-25. TTL/MSI 9304 dual full adder: (a) logic symbol; (b) logic diagram; (c) loading factors. (Courtesy of Fairchild Semiconductor.)

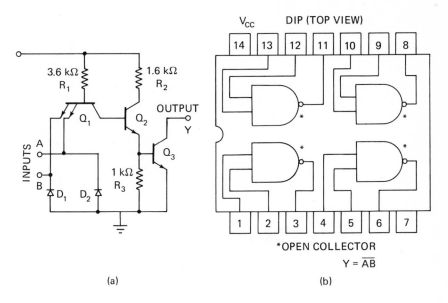

Figure 5-26. TTL/SSI 7403 quad 2-input NAND gate with open collector output: (a) schematic diagram; (b) logic diagram. (Courtesy of Fairchild Semiconductor.)

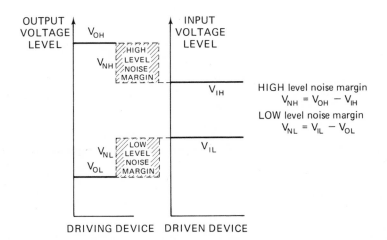

Figure 5-27. Graph defining noise margin. (Courtesy of Fairchild Semiconductor.)

5.10   NOISE MARGIN

Noise margin is defined in Figure 5-27 as the difference between the worst-case output voltage and the worst-case input voltage. Notice that there is a defined noise margin for the HIGH state (V_{NH}) and another for the LOW state (V_{NL}). The noise margin allows for variations in logic levels due to switching transients and/or voltage drops on lines that interconnect gates. The voltage transfer characteristics for a standard TTL gate are shown in Figure 5-28. Notice how the noise margin guarantees are obtained.

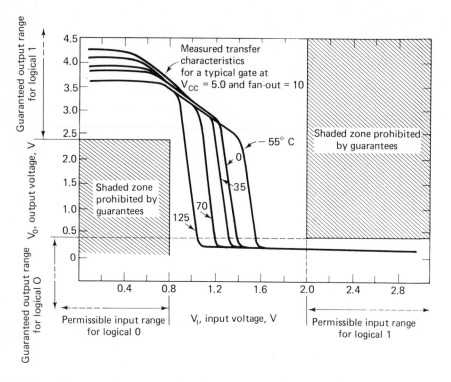

Figure 5-28.   Voltage transfer characteristic for a standard TTL 7400 gate. (Courtesy of Texas Instruments, Inc.)

5.11   UNUSED INPUTS

To minimize noise sensitivity and optimize switching times, unused inputs of all TTL circuits should be held between 2.4 volts and the absolute maximum of 5.5 volts. This eliminates the effect of distributed capacitance associated with the floating input and insures that no degradation will occur in the switching times. This may be accomplished by providing a pull-up resistor for these unused inputs or by tying them to the outputs of unused gates that are in the HIGH state.

5.12 PROPAGATION DELAY

Switching speeds in logic gates are usually expressed in terms of propagation delay times (t_{PD}) rather than rise and fall times. This is because it is important to know how long it takes to produce a function through a gate or series of gates. Two parameters are usually provided in specification sheets: (1) the propagation delay time t_{PHL} is the time required for a transition from the HIGH state to the LOW state; and (2) the delay time required for a transition from the LOW state to the HIGH state is called t_{PLH}. These output transitions are measured with respect to an input pulse. The manufacturer also specifies test conditions such as V_{CC}, temperature, loading factors, capacitance, etc. Figure 5-29 shows a test

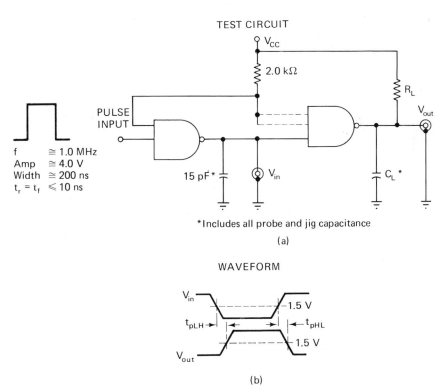

SWITCHING CHARACTERISTICS

TEST CIRCUIT

*Includes all probe and jig capacitance

(a)

WAVEFORM

(b)

Figure 5-29. Test circuit used to measure propagation delay: (a) test circuit; (b) waveform. (Courtesy of Fairchild Semiconductor.)

setup used by Fairchild Semiconductor to test the propagation delay of 9002 NAND gate. Average propagation delay (t_{PD}) is the average of t_{PLH} and t_{PHL} and is more generally used because every gate inverts. For example, the propagation delay for the two gates shown in Figure 5-30 is the sum of t_{PHL} and t_{PLH}.

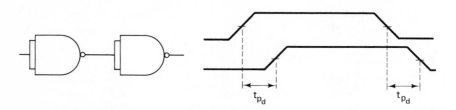

Figure 5-30.   Propagation delay for two TTL gates.

QUESTIONS AND PROBLEMS—CHAPTER 5

1. Explain the three different switching modes.

2. In Figure 5-2, when Q_1 is ON, what is the state of the output of Q_2?

3. In Figure 5-3, if V_{in} is greater than V_{BB}, through which transistor is the current path? In this situation, what is the output level?

4. In Figure 5-5, if input A is HIGH level, what is the output level (X)?

5. What switching mode do RTL gates utilize?

6. Define: (a) *fan-out;* (b) *logical 0;* (c) *logical 1.*

7. State the relative differences between RTL, DTL, and TTL, in terms of fan-out, switching speeds, power requirements, etc.

8. Which logic function is performed by: (a) DTL gates; (b) TTL gates; and (c) ECL gates?

9. Which switching mode is utilized by TTL gates?

10. Which logic function is performed by RTL gates?

11. In Figure 5-15, if input A is HIGH level, what is the output level?

12. In Figure 5-16, what condition(s) would be required to produce a LOW level output?

13. Which switching mode is used in ECL?

14. Define *propagation delay* and *noise margin.*

15. Draw the DeMorgan's equivalent circuit and construct the truth table for a 4-input NOR gate.

16. Draw the DeMorgan's equivalent circuit and construct the truth table for a 4-input NAND gate.

17. In Figure 5-13, does: (a) Q_1 saturate? (b) Q_2 saturate?

18. What is the advantage of an non-saturating transistor?

19. Of all the bipolar gates discussed in this chapter, which family is the fastest? Why?

20. Is it possible to wire-OR the outputs of:
    (a) Standard TTL gates? Why?
    (b) Open collector TTL gates? Why?

21. Using the waveforms in Figure 5-31, draw the waveforms for the output of the gate in Figure 5-32.

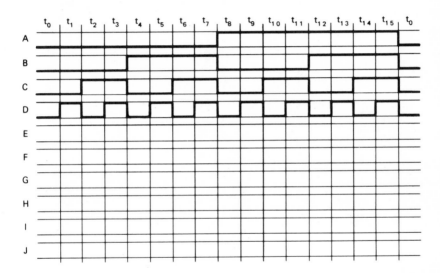

Figure 5-31. Waveforms for problems 21 through 26.

Figure 5-32.

22. Using the waveforms in Figure 5-31, draw the waveforms for the output of the gate in Figure 5-33.

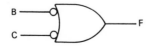

Figure 5-33.

23. Using the waveforms in Figure 5-31, draw the waveforms for the output of the gate in Figure 5-34.

Figure 5-34.

24. Using the waveforms in Figure 5-31, draw the waveforms for the output of the gate in Figure 5-35.

Figure 5-35.

25. Using the waveforms in Figure 5-31, draw the waveforms for the output (I) of the logic circuit in Figure 5-36.

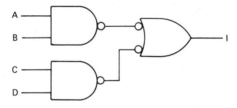

Figure 5-36.

26. Using the waveforms in Figure 5-31, draw the waveforms for the output (J) of the logic circuit in Figure 5-37.

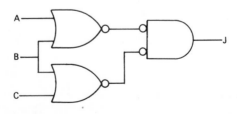

Figure 5-37.

6

MOS Logic Gates

Metal oxide silicon (MOS) devices offer the highest complexity of large-scale integrated circuits (LSI) because they are simple to fabricate, are small, and consume very little power. Approximately one-third of the process steps needed for standard bipolar ICs are required for MOS devices. Their most significant feature, however, is the large number of circuit elements that can be put on a single chip. More than 6000 devices may be placed on a chip of silicon only 150×150 mils. A MOS transistor requires approximately 1 square mil of chip area, as compared with roughly 50 square mils for a bipolar transistor. An even greater saving in chip area is achieved by the use of MOS transistors in place of the load resistor. In bipolar integrated circuits, the greatest part of chip area is for passive resistance elements.

The topics covered in this chapter are:

6.1   BASIC MOS INVERTER

A basic premise of MOS logic circuits is that all load resistances are provided by metal oxide silicon transistors (MOSFETs). Figure 6-1(a)
shows the basic MOS inverter, with Q_1 as the load for the active switch

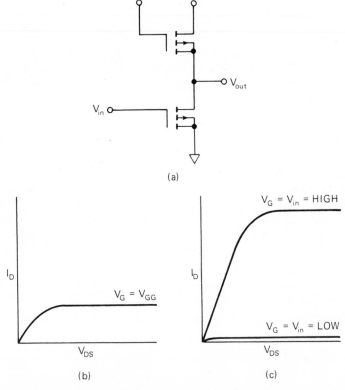

Figure 6-1.   Basic MOS inverter: (a) schematic diagram; (b) Q_1 drain
characteristics; (c) Q_2 drain characteristics.

Q_2. The gate of Q_1 is connected to a supply voltage (V_{GG}), which is somewhat less than V_{DD}. Because P channel devices are employed in this circuit, negative supply potentials are used. The purpose of V_{GG} is to lower
the transconductance (G_{FS}) of Q_1 as compared to the transconductance
of Q_2, as shown in Figure 6-1(b) and (c). The lower G_{FS} of Q_1 means
that the DC resistance of Q_1 is many times greater than that of Q_2.

You will recall that for a load line analysis, the active device characteristics are plotted as a positive slope and the load characteristics as
a negative slope, as illustrated in Figure 6-2. The drain characteristics of
Q_2 are plotted against the load characteristics of Q_1 in Figure 6-3. Note

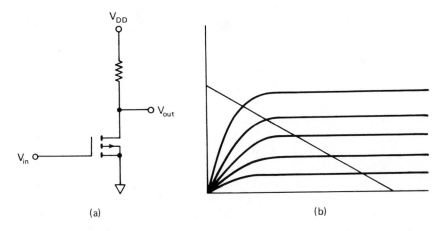

(a)   (b)

Figure 6-2.   MOSFET: (a) schematic diagram; (b) drain and load
characteristics.

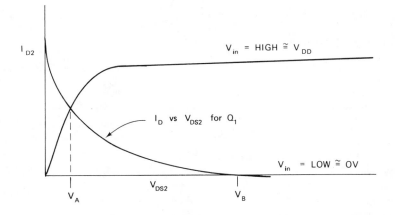

Figure 6-3.   Load characteristics of Q_1 versus drain characteristics of Q_2.

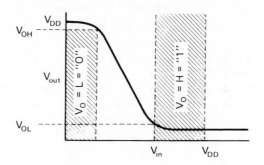

Figure 6-4.   Transfer characteristics of Q_1.

that because V_{GG} limits the current through Q_1, its characteristics tend to be those of a constant current load. Notice also that when V_{IN} is HIGH, the output voltage is LOW (V_A) and can be interpreted as the intercept of the two curves. When V_{IN} is LOW, Q_1 is turned off and the current is reduced to zero (V_B). The maximum current can be adjusted by selecting the value of V_{GG} on the load transistor Q_1.

Because V_{GG} limits the current, the load characteristics tend to be constant current. This improves the transfer characteristics and, in particular, the noise immunity as shown in Figure 6-4. The ideal switchover characteristic on the transfer curve occurs when V_{IN} equals approximately 50 percent of V_{DD}. In some of the latest MOS ICs, the gate supply V_{GG} has been eliminated. This is accomplished by changing the geometry of Q_1 so that its G_{FS} is much less than that of Q_2. When this is done, the gate of Q_1 is internally tied to V_{DD}.

6.2   LOGIC LEVELS

At this point, it is necessary to discuss logic levels. What is meant by HIGH level? Does it mean the most positive voltage or the voltage level furthest from zero? MIL-STD-806C allows for either definition. In a circuit using minus supply voltages, confusion can exist in determining the function of logic circuits. For example, some manufacturers state that the more negative potential is LOW level and equals a logical 1, and the more positive potential, which might be 0 volts, is HIGH level and equals a logical 0. This definition is inconsistent when applied to the logic symbols as we have studied them. Many manufacturers agree on the following definition, which is the one we will use in this text:

> *HIGH level* is defined as the greatest potential, whether positive or negative, from ground potential. *LOW level* is the potential nearest ground, whether positive or negative.

6.3 NOR GATE

If another transistor is added in parallel with Q_2 in the basic inverter circuit, the NOR function can be realized. Figure 6-5 shows a two-input NOR gate, its symbol, and its truth table. If A OR B is HIGH level, Q_2 OR Q_3 is turned on, and the output voltage is LOW level (0 volts). The equivalent function occurs when both A AND B are LOW level. This turns off both Q_2 AND Q_3, resulting in a HIGH level output.

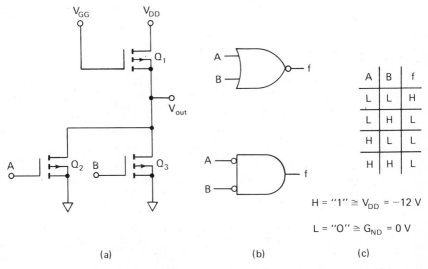

A	B	f
L	L	H
L	H	L
H	L	L
H	H	L

$$H = \text{"1"} \cong V_{DD} = -12 \text{ V}$$

$$L = \text{"0"} \cong G_{ND} = 0 \text{ V}$$

(a)　　　　　(b)　　　　　(c)

Figure 6-5. Two-input NOR gate: (a) schematic diagram; (b) symbols; (c) truth table.

6.4 NAND GATE

The basic inverter circuit can be modified to produce the NAND function by placing an additional transistor (Q_3) in series with Q_2, as shown in Figure 6-6. If A AND B are HIGH level (−12 volts), Q_2 AND Q_3 are turned on and the output voltage is LOW level (0 volts). The equivalent function occurs when A OR B is LOW level. If either Q_2 OR Q_3 is turned off, the resulting current is zero and the output will be HIGH level.

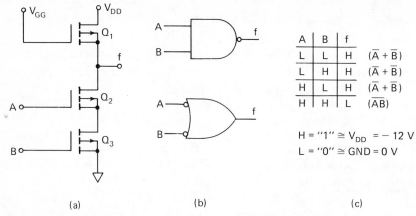

Figure 6-6. Two-input NAND gate: (a) schematic diagram; (b) symbols; (c) truth table.

6.5 DYNAMIC LOGIC GATES

Low power dissipation is a major advantage of MOS circuitry. However, low power dissipation requires low current levels, which in turn can be interpreted as high resistance circuits. High resistances, when coupled with interelectrode capacitances, mean long time constants and result in slower switching speeds. One method of limiting power requirements and still maintaining reasonably high switching speeds is the use of *dynamic logic*. The major difference between static logic and dynamic logic is that the latter samples data periodically. Figure 6-7 shows the conventional static MOS inverter. In this circuit, Q_1 is biased ON and serves as

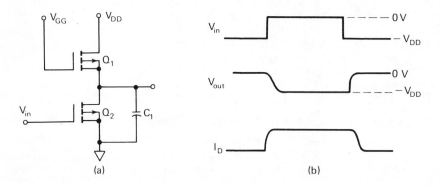

Figure 6-7. Ratio inverter: (a) schematic diagram; (b) input and output waveforms.

a fixed load resistance. Because of the value of V_{GG}, its resistance is many times that of the ON-resistance of Q_2 so that the output voltage can approach zero when Q_2 is turned ON. Because of the high resistance ratio between Q_1 and Q_2, it is called a *ratio inverter*.

When Q_2 is ON, the resistance ratio of Q_1 to Q_2 is high, resulting in a LOW level ouput. When Q_2 is OFF, the resistance ratio of Q_1 to Q_2 is low, resulting in a HIGH level output. C_1 is the output interelectrode capacitance of the inverter. When Q_2 is cut off, C_1 charges rather slowly through Q_1 because the resistance of Q_1 is relatively high. However, C_1 discharges more quickly when Q_2 is ON because of the lower resistance of Q_2.

Figure 6-8 is a dynamic two-phase ratioless inverter with associated waveforms. The gates of Q_1 and Q_3 are pulsed periodically so that their ON time is short, resulting in considerable power reduction. These pulses have two phases and are labeled ϕ_1 and ϕ_2. As shown in the waveforms, V_{IN} goes to 0 volts at t_0, turning OFF Q_2. Simultaneously, ϕ_1 goes negative, turning on Q_1 and causing C_1 to charge to V_{DD}. ϕ_2 turns ON Q_3 at t_1. This causes the charge on C_1 to be transferred to C_2. The charge transfer can occur without reduction of the output voltage because C_1 is made much larger than C_2.

Q_2 is turned ON at t_2 by V_{IN}, causing C_1 to partially discharge. The discharge does not influence C_2 at this time because Q_3 is turned OFF. When ϕ_1 turns Q_1 OFF at t_3, a partial charge still remains on C_1. However, because Q_2 is still held ON by V_{IN}, C_1 is discharged. Then, at t_4, ϕ_2 turns on Q_3, and C_2 discharges to ground through Q_2.

In one clock cycle (t_0 through t_4), drain current occurs only between t_2 and t_3 when Q_1 and Q_2 are ON simultaneously and when C_1

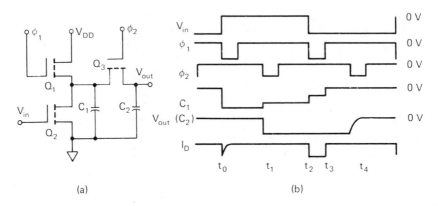

(a)   (b)

Figure 6-8. Two-phase ratioless inverter: (a) schematic diagram; (b) input and output waveforms.

charges at t_0. You will notice that drain current exists only when the ϕ_1 pulse is present. This circuit is called *ratioless* because the output voltage does not rely on the resistance ratio of Q_1 and Q_2.

6.6   DYNAMIC MOS GATES

Both NAND and NOR functions can be implemented from the basic dynamic inverter, as shown in Figure 6-9. The NOR function is produced by

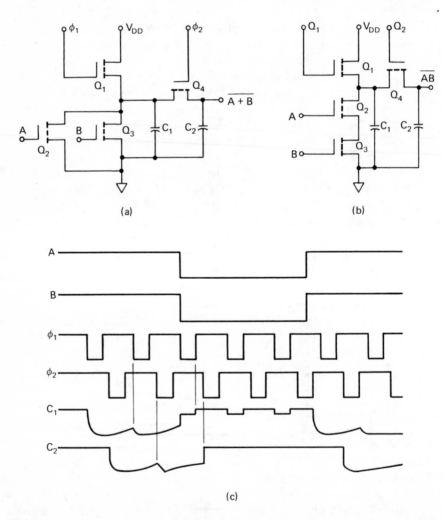

Figure 6-9.   Dynamic ratioless 2-phase gates: (a) NOR gate; (b) NAND gate; (c) waveforms.

adding Q_3 in parallel with Q_2. Because voltage levels are stored on capacitances, the clock pulses, ϕ_1 and ϕ_2, must be continually cycled in order to "refresh" these voltage levels.

The two-phase ratioless dynamic gate consumes DC power only for the duration of the ϕ_1 pulse. However, there is another dynamic circuit which dissipates no DC power at all. This circuit, called *ratioless-powerless*, dissipates only reactive power, which is provided by the ϕ_1 and ϕ_2 pulses. This circuit is illustrated in Figure 6-10. Note that no DC power supply is required.

To start, assume that at t_0, C_1 is charged, C_2 and C_3 are discharged, and V_{IN} is negative or in the 1 state. V_{IN} goes to 0 volts (0 state) at t_1.

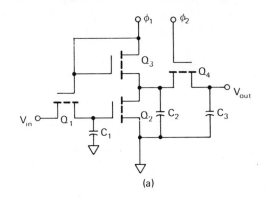

(a)

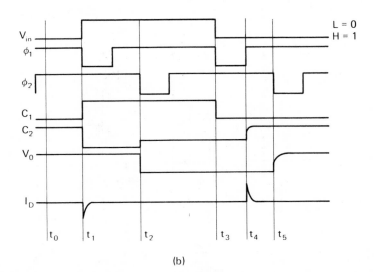

(b)

Figure 6-10. Dynamic 2-phase ratioless-powerless inverter: (a) schematic diagram; (b) waveforms.

Simultaneously, ϕ_1, appears to turn ON Q_1 and Q_3. This causes C_1 to discharge through Q_1, which turns OFF Q_2. Because Q_2 is OFF and Q_3 is ON, C_2 charges. This charge is transferred to C_3 at t_2 when ϕ_2 turns ON Q_4. At t_3 V_{IN} goes HIGH (1 state) and ϕ_1, appears, turning ON Q_1 and Q_3. This causes C_1 to charge which turns Q_2 ON. At this point, both Q_2 and Q_3 are ON, but because the source of Q_2 and the drain of Q_3 are at the same potential, C_2 cannot discharge.

Q_1 turns OFF at t_4, which is the end of the ϕ_1 pulse. At this time, C_1 remains charged, Q_3 is OFF, and Q_2 is ON. The result is that C_2 discharges through Q_2 to ground. Then, at t_5, Q_4 is turned on by ϕ_2, and C_3 also discharges through Q_2 to ground.

Other dynamic logic circuits exist requiring up to four clock pulse phases. Although dynamic logic circuits provide great savings in power dissipation, considerable complexity has been added to circuit design because of the two clock pulse phases. However, because of the low power dissipation, many circuits can be integrated on one chip.

6.7   MOS/BIPOLAR INTERFACE

A basic problem encountered when using MOS circuits with bipolar circuits involves the translation of voltage levels. For example, most MOS circuits use negative power supplies and voltage swings varying from —5 to —30 volts. (As yet, MOS circuits are not standardized throughout the industry.) Recall that TTL devices use positive supplies with swings from 0 to +5 volts.

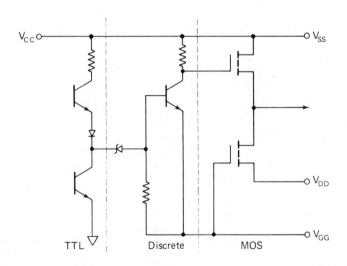

Figure 6-11.   TTL to MOS interface.

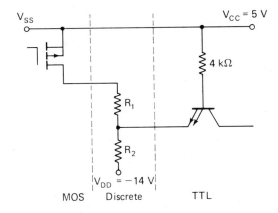

Figure 6-12. MOS to TTL interface.

Figure 6-11 shows how a TTL to MOS interface may be accomplished. The interface uses discrete components. One problem with this type of interface is that the two circuits cannot have a common ground. Notice that V_{SS} (which is usually GND) is connected to V_{CC}. Figure 6-12 shows how a MOS circuit might be connected to the input of a TTL circuit. Again, the problem of "common grounds" exists.

Although many types of interface circuits can be devised, the trend is toward MOS circuits that are compatible with TTL circuits. Many of these MOS circuits require a +5-volt supply and a —12-volt supply. The input and output logic levels are from 0 to +5 volts. Since the circuits are easily integrated, the interface is provided internally.

6.8 THRESHOLD VOLTAGES

For PMOS transistors, *threshold voltage* (V_T) is defined as the minimum gate voltage required to cause the silicon just underneath the gate electrode to invert from N-type to P-type material. This is the point at which the channel becomes just conductive. The same definition is true for NMOS transistors. However, the inversion is from P-type to N-type material. For typical MOS transistors, the threshold voltage is rather high, approximately 3.5 to 4.5 volts.

In Figure 6-13, typical transfer characteristics for high threshold devices are compared with those of typical saturated bipolar logic circuits, such as DTL and TTL. It should be quite obvious that bipolar logic circuits cannot turn on high threshold devices. It is also true that the logic levels of the high threshold devices are too great for the bipolar devices.

The principal advantage of high threshold devices is that of high noise immunity. These devices cannot easily be tripped on by noise spikes

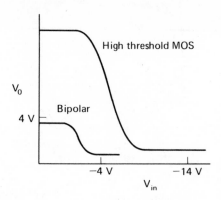

Figure 6-13.   Typical transfer characteristics for bipolar and high threshold MOS devices.

of a couple of volts when approximately 3.5 volts are required to turn on the transistors.

High threshold devices are generally slower than low threshold devices. High threshold devices also require more expensive supplies because of their higher voltage requirements. Perhaps the greatest disadvantage, however, is that they are not compatible with DTL and TTL devices.

There are a number of technological advances currently being used to reduce the threshold voltage. The most popular method is the silicon-gate-process. This type of MOS transistor uses a layer of deposited silicon as its gate electrode. This silicon layer is doped with impurities providing a lower work function than the metal gate process. This lower work function produces threshold voltages in the range of 0.5 and 2.5 volts. This process also reduces the gate to drain capacitance, thus increasing circuit speed. The silicon gate process also improves space utilization, thereby increasing circuit density.

6.9   CMOS

A MOS circuit that holds great promise of competition with other MOS circuits and with bipolar devices is the complementary MOS (CMOS). This circuit is compatible with bipolar devices because it can use a single positive supply of 5 volts. However, supply potentials as high as 18 volts may be used. The DC power dissipation is extremely low, approximately 2.5 nanowatts per gate. The circuit also features high fan-out, high noise immunity, and speeds that compete with TTL.

Figure 6-14 shows the basic CMOS inverter. Both active pull-up

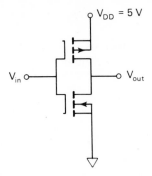

Figure 6-14. CMOS inverter.

and pull-down are provided by the complementary N and P channel transistors. This circuit is termed *ratioless* because the two transistors are not ON simultaneously. When V_{IN} is LOW, Q_1 is turned ON and Q_2 is turned OFF, resulting in a HIGH output.

The transistors are designed with a threshold voltage of approximately 2 volts. This causes the switching action to be sharp and to occur when V_{IN} equals about half of V_{DD}, as shown in the voltage transfer characteristic curve in Figure 6-15. The curve also reveals that the noise immunity is high, about 40 percent of V_{DD} (when $V_{DD} = 5V$, $V_{NH} = V_{NL} = 2V$). To fully understand this, one must realize that the output voltage swing equals V_{DD}. This is true because there is virtually no DC drain current and no DC output current. The input resistance of these devices is typically 10^{12} ohms; therefore, the logic swing is independent of fan-out. The only significant current occurs during transients when the various circuit capacitances are charged or discharged. For this

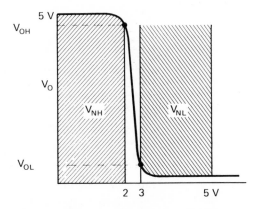

Figure 6-15. CMOS voltage transfer characteristic curve.

reason, the power dissipation increases in direct proportion to the frequency of these transitions.

Because one of the two transistors is always saturated, the output resistance is low and the fan-out is high. Fan-out capabilities greater than 50 are typical.

6.10   CMOS GATES

Both NAND and NOR gates are implemented from the basic CMOS inverter. Figure 6-16 shows a 3-input NAND gate, including its logic symbols. Q_1, Q_2, and Q_3 are the P channel pull-up transistors. Q_4, Q_5, and Q_6 are N channel pull-down transistors. When A, B, and C are all HIGH, Q_1, Q_2, and Q_3 are all OFF and Q_4, Q_5, and Q_6 are all ON. This results in a LOW level output as indicated in the NAND symbol. Because Q_4, Q_5, and Q_6 are in series and Q_1, Q_2, and Q_3 are in parallel, if only one input is LOW the output will be HIGH. Only one of the P channel transistors has to be ON and only one of the N channel transistors has to be OFF for this to occur. This makes the DeMorgan's equivalent function active as indicated in the logic symbol.

The converse of the above circuit configuration is the CMOS NOR gate, which along with its logic symbols appears in Figure 6-17. In this

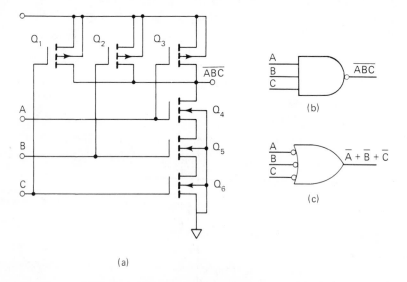

(a)

Figure 6-16.   CMOS NAND gate: (a) schematic diagram; (b) NAND symbol; (c) DeMorgan's equivalent symbol.

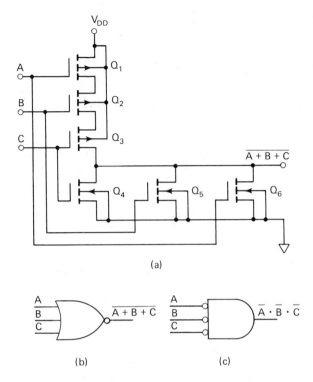

Figure 6-17. CMOS NOR gate: (a) schematic diagram; (b) NOR symbol; (c) DeMorgan's equivalent symbol.

circuit, the pull-up transistors Q_1, Q_2, and Q_3 are in series, while the pull-down transistors Q_4, Q_5, and Q_6 are in parallel. To make the OR function active, only one input need be HIGH. This causes the pull-up circuit to be open (Q_1, Q_2, or Q_3 is OFF) and the pull-down circuit to be closed (Q_4, Q_5, or Q_6 is ON). This is indicated by the NOR symbol in Figure 6-17(b).

The DeMorgan's equivalent function becomes active when all inputs are LOW. This causes the pull-up circuit to be closed (Q_1, Q_2, and Q_3 are all ON) and the pull-down circuit to be open (Q_4, Q_5, and Q_6 are OFF). The result is a HIGH output, as indicated by the symbol in Figure 6-17(c).

A quad 2-input NOR gate is shown in Figure 6-18. The additional diodes in the diagram are to protect the inputs from negative transients. Figure 6-19 shows a quad 2-input NAND gate. Full lines of CMOS devices similar to these are being produced by a number of manufacturers.

The MC14011AL/CL quad 2-input NAND gate is constructed with MOS P-channel and N-channel enhancement mode devices in a single monolithic structure. These complementary MOS logic gates find primary use where low power dissipation and/or high noise immunity is desired.

- Quiescent Power Dissipation = 10 nW/package typical
- Noise Immunity = 45% of V_{DD} typical
- Diode Protection on All Inputs
- Supply Voltage Range = 3.0 Vdc to 18 Vdc (MC14011AL)
  3.0 Vdc to 16 Vdc (MC14011CL)
- Single Supply Operation — Positive or Negative
- High Fan-out — > 50
- Input Impedance = 10^{12} ohms typical
- Logic Swing Independent of Fan-out
- Symmetrical Output Resistance — 750 ohms typical

(a)

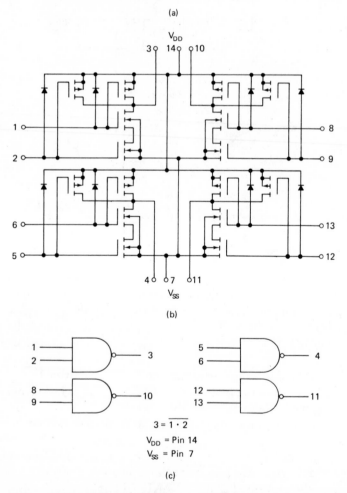

(b)

$3 = \overline{1 \cdot 2}$

V_{DD} = Pin 14

V_{SS} = Pin 7

(c)

Figure 6-18. MC 14001 quad 2-input NOR gate: (a) characteristics; (b) schematic diagram; (c) logic symbol. (Courtesy of Motorola Semiconductors.)

154

The MC 14001AL/CL quad 2-input NOR gate is constructed with MOS P-channel and N-channel enhancement mode devices in a single monolithic structure. These complementary MOS logic gates find primary use where low power dissipation and/or high noise immunity is desired.

- Quiescent Power Dissipation = 10 nW/package typical
- Noise Immunity = 45% of V_{DD} typical
- Diode Protection on All Inputs
- Supply Voltage Range = 3.0 Vdc to 18 Vdc (MC14001AL)
                       = 3.0 Vdc to 16 Vdc (MC14001CL)
- Single Supply Operation — Positive or Negative
- High Fan-out — > 50
- Input Impedance = 10^{12} ohms typical
- Logic Swing Independent of Fan-out
- Symmetrical Output Resistance — 750 ohms typical

(a)

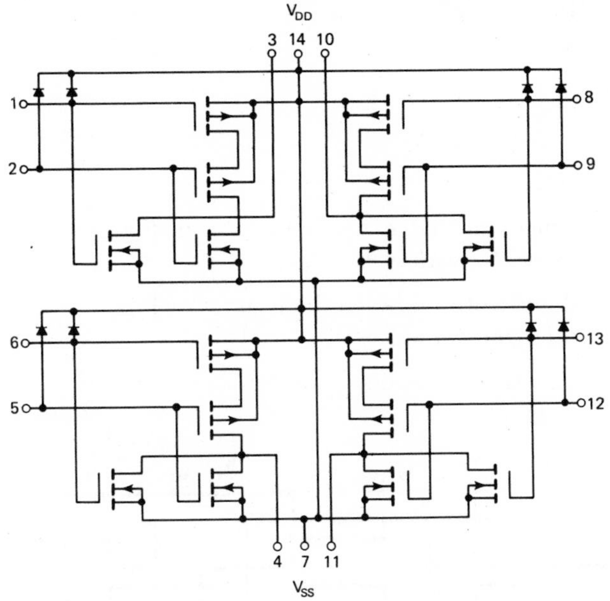

(b)

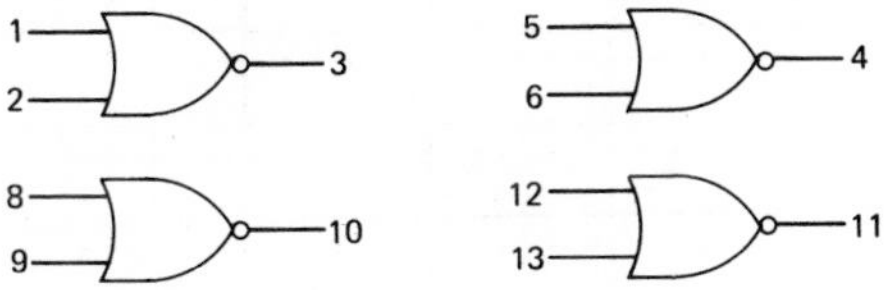

$$3 = \overline{1 + 2}$$

V_{DD} = Pin 14
V_{SS} = Pin 7

(c)

Figure 6-19. MC 14011 quad 2-input NAND gate: (a) characteristics; (b) schematic diagram; (c) logic symbol. (Courtesy of Motorola Semiconductors.)

QUESTIONS AND PROBLEMS—CHAPTER 6

1. What are the advantages of MOS over bipolar integrated circuits?

2. Define G_{FS}.

3. Define V_{DD}.

4. Define V_G.

5. Define V_{GG}.

6. Define "threshold voltage."

7. For each pair of values, determine which is LOW level and which is HIGH level.

   (a) 5 V; 0 V

   (b) —5 V; 0 V

   (c) —5 V; —10 V

   (d) —10 V; 5 V

8. Construct truth tables for both gates in Figure 6-5, substituting 1 and 0 for the variables.

9. Construct truth tables for both gates in Figure 6-6, substituting 1 and 0 for the variables.

10. Why do MOS gates dissipate very low power?

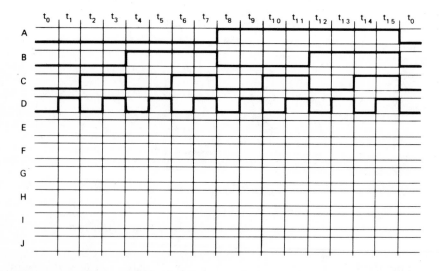

Figure 6-20.   Waveforms for problems 17 through 22.

11. Describe "ratio inverter."

12. Describe "ratioless inverter."

13. Describe "ratioless-powerless inverter."

14. How can the "interface" problem be solved between MOS and bipolar ICs?

15 Compare threshold voltages and noise immunity for both MOS and bipolar devices. Describe the effect of the silicon-gate process.

16. What is the primary advantage of CMOS over other MOS devices?

17. Using the waveforms in Figure 6-20, draw the waveforms for the output (E) of the circuit in Figure 6-21. (Refer to gate and truth table in Figure 6-5.)

Figure 6-21.

18. Using the waveforms in Figure 6-20, draw the waveforms for the output (F) of the circuit in Figure 6-22. (Refer to gate and truth table in Figure 6-5.)

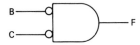

Figure 6-22.

19. Using the waveforms in Figure 6-20, draw the waveforms for the output (G) of the circuit in Figure 6-23. (Refer to gate and truth table in Figure 6-5.)

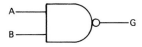

Figure 6-23.

20. Using the waveforms in Figure 6-20, draw the waveforms for the output (H) of the circuit in Figure 6-24. (Refer to gate and truth table in Figure 6-5.)

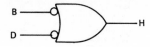

Figure 6-24.

21. Using the waveforms in Figure 6-20, draw the waveforms for the output (I) of the circuit in Figure 6-25. (Refer to gate and truth table in Figure 6-16.)

Figure 6-25.

22. Using the waveforms in Figure 6-20, draw the waveforms for the output (J) of the circuit in Figure 6-26. (Refer to gate and truth table in Figure 6-17.)

Figure 6-26.

Flip-Flops

We have learned that the three basic logic functions are AND, OR, and NOT. *All of the computer's decision-making and arithmetic circuits are made up of these three functions.* However, in order to make a decision or perform an arithmetic operation, the computer must have a place to store the input and output information (*data*). For example, if we are to find the sum of 3 and 6, we must write these numbers down on a piece of paper (*store* them), or use some of the cells of our brain to remember them. After the sum has been obtained, it is again necessary to store or remember that sum—or its value, 9, will be lost. In computers, this very important function of remembering or storing bits of data is performed by various types of *flip-flops*.

The topics covered in this chapter are:

7.1   RS FLIP-FLOP

The simplest method of storing a bit of information is with an RS flip-flop. The RS flip-flop is bistable; that is, it has two states. When it is storing a 1-bit, it is said to be in the SET state. Conversely, when it is in the RESET state, a 0-bit is stored. Such a flip-flop has two inputs: the SET input (S) and the RESET input (R); and two outputs: the SET output (Q) and the RESET output ($\overline{Q}$). A bit of information may be "erased" by making the R input active. If the flip-flop is SET, the set output (Q) is active. If the flip-flop is RESET, the reset output ($\overline{Q}$) is active. The symbol for the RS flip-flop appears in Figure 7-1(b).

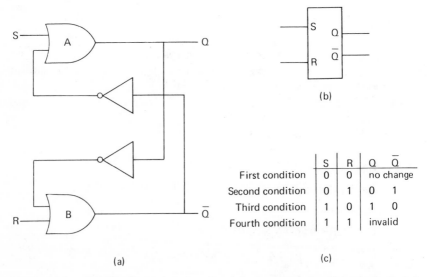

	S	R	Q	$\overline{Q}$
First condition	0	0	no change	
Second condition	0	1	0	1
Third condition	1	0	1	0
Fourth condition	1	1	invalid	

(a)

(c)

Figure 7-1.   RS flip-flop: (a) first-level logic diagram; (b) symbol; (c) truth table.

The function of the RS flip-flop may be described by using the first-level logic diagram in Figure 7-1(a). The flip-flop consists of two OR gates and two inverters.

If the S input goes to 1 and the R input remains at 0, the output of gate A will be 1. The output from gate A is fed back through an inverter to gate B, causing both inputs to gate B to be 0, and its output to be 0. The output of gate B is also fed back through an inverter to an input of gate A. This input will be a 1, causing gate A to continue to be active even when the S input returns to 0. With both S and R inputs at 0, the flip-flop will remain in the 1 state (SET).

In order to reset the flip-flop, the R input is made active by providing it with a 1. Gate B becomes active, and a 1 appears at its output. The output of gate B is fed back through an inverter to the input of gate A, causing it to become inactive. A 0 will now appear at the output of gate A, which is inverted and becomes a 1 at the input of gate B, holding it active. When the R input returns to 0, the flip-flop remains in the RESET, or 0, state.

Because the effect of holding one of the two gates active is sometimes called "latching," the RS flip-flop is often referred to as a *latch*.

A truth table may be constructed for the RS flip-flop, like the one in Figure 7-1(c). The two input variables, S and R, appear in the first two columns of the truth table.

The first input condition in the truth table shows both the S and R inputs inactive (both inputs have a 0 present). In this condition, the flip-flop remains in its last state.

The second input condition shows the reset input active (a 1 is present). This causes the flip-flop to reset. When the flip-flop is reset, it is in the 0 state. The Q output is inactive (0), and the $\overline{Q}$ output is active (1).

In the third condition, the flip-flop is set because a 1 is present at the set input. When the flip-flop is set, it is in the 1 state. The Q output is active (1), and the $\overline{Q}$ output is inactive (0).

The fourth input condition shown in the truth table should be avoided because we cannot predict the state of the flip-flop if both inputs return to zero simultaneously. This can be thought of as asking the flip-flop to set and reset at the same time, which is invalid.

A 1 is stored in the flip-flop when the set input is made active momentarily and then returned to 0. A 0 is stored in the flip-flop when the reset input is made active momentarily and then returned to 0. Remember, the primary purpose of the flip-flop is to store a bit of information. Therefore, both inputs are normally 0, causing the flip-flop to remember its last state. If the set input was activated last, the flip-flop will remember that it is in the 1 state so long as both inputs remain in the 0 state, and vice versa.

7.2   IMPLEMENTING THE RS FLIP-FLOP USING NOR GATES

The student is reminded that an implemented version of a logic diagram indicates levels—either high or low—rather than 1 or 0. A HIGH level in this text is assigned a value of 1; a LOW level, a value of 0.

In order to set the flip-flop in Figure 7-2, the set input is made high, causing gate A to become active and providing a low level at the set output. This low level is fed back to the input of gate B, causing it to become inactive. The output of gate B will then be high level holding gate A active. The effect is that gate A disables gate B and gate B enables gate A. This condition will remain even when the S input returns to low level.

In order to reset the flip-flop, the S input is made low and the R input is made high, causing the flip-flop to change state. In this condition, gate A enables gate B and gate B disables gate A.

The symbol for the RS flip-flop is shown in Figure 7-2(b). As in Figure 7-1, the set and reset inputs are identified by S and R. The set output is Q, and the reset output is $\overline{Q}$. The low level indicators present at the output of the symbol imply inversion. Therefore, the output variable will be opposite on either side of the implied inversion; that is, Q becomes $\overline{Q}$ and $\overline{Q}$ becomes Q. This concept should be clearly understood; it is further clarified by Figure 7-3.

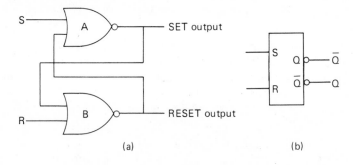

(a)   (b)

Input		Command	Output	
S	R		Q	$\overline{Q}$
L	L	Remember	No change	
L	H	Reset	L	H
H	L	Set	H	L
H	H	Invalid	L	L

(c)

Figure 7-2.   RS flip-flop implemented using NOR gates: (a) logic diagram; (b) symbol; (c) truth table.

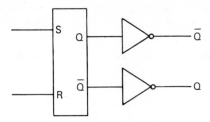

Figure 7-3. Figure 7-2(b) clarified.

RS flip-flops may be implemented with NOR gates from other logic families—such as TTL, ECL, and CMOS—in the same manner as in Figure 7-2. However, the electrical characteristics may differ.

7.3 IMPLEMENTING THE RS FLIP-FLOP USING NAND GATES

The RS flip-flop may be implemented by using NAND gates, such as in Figure 7-4. The DeMorgan's equivalent symbols for the NAND gates are used to clarify the active levels of the RS flip-flop. The function of this flip-flop may be described in the same manner as was done with NOR gates. However, the input active levels are low, and the output active levels are high. The symbol for the RS flip-flop and its truth table appear in Figure 7-4. It should be noted that the S and R inputs for Figure

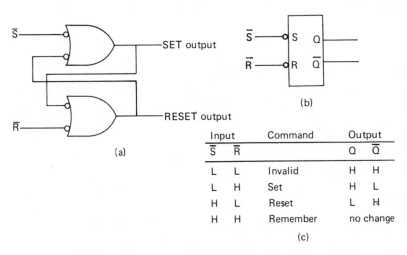

Input		Command	Output	
$\overline{S}$	$\overline{R}$		Q	$\overline{Q}$
L	L	Invalid	H	H
L	H	Set	H	L
H	L	Reset	L	H
H	H	Remember	no change	

(c)

Figure 7-4. RS flip-flop using DeMorgan's equivalent symbols for TTL 7400 NAND gates: (a) logic diagram; (b) symbol; (c) truth table.

7-4 both have low level indicators, which define the levels required to set or reset the flip-flops. In addition, the set and reset outputs of the flip-flops also agree with the active high levels.

RS flip-flops may be implemented using NAND gates from other logic families—such as DTL, ECL, and CMOS—in the same manner as the TTL 7400.

7.4   *GATED RS FLIP-FLOP (RST)*

When flip-flops are used, it is often desirable to enable or condition the S and R inputs with their required levels *before* the flip-flop is to set or reset. The new state of the flip-flop will now depend on a third input—possibly a timing pulse. The input information to the flip-flop gives instructions as to *what* to do (set, reset, or remember) and *when* to do it (timing pulse). Figure 7-5 shows the logic diagram for a gated RS flip-flop that uses RTL 9914 NOR gates.

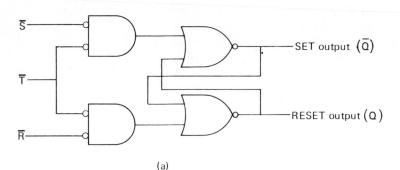

(a)

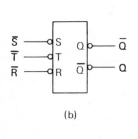

(b)

Input			Command	Output	
$\overline{S}$	$\overline{R}$	$\overline{T}$		Q	$\overline{Q}$
L	L	L	Invalid	L	L
L	L	H		NC	NC
L	H	L	Set	H	L
L	H	H		NC	NC
H	L	L	Reset	L	H
H	L	H		NC	NC
H	H	L	Remember	NC	NC
H	H	H		NC	NC

(c)

Figure 7-5.   Gated RS flip-flop: (a) logic diagram; (b) symbol; (c) truth table (NC = no change).

In order to set the flip-flop, the S input must be low level and the R input must be high. The flip-flop will not set until a low level appears at the T input. With the flip-flop in the set state, a low level appears at the set output and a high level at the reset output.

In order to reset the flip-flop, the R input must be low and the S input high. The flip-flop will reset when the T input goes low level.

The symbol for the gated RS flip-flop and its truth table appear in Figure 7-5(b) and (c). Note that the active levels for Figure 7-5(a) and (b) agree.

The first condition in the truth table should be avoided because there is no way to predict the state of the flip-flop after the T input returns to high level. Conditions 2, 4, 6, 7, and 8 cannot affect the state of the flip-flop because the T input is high level. Therefore, neither of the two input gates are enabled. Conditions 3 and 5 will set and reset the flip-flop, respectively.

7.5 GATED RS FLIP-FLOP USING NAND GATES

The RST flip-flop may also be implemented using NAND gates. The RST flip-flop in Figure 7-6 uses TTL 7400 NAND gates. The function of this flip-flop may be described in the same manner as before, except that all inputs and outputs are active high level. Figure 7-6(b) and (c) shows the symbol and truth table for the NAND gate gated RS flip-flop.

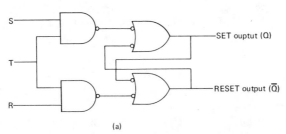

(a)

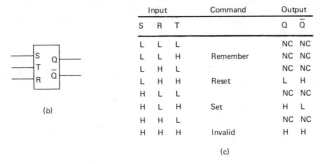

Input			Command	Output	
S	R	T		Q	Q̄
L	L	L		NC	NC
L	L	H	Remember	NC	NC
L	H	L		NC	NC
L	H	H	Reset	L	H
H	L	L		NC	NC
H	L	H	Set	H	L
H	H	L		NC	NC
H	H	H	Invalid	H	H

(c)

Figure 7-6. Gated RS flip-flop using TTL 7400 NAND gates: (a) logic diagram; (b) symbol; (c) truth table.

7.6   MASTER-SLAVE RST FLIP-FLOP

In many applications, the gated RS flip-flop has a serious problem. If the input conditions change at the time the state of the flip-flop is to change, the desired state may not be achieved. This problem is referred to as *race*. This problem may occur when flip-flops are used in counters, shift registers, and other applications. As an example, in counters, certain conditions (counts) may or may not appear; this results in erroneous counting. Therefore, race problems must be prevented.

The race problem is solved by using a master-slave flip-flop. The input variables of this flip-flop are disconnected at the instant the timing or clock pulse first appears. The new state of the flip-flop will then become dependent upon the input levels as they appeared just prior to the clock pulse. This is achieved by using two gated RS flip-flops, as shown in Figure 7-7.

Note that an inverter appears between the T inputs of the two flip-flops. This means that when the T input is high level, the first flip-flop (master) is enabled and will set or reset according to the input (S and R) levels. When the T input goes low level, the master input is disabled, and the second flip-flop (slave) is enabled and will set or reset in accordance with the state of the master flip-flop.

It should be apparent that after the clock input goes low level, the slave flip-flop cannot be affected by any change of the S and R input levels in the master flip-flop.

The master-slave RST flip-flop is said to be *cocked* when the master sets; it is *triggered* when the slave flip-flop sets. When the slave sets, it will be necessary for the clock pulse to return to high level (cocked) and then to low level (triggered) again before the slave can change state. Hence, the problem has been solved. The symbol for the master-slave RST flip-flop and its truth table appear in Figure 7-7(b) and (c).

The master-slave RST flip-flop can only change state with a low-going transient at the T input (clock pulse). This leaves only the S and R input variables, as is indicated in the truth table of Figure 7-7(c). The output variables, Q and $\overline{Q}$, are shown as they would appear *after* the clock pulse.

7.7   DIRECT SET AND RESET INPUTS

Figure 7-8 shows the symbol for an RST flip-flop having two additional inputs: direct set (S_D) and direct reset (R_D). These two inputs are gated directly into the slave flip-flop; as a result, they have priority over all of

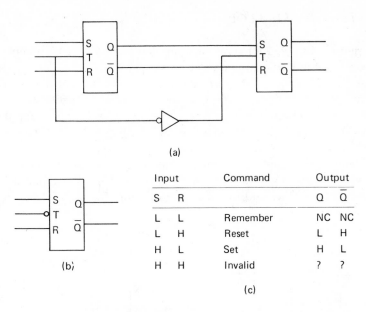

(a)

(b)

Input		Command	Output	
S	R		Q	Q̄
L	L	Remember	NC	NC
L	H	Reset	L	H
H	L	Set	H	L
H	H	Invalid	?	?

(c)

Figure 7-7. Two gated RS flip-flops connected together to form a master-slave flip-flop: (a) logic diagram; (b) symbol; (c) truth table.

the other inputs. Whenever the S_D input is made active (high level), the flip-flop will set regardless of the state of the master flip-flop and the T input. It would be illogical and therefore invalid to activate both the S_D and R_D inputs simultaneously. These inputs are typically used to preset the state of the flip-flop.

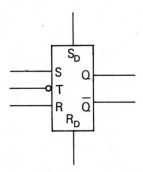

Figure 7-8. Direct set and direct reset inputs for the RST flip-flop.

7.8   TYPE D FLIP-FLOP

A variation of the master-slave RST flip-flop is the type D flip-flop, which is shown in Figure 7-9. Figure 7-9(a) shows an RST flip-flop and an inverter wired to form a type D flip-flop. Its symbol appears in Figure 7-9(b). Note that there is only one input variable—the D (data) input. Whenever the D input is active, the flip-flop will set with an input clock pulse. When the D input is inactive, the flip-flop will reset with a clock pulse.

This type of input can simplify the handling of binary data, such as 1–0, yes–no, go–no-go, etc. Like some RST flip-flops, the D flip-flop may have direct set (S_D) and direct reset (R_D) inputs.

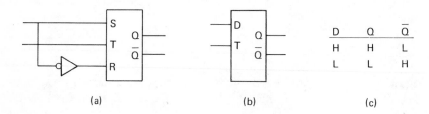

	D	Q	$\overline{Q}$
	H	H	L
	L	L	H

(a)   (b)   (c)

Figure 7-9.   Type D flip-flop: (a) RST flip-flop and inverter wired to form a type D flip-flop; (b) prewired type D flip-flop; (c) truth table.

7.9   COMPLEMENTING FLIP-FLOPS

In some cases, as when counting, it is necessary to have a flip-flop that will change state with each input clock pulse. Such a flip-flop is called a *complementing flip-flop*. The RST master-slave flip-flop may be wired as in Figure 7-10 to accomplish this function. Note that the output is cross-connected back to the input so that the flip-flop is conditioned to its opposite state. For example, if the flip-flop is in the reset state, the S input will have a high level present and the R input will be low level. This conditions the flip-flop to set with the next input pulse. The flip-flop will then set and become conditioned to reset. An RST flip-flop, when connected in this manner, is sometimes called a *T flip-flop*.

Two complete input cycles are required for each output cycle, resulting in a frequency divider of 2:1. Two such flip-flops connected in series will result in a frequency divider of 4:1. Frequency dividers and counters will be covered in Chapter 11.

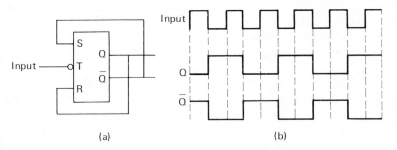

(a)   (b)

Figure 7-10.   Complementing flip-flop: (a) RST flip-flop wired to complement; (b) input and output waveforms.

7.10   THE J-K FLIP-FLOP

Perhaps the most useful flip-flop is called the *J-K flip-flop*. The two most important characteristics of the J-K flip-flop are that: (1) it has no invalid input; and (2) it can complement. Its symbol and truth table appear in Figure 7-11(b) and (c).

Since there are two input variables, there are four input conditions. If both J and K are inactive, the flip-flop is instructed to remain in its present state with the next clock pulse. The second condition is where K is active and J is inactive, instructing the flip-flop to reset. If J is active and K inactive, the flip-flop is instructed to set.

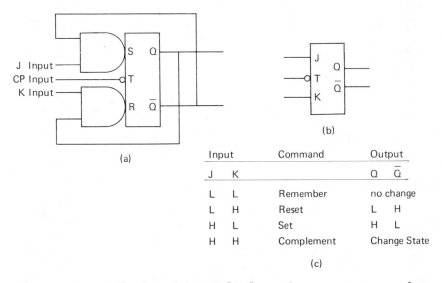

(a)   (b)

Input		Command	Output	
J	K		Q	$\overline{Q}$
L	L	Remember	no change	
L	H	Reset	L	H
H	L	Set	H	L
H	H	Complement	Change State	

(c)

Figure 7-11.   J-K flip-flop: (a) RST flip-flop and two gates connected to form a J-K flip-flop; (b) symbol; (c) truth table.

The uniqueness of the J-K flip-flop is exemplified by the fourth condition. Both J and K are active, instructing the flip-flop to change state (*complement*). It is important to note that all four input conditions are valid!

The J-K flip-flop can be thought of as an RST flip-flop connected to complement, but with two additional gates at the input. These two gates allow for the input variables J and K, as shown in Figure 7-11(a).

An understanding of how the J-K flip-flop works can be gained by studying how it is internally wired. The diagram in Figure 7-12 shows a J-K flip-flop wired using NAND gates from the DTL or TTL families. The example shown here is for the TTL 7476. Notice how the S_D and R_D inputs are gated into the slave flip-flop. This allows the slave to be directly set or reset, regardless of the other inputs.

Let's examine the T input. Gates A and B are enabled when the T input is high level, which sets or resets the master flip-flop. When the T input goes low level, the J and K inputs are disabled and gates C and D

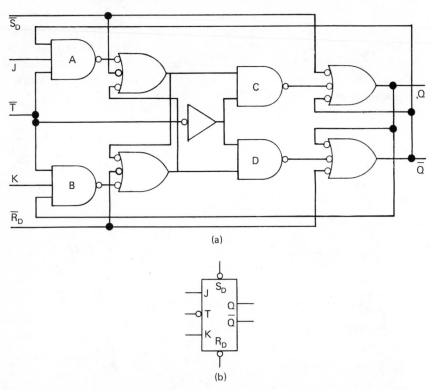

(a)

(b)

Figure 7-12.   TTL 7476 NAND gates wired to form a J-K flip-flop; (a) logic diagram; (b) symbol.

are enabled. This allows the slave to set to the state of the master flip-flop. The timing of these functions is described in Figure 7-13.

Figure 7-14 shows how NOR gates, such as those from the RTL

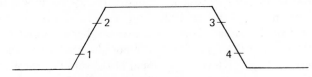

1. Isolate slave from master.
2. Enable J and K inputs to set master.
3. Disable J and K inputs.
4. Transfer data from master to slave.

Figure 7-13.   Timing pulse for the T input of the J-K flip-flop.

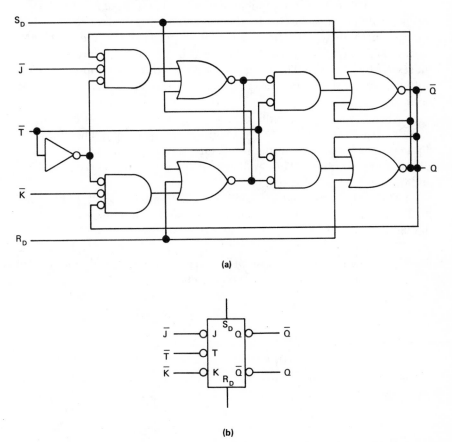

(a)

(b)

Figure 7-14.   RTL 926 NOR gates wired to form a J-K flip-flop: (a) logic diagram; (b) symbol.

family, are wired internally to form a J-K flip-flop. The example shown here is for the RTL 926.

The student should be aware of the difference between the active levels for the two examples given. Compare the two symbols in Figure 7-15. Note that for the DTL or TTL flip-flop, a high level is required to activate the J and K inputs, whereas a low level is required to activate the J and K inputs of the RTL flip-flop. Note also that when the DTL or TTL flip-flop is set, a high level will appear at the set (Q) output, whereas a low level will appear at the set output of the RTL flip-flop when it is set.

Remember that an implied inversion exists whenever a low-level indicator is used. Therefore, one could think of the output variable as being true inside the block symbol and false ($\overline{Q}$) outside the symbol.

In order to help the student understand the functions of the J-K flip-flop as well as the active levels associated with them, a timing chart appears in Figure 7-16. This chart relates specifically to the DTL 9097 or the TTL 7476 flip-flops. Across the top of the chart are the clock pulses t_1, through t_{10}. These are followed by the other inputs and the outputs. Across the bottom are the functions the flip-flop is instructed to perform.

At time zero (t_0), the direct reset is active (a low level is present), and the direct set is not active (a high level is present). Therefore, the flip-flop must be initially in the reset state because these two inputs have priority over all others. The result is that the set output (Q) is low level and the reset output ($\overline{Q}$) is high level.

At time t_1, both J and K are inactive (both are low level). Therefore, the flip-flop is instructed to remain in its present state (remember). Since the flip-flop is in the reset state, it will remain there.

Before t_2, K becomes active (high), and J is inactive (low). The flip-flop is now instructed to reset. However, since it is already in the reset state, it will remain there. Q and $\overline{Q}$ do not change.

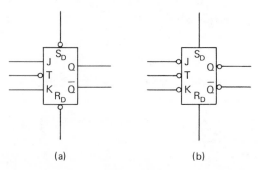

(a)   (b)

Figure 7-15.   Comparison of logic families used to form J-K flip-flops: (a) DTL 9097 or TTL 7476; (b) RTL 926.

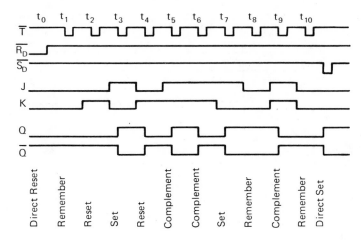

Figure 7-16. Timing chart for DTL 9097 or TTL 7476.

At t_3, J is active (high), and K is inactive (low). This instructs the flip-flop to set. As a result, Q becomes high level, and $\overline{Q}$ becomes low level.

At t_4, the input variables J and K are reversed. J is inactive and K is active. The instruction is to reset. Q becomes low level, and $\overline{Q}$ becomes high level.

At t_5, both J and K are active. This is the instruction to change state (complement). Since the flip-flop is in the reset state, it will set.

At t_6, J and K are still both active. The effect is that the flip-flop again changes state (resets) with the clock pulse.

At t_7, J is active, and K is inactive. This is another instruction to set. Q goes high, and $\overline{Q}$ goes low.

At t_8, the flip-flop is again instructed to remember. There is no change in the output.

Clock pulse t_9 causes the flip-flop to change state because both J and K are active. Q goes low, and $\overline{Q}$ goes high.

The state of the flip-flop remains unchanged at t_{10} because both J and K are inactive. However, following t_{10} the set input becomes active, causing the flip-flop to set.

One should remember that the J and K inputs are used to tell the flip-flop what to do, and the T input is used to tell it when to do it. Keep in mind however that the direct set and direct reset inputs do not require a clock pulse.

Figure 7-17 is a timing chart for the RTL 926 J-K flip-flop. The significant difference lies in the fact that the J and K inputs are active when they are low level, and the direct set and reset inputs are active when they are high level. Note that the outputs Q and $\overline{Q}$ are active when they are

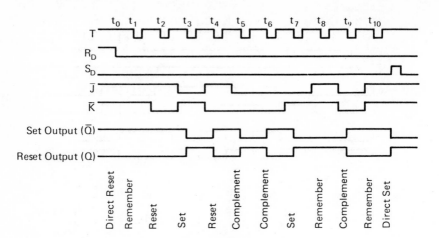

Figure 7-17.   Timing chart for RTL 926 flip-flops.

low. Remember, an implied inversion exists with the use of the low-level indicator.

7.11   MONOSTABLE MULTIVIBRATORS

The *monostable multivibrator* is a flip-flop used to produce pulses with a duration that can be independent of the input frequency. The flip-flop is triggered with an input transient, causing it to flip to its unstable state for a time determined by an RC time constant. Then it returns to its original stable state. Because of this characteristic, it is sometimes called a *one-shot*. Short pulses are sometimes desirable. For example, the pulse used to trigger the J-K flip-flop is often very short so that ambiguity in the triggering time is removed. Sometimes, where high noise conditions exist, it is desirable to allow as short a time as possible to cock the J-K flip-flop. Only while the flip-flop is cocked, is it ready to be triggered. Therefore, the possibility of accidentally triggering the flip-flop is lessened.

The 9601 monostable multivibrator shown in Figure 7-18 has an output pulse whose accuracy and duration are functions of external timing components. These components are C_x and R_x. The output is either a positive or a negative pulse. Its width is defined as follows:

$$t = 0.32\ R_x C_x \left(1 + \frac{0.7}{R_x}\right)$$

where R_x is in k ohms, C_x is in pF, and t is in nanoseconds.

Four inputs are provided, two active HIGH and two active LOW. This allows for either leading-edge or trailing-edge triggering. Figure 7-19 shows how a square wave is used to produce pulses whose widths

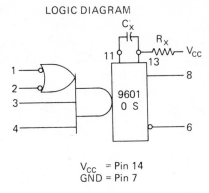

Figure 7-18. TTL/9601 monostable multivibrator. (Courtesy of Fairchild Semiconductor.)

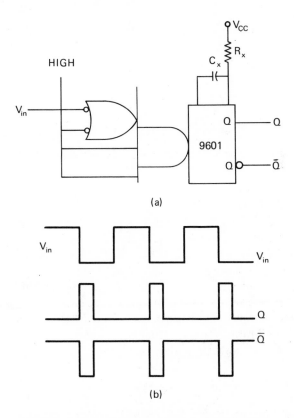

Figure 7-19. Circuit used to generate short pulse width: (a) logic diagram; (b) waveforms.

are determined by R_x and C_x. Notice that the pulse is initiated with the LOW-going transient of the square wave because an active LOW input is used. The pulse width is independent of the square wave. An input period shorter than the output cycle time will retrigger the 9601, causing a continuous true output. Retriggering may be inhibited by tying the negated output to an active LOW input.

A dual monostable multivibrator is shown in Figure 7-20. It is the 9602 from the TTL family, similar to the 9601. Notice the two inputs allow for either HIGH-going or LOW-going triggering.

LOGIC DIAGRAM

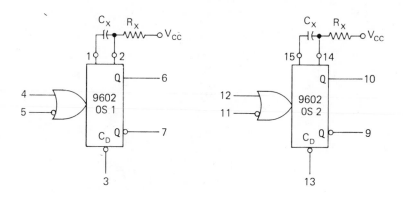

Figure 7-20.   TTL/9602 multivibrator. (Courtesy of Fairchild Semiconductor.)

7.12   ASTABLE MUTIVIBRATORS

The 9602 may be connected as in Figure 7-21 to form an astable multivibrator, which is an oscillator. When the first one-shot (A) is triggered, an output pulse, whose width (t_1) is determined by C_{x1} and R_{x1}, is produced. The trailing edge of this pulse triggers the second one-shot (B) by producing a pulse whose width (t_2) is determined by C_{x2} and R_{x2}. The trailing edge of this pulse triggers the A one-shot. The result is a free-running multivibrator having a frequency equal to

$$\frac{1}{t_1 + t_2}$$

If R_{x1} and R_{x2} are equal and C_{x1} and C_{x2} are equal, a symmetrical square wave results.

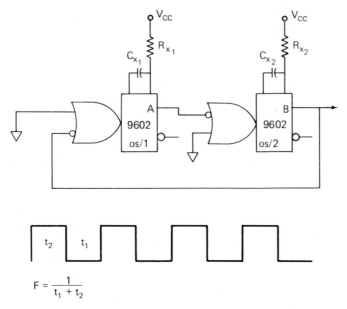

Figure 7-21. 9602 connected to form an astable multivibrator.

A very versatile timing device is shown in Figure 7-22. The NE/SE 555 monolithic timing circuit is a highly stable controller capable of producing accurate time delays or oscillation. Additional terminals are provided for triggering or resetting if desired. In the time delay mode of operation, the time is precisely controlled by one external resistor and one capacitor. For a stable operation as an oscillator, the free-running frequency and the duty cycle are both accurately controlled with two external resistors and one capacitor. The circuit may be triggered and reset on LOW-going transients, and the output structure can source or sink up to 200 mA or drive TTL circuits.

The circuit in Figure 7-23 illustrates the monostable mode of operation. The external capacitor is initially held discharged by a transistor inside the timer. Upon application of a negative trigger pulse to pin 2, the flip-flop is set; this releases the short circuit across the external capacitor and drives the output high. The voltage across the capacitor then increases exponentially with the time constant $t = R_A C$. When the voltage across the capacitor equals $\frac{2}{3}V_{CC}$, the comparator resets the flip-flop, which in turn discharges the capacitor rapidly and drives the output to its low state. Figure 7-24 shows the waveforms generated in this mode of operation.

The circuit triggers on a negative-going input signal when the level reaches $\frac{2}{3}V_{CC}$. Once triggered, the circuit will remain in this state until

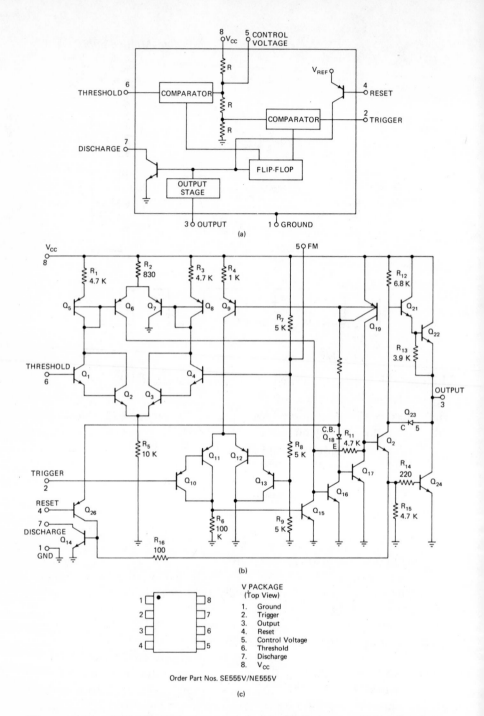

Figure 7-22. NE/SE 555 timer: (a) block diagram; (b) schematic diagram; (c) pin connections. (Courtesy of Signetics Corporation.)

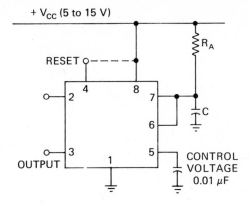

Figure 7-23.   NE/SE 555 timer connected in the monostable mode. (Courtesy of Signetics Corporation.)

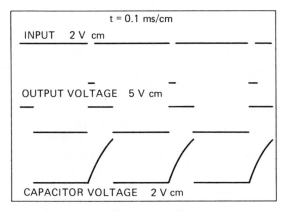

R_A = 9.1 KΩ, C = 0.01 μF, R$_L$ = 1 KΩ

Figure 7-24.   Waveforms for the monostable multivibrator. (Courtesy of Signetics Corporation.)

the set time is elapsed, even if it is triggered again during this interval. The time that the output is in the high state is given by $t = 1.1 \, R_A C$ and can easily be determined by Figure 7-25. Notice that since the charge rate and the threshold level of the comparator are both directly proportional to supply voltage, the timing interval is independent of supply. Applying a negative pulse simultaneously to the reset terminal (pin 4) and the trigger terminal (pin 2), during the timing cycle, discharges the external capacitor and causes the cycle to start over again. The timing cycle will now commence on the positive edge of the reset pulse. During the time the reset pulse is applied, the output is driven to its low state.

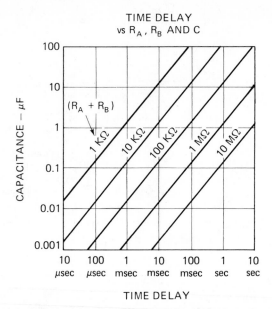

Figure 7-25. Time chart for the monostable multivibrator. (Courtesy of Signetics Corporation.)

When the reset function is not in use, it is recommended that it be connected to V_{CC} to avoid any possibility of false triggering .

If the circuit is connected as shown in Figure 7-26 (pins 2 and 6 connected), it will trigger itself and free run as a multivibrator. The external capacitor charges through R_A and R_B and discharges through R_B

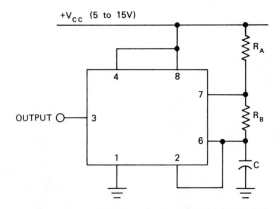

Figure 7-26. NE/SE 555 connected as a free-running oscillator. (Courtesy of Signetics Corporation.)

only. Thus, the duty cycle may be precisely set by the ratio of these two resistors. In this mode of operation, the capacitor charges and discharges between $\frac{1}{3}V_{CC}$ and $\frac{2}{3}V_{CC}$. As in the triggered mode, the charge and discharge times, and therefore the frequency, are independent of the supply voltage. Figure 7-27 shows waveforms generated in this mode of operation.

The charge time (output high) is given by

$$T_1 = 0.693(R_A + R_B)C$$

and the discharge time (ouput low) by

$$t_2 = 0.693(R_B)C$$

Thus, the total period is given by

$$T = t_1 + t_2 = 0.693(R_A + 2R_B)C$$

The frequency of oscillation is then

$$f = \frac{1}{T} = \frac{1.44}{(R_A + 2R_B)C}$$

and may easily be found by using Figure 7-28.

$$D = \frac{R_B}{R_A + 2R_B}$$

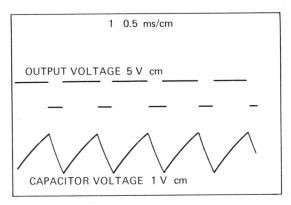

$$R_A = 4\ K\Omega,\ R_B = 3\ K\Omega,\ R_L = 1\ K\Omega$$

Figure 7-27.   Waveforms for the free-running multivibrator. (Courtesy of Signetics Corporation.)

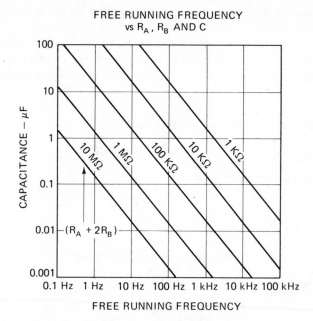

Figure 7-28.   Chart to determine frequency of the free-running multi-vibrator. (Courtesy of Signetics Corporation.)

QUESTIONS AND PROBLEMS—CHAPTER 7

1. Define each of the following.
   (a) bistable
   (b) astable
   (c) monostable
2. Using NOR gates, draw the logic diagram for a latch.
3. Draw the logic symbol for the above diagram using the appropriate level indicators.
4. Construct a truth table for the above.
5. Using NAND gates, draw a logic diagram for an RS flip-flop.
6. Draw the logic symbol for the above diagram using the appropriate level indicators.
7. Construct a truth table for the above.
8. What is the significant advantage of the gated RS flip-flop over the RS flip-flop?

9. What does *race* mean when referring to flip-flops?

10. How is the race problem solved?

11. What is the primary advantage of the master-slave flip-flop?

12. What do the terms *cocked* and *triggered* mean when applied to the master-slave flip-flop?

13. What is the disadvantage of the master-slave RST flip-flop?

14. What is the advantage of the J-K over the master-slave RST flip-flop?

15. Construct a truth table for the J-K flip-flop assuming all inputs and outputs are active HIGH. Include the direct SET (S_D) and RESET (R_D) inputs.

16. What does the term *complement* mean when referring to flip-flops?

17. How is the level indicator on the T input of a master-slave flip-flop interpreted?

18. What is the purpose of a monostable flip-flop?

19. What is the purpose of an astable flip-flop?

20. What determines the frequency of an astable multivibrator?

21. Using the waveforms in Figure 7-29, draw the output waveform (E) for the logic circuit in Figure 7-30. For the times when the results cannot be determined, leave waveform blank. (Refer to flip-flop in Figure 7-4.)

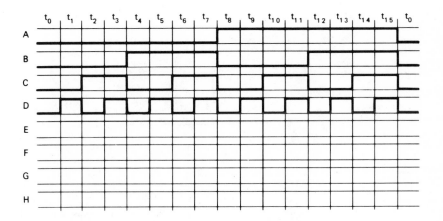

Figure 7-29.   Waveforms for problems 21 through 24.

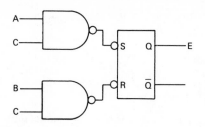

Figure 7-30.

22. Using the waveforms in Figure 7-29, draw the output waveform (F) for the logic circuit in Figure 7-31. For the times when the results cannot be determined, leave waveform blank. (Refer to flip-flop in Figure 7-5.)

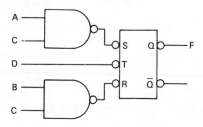

Figure 7-31.

23. Using the waveforms in Figure 7-29, draw the output waveform (G) for the logic circuit in Figure 7-32. For the times when the results cannot be determined, leave waveform blank. (Refer to flip-flop in Figure 7-7.)

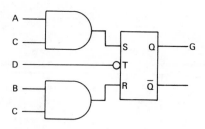

Figure 7-32.

24. Using the waveforms in Figure 7-29, draw the output waveform (H) for the logic circuit in Figure 7-33. For the times when the results cannot be determined, leave waveform blank. (Refer to flip-flop in Figure 7-11.)

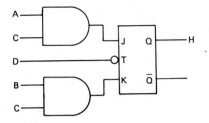

Figure 7-33.

8

Implementation of Logic Functions

Chapter 4 dealt with concepts. The logic symbols used referred only to concepts. Variable values were true (1) or false (0). Gates were discussed in terms of their functions and characteristics. No actual circuits were implied.

In this chapter you will learn how to implement the functions using real circuits. This chapter shows how logic expressions may be implemented using gates of the RTL, DTL, and TTL families. However, the approach to implementation used can be applied to all inverting and non-inverting gates. To assist the student in understanding the implementation process, a logic design problem is presented and solved.

Voltage levels (HIGH and LOW) are used to represent variable values. Active level indicators are introduced and MIL-STD-806C symbols are used.

The topics covered in this chapter are:

8.1 Applying DeMorgan's Theorem to Logic Gates

8.2 Interpreting Logic Diagrams

8.3 Implementing Logic Expressions

8.1 APPLYING DEMORGAN'S THEOREM TO LOGIC GATES

The NAND gate in Figure 8-1(a) tells the technician, "If both inputs are HIGH level, the output will be LOW level." Although the symbol did not say anything about the other possible conditions, the technician could assume by implication that if either input OR both inputs are LOW level, the output would be HIGH level. The gate, therefore, could be redrawn as in Figure 8-1(b).

This basic equivalency of gates is expressed by DeMorgan's theorem:

$$\overline{AB} = \overline{A} + \overline{B}$$

The redrawn version of the NAND gate is called *DeMorgan's equivalent* gate and may be read as, "If either OR both inputs are LOW level, the output will be HIGH level."

The technician would read the NOR symbol in Figure 8-2(a) as, "If either input OR both inputs are HIGH level, the output will be LOW

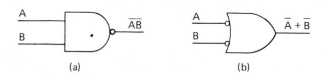

Figure 8-1.   (a) NAND gate; (b) DeMorgan's equivalent gate.

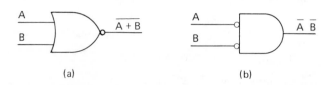

Figure 8-2.   (a) NOR gate; (b) DeMorgan's equivalent gate.

level." The DeMorgan's equivalent gate appears in Figure 8-2(b) and is read, "If both inputs are low level, the output will be high level."

Again, this basic equivalency of gates is expressed by DeMorgan's theorem:

$$\overline{A + B} = \overline{A} \cdot \overline{B}$$

A very simple rule may be applied to all gates when the DeMorgan's equivalent is desired: *The AND and OR symbols may be interchanged if ALL of the level indicators are interchanged.* Figure 8-3 shows a number of examples of this rule. For reasons of economy, families of logic

GATES AND DEMORGAN'S EQUIVALENTS	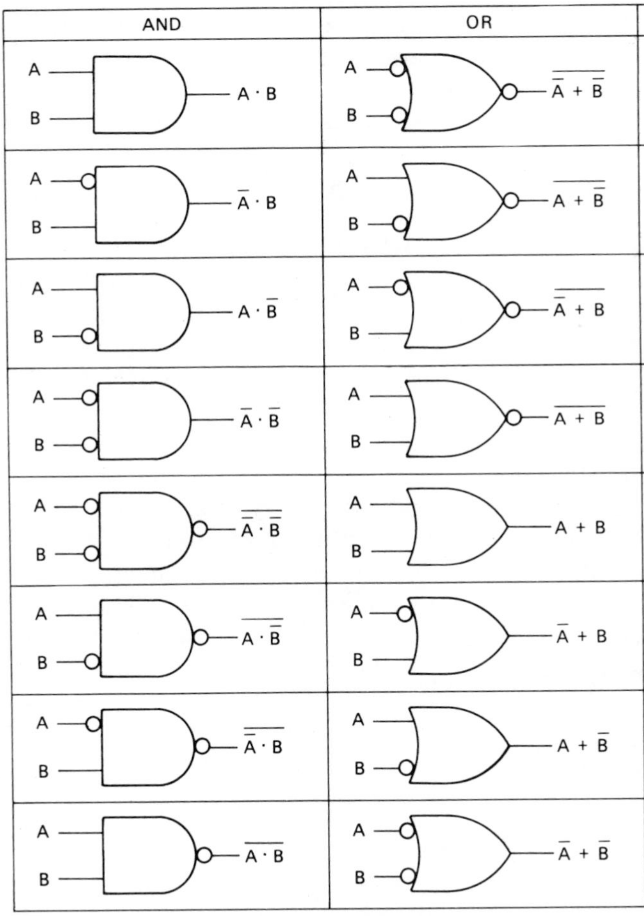	TABLE OF COMBINATIONS		
AND	OR	A	B	F
$A \cdot B$	$\overline{\overline{A} + \overline{B}}$	H H L L	H L H L	H L L L
$\overline{A} \cdot B$	$\overline{A + \overline{B}}$	H H L L	H L H L	L L H L
$A \cdot \overline{B}$	$\overline{\overline{A} + B}$	H H L L	H L H L	L H L L
$\overline{A} \cdot \overline{B}$	$\overline{A + B}$	H H L L	H L H L	L L L H
$\overline{A \cdot B}$	$A + B$	H H L L	H L H L	H H H L
$\overline{A \cdot \overline{B}}$	$\overline{A} + B$	H H L L	H L H L	H L H H
$\overline{\overline{A} \cdot B}$	$A + \overline{B}$	H H L L	H L H L	H H L H
$\overline{\overline{A} \cdot \overline{B}}$	$\overline{A} + \overline{B}$	H H L L	H L H L	L H H H

Figure 8-3. Gates and their DeMorgan's equivalents.

are often manufactured so that all functions are performed with one basic circuit. For example, if the designer has chosen to use the RTL family, he is restricted to only NOR gates. This means that if an AND function is required, the DeMorgan's equivalent gate will be used as illustrated in Figure 8-2.

If the designer has chosen to use the DTL family, only NAND gates are available. When the OR function is required, the DeMorgan's equivalent gate is used as illustrated in Figure 8-1. The student should realize that all functions (AND, OR, and NOT) are inherent in every inverting gate.

8.2   INTERPRETING LOGIC DIAGRAMS

Logic diagrams should be drawn so that the reader understands the logic function intended as well as the levels required to activate these functions. Remember that for purposes of this text, high level is associated with "true," "yes," or "one," and low level is associated with "false," "no," or "zero."

For example, Figure 8-4 demonstrates that the logic symbol for the NOT function can be drawn two ways. The symbol in Figure 8-4(a) should be used to indicate that a high level at its input will produce a low level at its output. In other words, if A is true or equal to 1 at the input, A will be false and will be equal to 0 at the output. Figure 8-4(b) should be used to indicate that a low level at the input will produce a high level at the output. In other words, if A is false or equal to 0 at the input, the output ($\overline{A}$) will be true or equal to 1. This distinction is important for the technician because it tells him the level that will exist for the function intended.

There are only two input conditions. The symbol should be drawn to indicate to the technician which condition is significant. The symbol in Figure 8-4(b) tells the technician that the significant input condition is LOW and the significant output condition is HIGH. The symbol in Figure 8-4(a) tells the technician that the significant input condition is HIGH and the significant output condition is LOW.

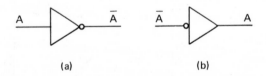

(a)   (b)

Figure 8-4.   (a) Inverter with active high level input; (b) Inverter with active low level input.

Consider the diagram in Figure 8-5. There are four possible input conditions. Because of the AND function, only one of the four is significant. Only when A is true and B is false do we get the indicated output. The diagram tells the technician what the significant input condition is when A is HIGH (no low-level indicator) and B is LOW (low-level indicator on input side of the inverter). The technician should know that this condition is significant even though he has not yet read all the logic that follows. The diagram also tells him that the significant output condition is high level because there is no low-level indicator at the output.

The logic diagram in Figure 8-6(a) is poorly drawn because it does

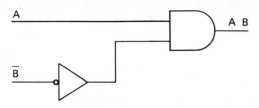

Figure 8-5.   Circuit indicating significant input levels.

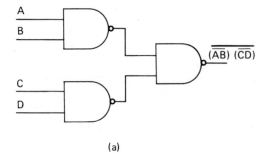

(a)

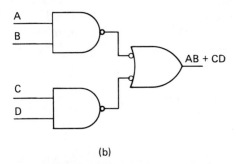

(b)

Figure 8-6.   (a) Poorly drawn logic diagram; (b) DeMorgan's theorem applied to simplify meaning of logic.

not clearly indicate the function performed by the circuit. If we apply DeMorgan's theorem to the output expression, we simplify it to AB + CD. DeMorgan's theorem may also be applied to the diagram itself. The output gate is redrawn so that the AND symbol is changed to OR and the level indicators are all changed. The resulting diagram in Figure 8-6(b) may now be read as

$$\overline{\overline{AB} + \overline{CD}}$$

Logically, whenever a low-level indicator appears at both ends of a line, as they do here, between gates, they may be ignored. The expression then reads

$$AB + CD$$

The technician can interpret this as meaning that whenever A and B are high level or whenever C and D are high level (the two significant input conditions), the output will be high level. Since there is no low-level indicator at the output, the indication is that the significant level is high.

The way that the diagram is drawn can greatly simplify the technician's job of troubleshooting a circuit. Consider this: The logic diagram in Figure 8-6 has a total of 16 possible input conditions, but the technician need only look for two of the 16!

8.3   IMPLEMENTING LOGIC EXPRESSIONS

The process of changing a logic expression into a logic diagram by using a specific family of logic gates is called *implementation*.

Let's implement the simple expression A · B using NAND gates from the DTL family. Since the AND gate does not exist in the DTL family, the NAND gate must be used as shown in Figure 8-7. Obviously, the NAND gate could not be used by itself because its output is opposite that desired. Therefore, an inverter is required.

When using a NAND gate to perform the OR function, the De-Morgan's equivalent symbol is used as in Figure 8-8. Since the NAND gate "ORs" only low levels, two inverters are required at the input.

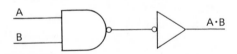

Figure 8-7.   Implementing the expression AB using a NAND gate and an inverter.

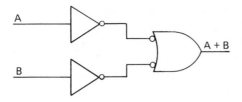

Figure 8-8.   Implementing the OR function using a NAND gate.

The AND and OR functions can be accomplished by using NOR gates as in Figure 8-9.

The student will note that inverters must be used for each example above. However, gates are usually connected in such a sequence that the inversions tend to cancel themselves.

The rules given for analyzing logic circuits are also useful for determining how a logic expression may be implemented. In general, a logic expression can be implemented in various ways. The implementation selected depends on considerations like the number of gates, type of gates, signal levels available, and signal levels required.

In Chapter 4, the Boolean laws and theorems were discussed. These expressions are implemented using NOR and NAND gates as in Figure 8-10 (see inside front and back covers for artwork).

Expressions 1, 3, 6, and 7 all have a single variable at the input (x), and all provide the same variable at the output (x). It should be recognized that a fallacy exists when attempting to implement expressions of

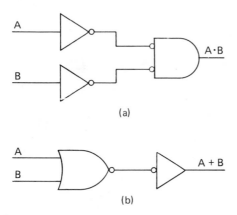

Figure 8-9.   (a) Implementing the AND function using a NOR gate and two inverters; (b) implementing the OR function using a NOR gate and one inverter.

this nature because no logic function is performed. Such circuits can be replaced with a single conductor.

When implemented, expressions 2 and 4 will always have a high level at their outputs. These "always" gates should be avoided because no logic function is performed. A high level may be obtained by simply connecting to V_{CC}.

Expressions 5 and 8 always provide a low level at their outputs. These "never" gates should be avoided because no logic function is performed. A low level may be obtained by connecting to common.

When implementing a logic expression, a systematic approach should be followed. First, simplify the expression as much as possible using Boolean algebra. Second, draw a first-level logic diagram. Third, replace gates in drawings with gates available from the family you have selected. Finally, make modifications as required. For example, if we were to implement the expression AB + CD + AB using the DTL family of NAND gates, our first step would be to use Boolean algebra to simplify the expression to

$$AB + CD \qquad\qquad (AB + AB = AB)$$

At this time, such a simplification may seem obvious. However, it may not have been so at the time the problem was first expressed. Next, the first-level logic diagram is drawn—as in Figure 8-11. The third step is to redraw the diagram using gate symbols from the DTL NAND family as in Figure 8-12. The fourth step is not necessary since modifications are not required.

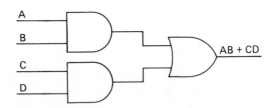

Figure 8-11.   First-level logic diagram of the expression AB + CD.

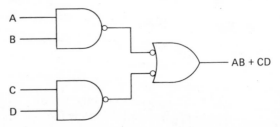

Figure 8-12.   Redrawn diagram using symbols from DTL NAND family.

Let us now implement the above expression using the RTL family of NOR gates. Steps 1 and 2 would be the same as above. Step 3 produces the diagram shown in Figure 8-13.

You will note that low level indicators appear at the inputs and the output. Therefore, modification is required. In order to obtain a true output (high level), an inverter is required as shown in Figure 8-14.

The variables are usually stored in flip-flops. Therefore, both the variable and its complement are available. Because the inputs to the circuit of Figure 8-14 require the complement form (low level active), they should be connected to the set output of the appropriate flip-flop. Figure 8-15 shows how the variable A is connected.

Let's implement the expression AB + AC using the RTL family of logic. The first step is to simplify the expression, if that is possible, by using Boolean algebra as follows:

1. AB + AC

2. A(B+ C)   (distributive property)

The next step is to draw the first-level logic diagram as shown in Figure 8-16.

In the third step, the gates in the drawing are replaced by the gates available from the RTL family as shown in Figure 8-17. Note that the

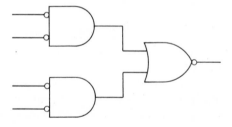

Figure 8-13. Redrawn diagram using symbols from RTL NOR family.

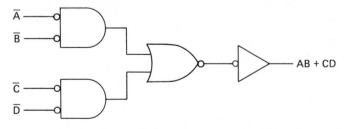

Figure 8-14. Redrawn logic diagram with an inverter at output and appropriate variables at input.

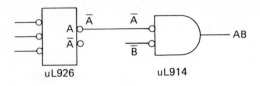

Figure 8-15.   Connection of variable A.

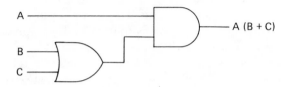

Figure 8-16.   First-level logic diagram of the expression A(B + C).

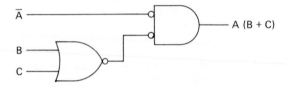

Figure 8-17.   Implementation of the expression A(B + C) using NOR
gates.

complement of A is required at the input of the equivalent gate. The
complement may be obtained from the appropriate output of a flip-flop
or by the use of an inverter.

   If we implemented the above expression using only NAND gates
of the TTL family, the first two steps of the implementation sequence
would be the same. The third step is to substitute gate symbols as shown
in Figure 8-18. Note that in order to obtain an active high level at the
output, an inverter is required. Also, the complements of B and C are
required at the input.

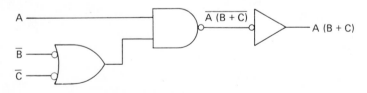

Figure 8-18.   Implementation of the expression A(B + C) using NAND
gates.

The original expression (AB + AC) may be implemented using TTL NAND gates as in Figure 8-19.

Implementing the expanded form of the expression is sometimes desirable, as in this case, where all inputs and the output are active high level.

The student should keep in mind that the input to the next logic function or to a flip-flop may require either a LOW level or a HIGH level to activate it. Therefore, the output level should be modified only if necessary.

It may also be helpful to understand that an expression with a bar extending completely across it relates directly to low-level active. For example, if the expression $\overline{A\overline{B} + \overline{A}B}$ appeared at the input of a logic gate, it should mean that the input is active when it is low level. On the other hand, if the expression $A\overline{B} + \overline{A}B$ appears at the input of a logic gate, that input should be active when it is high level.

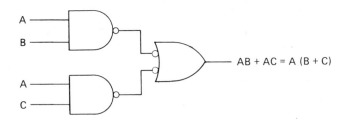

Figure 8-19. Implementation of the expression AB + AC using NAND gates.

8.4 THE COKE MACHINE

A coke-dispensing machine is a good example for a logic design problem. The purpose of this example is to show how an implemented-level logic diagram is developed. To start with, let's identify the input variables.

Item	Symbol
Power	P
Quarter	Q
Dime 1, 2	D_1, D_2
Nickel 1, 2, 3	N_1, N_2, N_3
"No Ice" button	No I

The input variable, Power (P), is used to determine if the machine is electrically connected; if not, a "Return" is made. Each coke costs 15 cents. Therefore, one quarter (Q), two dimes (D_1, D_2), one dime and a nickel (D_1, N_1), or three nickels (N_1, N_2, N_3) may be inserted. The machine must make the correct change. One of the features of the machine is that it has a "No Ice" button to accommodate those people who prefer no ice in their drink.

The output variables are

Item	Symbol
Cup	K
Ice	I
Coke	C
Return	R
Change 1, 2, 3	Ch_1, Ch_2, Ch_3

The "Return" is provided in order to return the coins inserted in case of an electrical power failure or in the absence of supplies or change. Three possible combinations of change are available: one dime or two nickels when a quarter is inserted, or one nickel when two dimes are inserted.

There are certain internal variables necessary to determine the status of the supplies and the remaining change. These variables are

Item	Symbol
Dime Sense	D_S
Nickel Sense	N_S
Coke Sense	C_S
Ice Sense	I_S
Cup Sense	K_S

The Dime Sense (D_S) is active if at least one dime is available for change. However, two nickels must be available to activate the Nickel Sense (N_S). Coke, Ice, and Cup Senses indicate whether these supplies are present.

The first step in our design problem is to write expressions for each output variable. The expression for Cup (K) would be the same as for the Coke (C) since one should not be dispensed without the other. Therefore, the expression for the Coke and Cup would be

$$C = K = P \cdot C_S (I_S + \text{No I}) K_S \; [(D_1 \cdot N_1 + N_1 \cdot N_2 \cdot N_3) + (Q \cdot D_S) + (Q \cdot N_S) +$$
$$(D_1 \cdot D_2 \cdot N_S)]$$

which says that Coke will be dispensed if;

> the power is on,
> AND Coke is available,
> AND there is ice, OR the No Ice Button is pressed,
> AND cups are available,
> AND one dime and one nickel have been inserted,
> OR three nickels have been inserted,
> OR a quarter has been inserted and there is a dime for change,
> OR a quarter has been inserted and there are nickels available for change,
> OR two dimes have been inserted and there are nickels available for change.

The expression for "Ice" would be

$$I = C \cdot \overline{No\ I}$$

that is, a Coke is being dispensed and the "No Ice" button has not been pressed. The expression would read: Ice equals Coke and not "No Ice." (Poor English, but valid logic.)

The "Return" expression would be

$$R = \overline{C_s} + \overline{(I_s \cdot No\ I)} + \overline{K_s} + \overline{P}$$

A return is made if there is no Coke available,

> OR there is no ice and the No Ice button has not been pressed,
> OR there are no cups available,
> OR the power is not on.

There are three expressions for change:

$$Ch_1 = C \cdot Q \cdot D_s = \text{one dime change}$$

$$Ch_2 = C \cdot Q \cdot \overline{D_s} \cdot N_s = \text{two nickels change}$$

$$Ch_3 = C \cdot D_1 \cdot D_2 \cdot N_s = \text{one nickel change}$$

A change of one dime will be made (Ch_1):

> when a Coke has been dispensed,
> AND a quarter was inserted,
> AND a dime is available for change.

A change of two nickels will be made (Ch_2):

> when a Coke has been dispensed,
> AND a quarter was inserted,
> AND no dimes are available for change,
> AND nickels are available for change.

A change of one nickel will be made (Ch_3):

> when a Coke has been dispensed,
> AND two dimes were inserted,
> AND there are nickels available for change.

As you remember, the first step in the implementation process is to try to simplify the expression by using Boolean algebra.

The expression for Coke and Cup may be simplified in two ways. Therefore, we have three possible expressions that may be implemented:

1. $C = K = P{\cdot}C_S{\cdot}(I_S + \text{No } I){\cdot}K_S{\cdot}[(D_1{\cdot}N_1 + N_1{\cdot}N_2{\cdot}N_3) + (Q{\cdot}D_S) + (Q{\cdot}N_S) + (D_1{\cdot}D_2{\cdot}N_S)]$

2. $C = K = P{\cdot}C_S{\cdot}(I_S + \text{No } I){\cdot}K_S{\cdot}[N_1{\cdot}(D_1 + N_2{\cdot}N_3) + Q(D_S + N_S) + (D_1{\cdot}D_2{\cdot}N_S)]$ (distributive property)

3. $C = K = P{\cdot}C_S{\cdot}(I_S + \text{No } I){\cdot}K_S{\cdot}[N_1 (D_1 + N_2{\cdot}N_3) + (Q{\cdot}D_S) + N_S {\cdot}(Q + D_1{\cdot}D_2)]$ (distributive property)

The expression for "Ice" appears to be in its simplest form:

1. $I = C{\cdot}\overline{\text{No } I}$

An attempt to simplify the "Return" expression reveals:

1. $R = \overline{C_S} + (\overline{I_S}{\cdot}\overline{\text{No } I}) + \overline{K_S} + \overline{P}$

2. $R = (\overline{I_S + \text{No } I}) + \overline{C_S{\cdot}K_S{\cdot}P}$ (DeMorgan's theorem)

However, one should realize that if for some reason no Coke is dispensed, a "Return" should be made; therefore:

3. $R = \overline{C}$

The three expressions for change (Ch_1, Ch_2, Ch_3) cannot be further simplified.

The second step in implementation is the drawing of the first-level logic diagram. Since it is not obvious which Coke expression we should implement, first-level diagrams will be drawn for each of the expressions. These appear in Figures 8-20, 8-21, and 8-22.

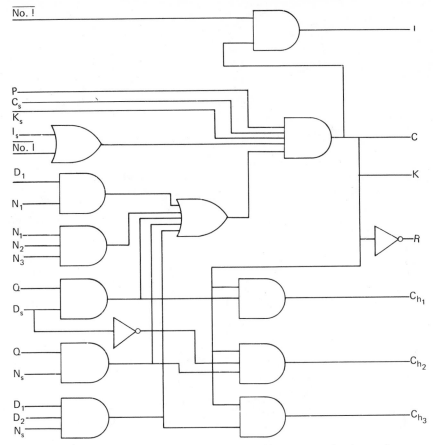

Figure 8-20. First-level logic diagram of expression #1 for coke machine.

Now a decision has to be made as to which of the three first-level diagrams we should use to continue the implementation process. You will note that parts of the change expressions also exist in Coke expression #1. These are available as $Q \cdot D_S$, $Q \cdot N_S$, and $D_1 \cdot D_2 \cdot N_S$. Because of this and because the Coke expression could not be significantly simplified, expression #1 is chosen.

The third step in the implementation sequence is the replacement of the symbols in the first-level drawing with symbols of the family of gates selected. The TTL 7400 series is chosen because

1. It is a very versatile family of logic.
2. It has a high noise immunity, which is required because of the noise transients produced by the relays and motors in the dispensing apparatus.
3. It is relatively inexpensive and readily available.

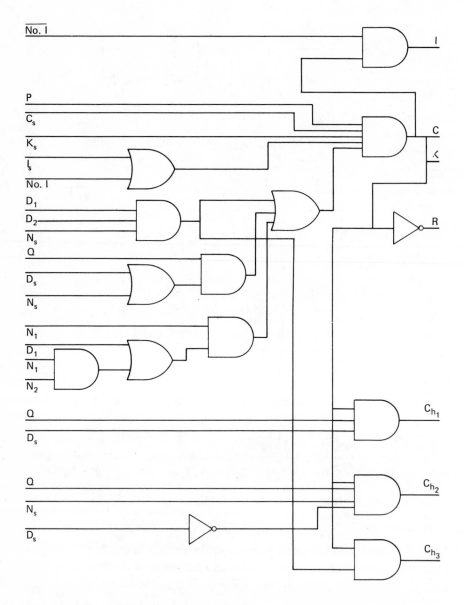

Figure 8-21.   First-level logic diagram of expression #2 for coke machine.

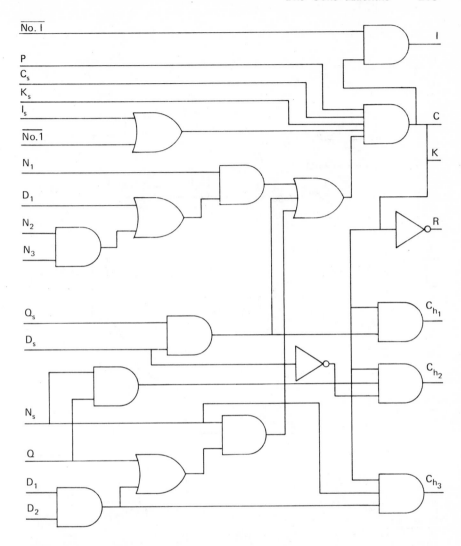

Figure 8-22. First-level logic diagram for expression #3 for coke machine.

The implemented logic drawing using the TTL 7400 family appears in Figure 8-23. Because gates having more than four inputs are very expensive, the implemented version was changed so that four-input gates could be used. These two gates are the 7420 package.

You will note that in some cases the active inputs are low level so as to facilitate the selection of the gates. The active levels of the input

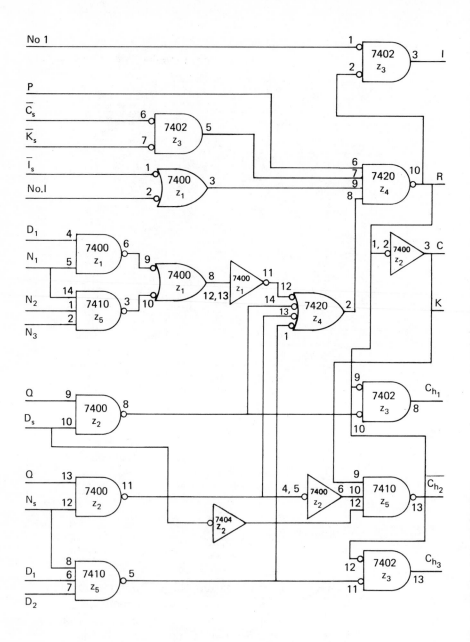

Figure 8-23.   Implemented logic diagram of coke machine.

variables C_S, K_S, I_S, and No I are all changed. This is achieved by appropriate returns to the sensing switches.

One of the outputs (Ch_2) is active low level. Mechanical functions may be activated by either active high or low levels, depending on the choice of hardware.

For this example, the sequence of events (decision making, dispensing sequence, etc.) called *timing* was not considered. Many logic problems are solved sequentially; such a solution requires special timing circuits and will be covered in Chapter 12.

8.5   COMPONENT REDUCTION

In the implementation of this problem, an attempt was made to keep the component count to a minimum by using all available gates in a given package. One method of doing this is to use two-input NAND gates as inverters by connecting the two inputs together. In the implemented logic diagram, these gates are symbolized as inverters because of the function they perform. (You will note that only five packages were used and that all gates in each package were utilized.)

8.6   SECOND-LEVEL LOGIC DIAGRAM

A great deal of information can be communicated in a properly drawn second-level (implemented) logic diagram. As was indicated previously, significant levels (those levels required to activate a function) should be carefully considered when drawing the diagram.

Note that in most cases both ends of a line between gates are terminated by the same level indicator; that is, the line connects either two low level indicators or two high level indicators. Note also that the symbols for the input and output variables are consistent with the active levels. "Not No I" may seem inconsistent. However, "No Ice" is active when high level. Normally, when ice is desired, the "No Ice" button is NOT pressed.

8.7   COMPONENT AND PIN-OUT IDENTIFICATION

The standard identification symbol for a digital component package is U or Z. In our example, the five packages are identified as Z_1 through Z_5. These identification symbols can appear on a layout diagram to identify physical location. The type of gate is also identified by a number inside the logic symbol, such as 7400 for the two-input NAND gate.

Pin numbers are identified with numbers on the outside of each gate symbol. The input to the inverters has two numbers indicating that two inputs of the NAND gate were tied together to form an inverter.

A well-organized, implemented logic diagram should concisely communicate the following information:

1. Wiring diagram

2. Logic diagram

3. Active levels

4. Significant conditions to activate the function

5. Pin connections

6. Physical locations of components

7. Mnemonics to easily recognize variables

When troubleshooting logic circuits, the technician will need all of this information, and it will be most convenient if it all appears together.

QUESTIONS AND PROBLEMS—CHAPTER 8

1. What functions are inherent in
   (a) NAND gates?
   (b) NOR gates?

2. Draw DeMorgan's equivalent of a three-input:
   (a) OR gate
   (b) AND gate
   (c) NOR gate
   (d) NAND gate

3. What is meant by the term *implementation?*

4. What type of gates are available from each of the following families?
   (a) DTL
   (b) TTL
   (c) ECL
   (d) RTL

5. What does the low level indicator say to a technician?

6. What is the significance of low level indicators appearing at both ends of a line?

7. Implement the circuit in Figure 8-11 by using:

(a) DTL gates

(b) RTL gates

(c) TTL gates

8. By constructing truth tables, verify expressions 15, 16, 20, and 22 in Figure 8-10.

9. Design the logic and the implementation for a candy dispensing machine according to the following specifications:

(a) The machines dispense 12 different candy bars.

(b) Eight selections cost 15¢ each; four selections cost 10¢ each.

(c) The largest coin the machine will accept is a quarter.

(d) The machine accepts nickels and dimes and will make the appropriate change when required.

For problems 10 through 19: (a) draw the first level and the implemented level logic diagram using only NAND gates and inverters if necessary; (b) draw the first level and implemented level logic diagrams using only NOR gates and inverters if necessary. (Use minimum number of gates.)

10. A B

11. A + B

12. $\overline{A\,B}$

13. $\overline{A + B}$

14. A B + A C

15. (A + B)(A + C)

16. (A + B) $\overline{A\,B}$

17. $\overline{A + CD}$

18. $\overline{D(A + C)}$

19. $\overline{(A + B)(C + D)}$

Using the waveforms in Figure 8-24, draw the output waveforms (E through M) for the following logic diagrams (Problems 20 through 28).

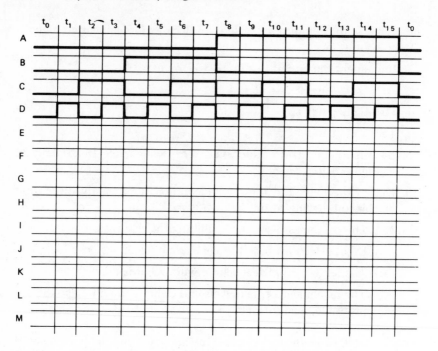

Figure 8-24.   Waveforms for problems 20 through 28.

20.

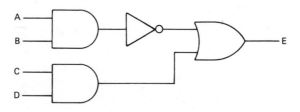

Figure 8-25.

21.

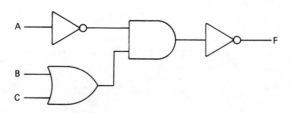

Figure 8-26.

22.

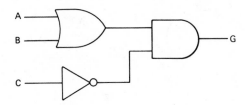

Figure 8-27.

23.

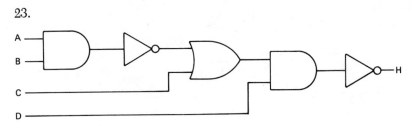

Figure 8-28.

24.

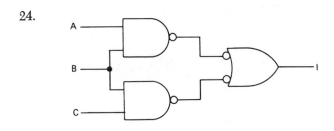

Figure 8-29.

25.

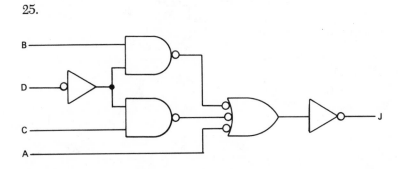

Figure 8-30.

26.

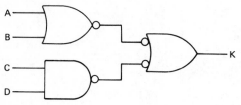

Figure 8-31.

27.

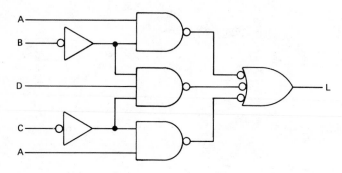

Figure 8-32.

28.

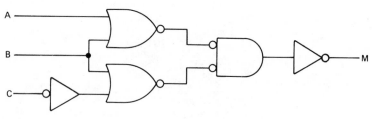

Figure 8-33.

9

Registers

Flip-flops are used to store bits of information. Binary numbers, binary-coded numbers, or binary-coded alphanumeric characters, etc., require many flip-flops. When flip-flops are organized to store multibit information, they are called *registers*.

Registers are classified according to the way information is entered and removed. For example, if all flip-flops are set simultaneously, the register is referred to as a *parallel register*. If data is entered and removed one bit at a time, the register is referred to as a *serial* or *shift register*.

The topics covered in this chapter are:

9.1   Parallel Entry

9.2   Jam Entry

9.3   Transfer Function

9.4   Shift Registers

9.5   Left-Right Shift Registers

9.6   Recirculating Shift Registers

9.7   MOS Dynamic Shift Registers

9.1   PARALLEL ENTRY

Figure 9-1 shows a 4-bit register using parallel entry. The flip-flops may be simple R-S flip-flops or latches. Before data may be entered, the contents of the register must be erased. This is done by activating the reset input, which resets all of the flip-flops. To be entered, the data must be present at the inputs of the gates. When the SET input is made active, only those flip-flops which store 1s are set. The stored data is available in parallel form at the outputs of the flip-flops.

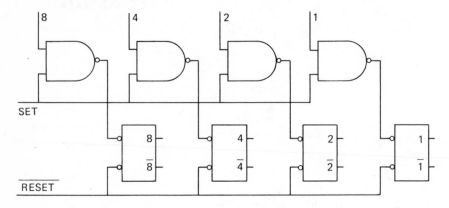

Figure 9-1.   Parallel entry.

9.2   JAM ENTRY

Jam entry is used to overcome the necessity of resetting the register before new data is entered. Figure 9-2 shows a 4-bit register with jam entry. The data is gated into the set input of each flip-flop, and the complement of the data is gated into the RESET input of each flip-flop. The result is that new data enters the register whenever an ENTER pulse appears without first having to reset all of the flip-flops. As before, the data stored is available in parallel form at the outputs of the flip-flops as indicated.

9.3   TRANSFER FUNCTION

It is often desirable to transfer data from one register to another. A simple way to do this is to use jam entry as in Figure 9-3. RST flip-flops are employed so that additional gates are not required. When the TRANSFER line is activated, the contents of the A register are duplicated in the B

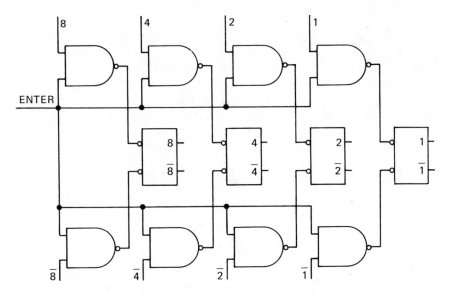

Figure 9-2. Jam entry.

register. Note that each time the TRANSFER line is activated, the old contents of the B register are destroyed.

Registers are commonly available in MSI circuits. Figure 9-4 shows the 9314 4-bit latch. Actually, it consists of four type D flip-flops. The inputs D_0, D_1, D_2, and D_3 are for the four bits of data. If the set inputs (S_0–S_3) are all active, the data will be jam entered into the register when the enable input (E) is made active. The stored data is available at the outputs, Q_0–Q_3. The four set inputs are available separately so that the 9314 may be used as four separate type D flip-flops.

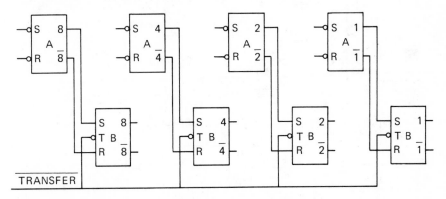

Figure 9-3. The transfer function using jam entry.

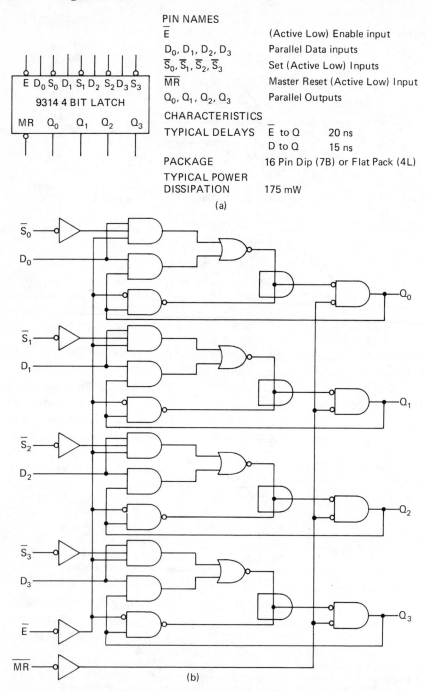

PIN NAMES

$\overline{E}$	(Active Low) Enable input
D_0, D_1, D_2, D_3	Parallel Data inputs
$\overline{S}_0, \overline{S}_1, \overline{S}_2, \overline{S}_3$	Set (Active Low) Inputs
$\overline{MR}$	Master Reset (Active Low) Input
Q_0, Q_1, Q_2, Q_3	Parallel Outputs

CHARACTERISTICS

TYPICAL DELAYS	$\overline{E}$ to Q	20 ns
	D to Q	15 ns
PACKAGE	16 Pin Dip (7B) or Flat Pack (4L)	
TYPICAL POWER DISSIPATION	175 mW	

(a)

(b)

Figure 9-4. TTL/MSI 9314 4-bit latch: (a) logic diagram; (b) symbol. (Courtesy of Fairchild Semiconductor.)

9.4 SHIFT REGISTERS

Occasionally, it will be preferable to handle data in serial. When this is the case, a shift register is used. Figure 9-5 shows a 4-bit shift register that uses JK flip-flops. With the exception of the first, the input of each flip-flop is conditioned by its preceding flip-flop. The result is that the data appearing in each flip-flop PRIOR to a clock pulse is shifted to the next flip-flop WITH the clock pulse. Because there are four flip-flops, four clock pulses are required to shift data through this register. As indicated in the diagram, any number of flip-flops may be used in a shift register. Data appearing at the input and data removed from the output must be synchronized with the clock pulses. Clock pulse generators are discussed in Chapter 12. Shift registers are available as MSI circuits and in various capacities (bit positions). Figure 9-6 shows a serial-in serial-out 8-bit shift register utilizing TTL circuits.

In logic diagrams, entire registers are symbolized by a single block symbol. Figure 9-7 shows the symbol for the shift register shown in Figure 9-6. Inputs for such blocks are at either the top or the left side. Outputs appear at the bottom or the right side.

It is sometimes desirable to change data from the serial mode to the parallel mode. Conversely, it may be desirable to change data from parallel to serial mode. Whenever this is the case, a register is used with both serial and parallel modes of operation. Figure 9-8 shows a 4-bit shift register with parallel inputs and parallel outputs. The direct SET and RESET inputs are used for jam entry of parallel data. The outputs of each flip-flop are available so that data may be obtained in the parallel mode.

Registers that are used between two systems where timing and/or mode variations exist are called *buffer registers*. Figure 9-9 shows an MSI buffer register from the TTL family. The 9300 is a synchronous 4-bit shift register with parallel outputs. Data entry is synchronous with the registers, changing state after each low-to-high transition of the clock when the parallel enable input is LOW, the parallel inputs determine the next condition of the shift register. When the parallel enable input is HIGH, the shift register performs a 1-bit shift to the right, with data entering the first-stage flip-flop through JK inputs. By tying the two inputs

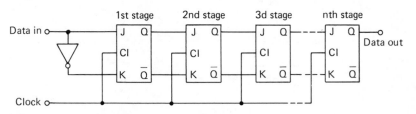

Figure 9-5. Four-bit shift register.

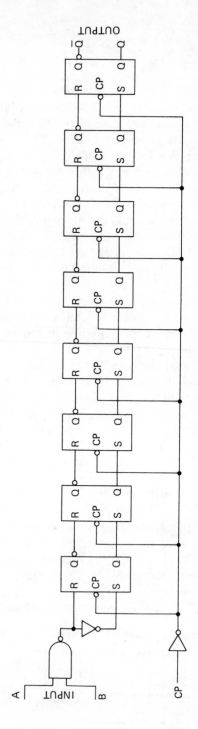

Figure 9-6. TTL/MSI 9391 8-bit shift register. (Courtesy of Fairchild Semiconductor.)

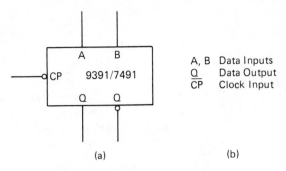

(a)   (b)

Figure 9-7. TTL/MSI 9391 8-bit shift register: (a) symbol; (b) pin names.

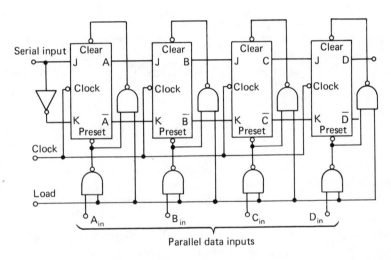

Figure 9-8. Four-bit serial-in serial-out, parallel-in parallel-out buffer register.

together, we can obtain D type entry. When the master RESET is activated, the register clears, regardless of the other inputs.

9.5 LEFT-RIGHT SHIFT REGISTERS

Shift registers normally shift data in only one direction. However, sometimes it is desirable for a register to be able to shift data left or right. An example of a left-right shift register appears in Figure 9-10. The output to each D type flip-flop is gated to the input of its following flip-flop to

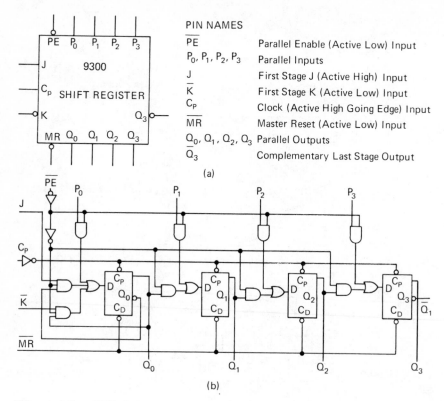

PIN NAMES

$\overline{PE}$	Parallel Enable (Active Low) Input
P_0, P_1, P_2, P_3	Parallel Inputs
J	First Stage J (Active High) Input
$\overline{K}$	First Stage K (Active Low) Input
C_P	Clock (Active High Going Edge) Input
$\overline{MR}$	Master Reset (Active Low) Input
Q_0, Q_1, Q_2, Q_3	Parallel Outputs
$\overline{Q_3}$	Complementary Last Stage Output

(a)

(b)

Figure 9-9. TTL/MSI 9300 buffer register: (a) symbol; (b) logic diagram. (Courtesy of Fairchild Semiconductor.)

perform a right shift. In order to accomplish a left shift, the output of each flip-flop is gated to the input of its preceding flip-flop.

The direction of the shift is determined by the MODE input. If the MODE input is low, the shift will be to the right; if the MODE input is high, the shift will be to the left.

An MSI example of the left-right shift register appears in Figure 9-11. When a logical 0 level is applied to the mode control input, the number 1 AND gates are enabled and the number 2 AND gates are inhibited. In this mode, the output of each flip-flop is coupled to the R-S inputs of the succeeding flip-flop and right-shift operation is performed by clocking at the clock 1 input. In this mode, serial data is entered at the serial input. Clock 2 and parallel inputs P_A through P_D are inhibited by the number 2 AND gates.

When a logical 1 level is applied to the mode control input, the number 1 AND gates are inhibited (decoupling the outputs from the

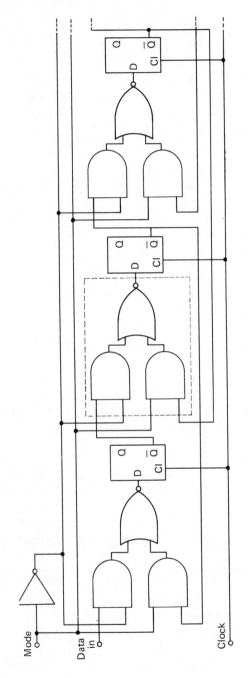

Figure 9-10. Left-right shift register.

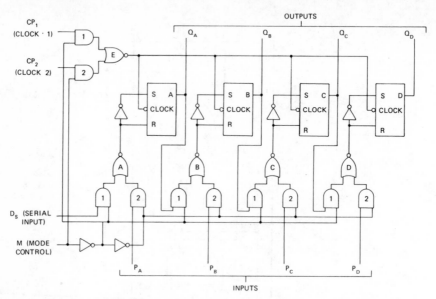

Figure 9-11.   9395/7495 4-bit left-right shift register. (Courtesy of Fairchild Semiconductor.)

succeeding R-S inputs to prevent right-shift), and the number 2 AND gates are enabled to allow entry of data through parallel inputs P_A through P_D and clock 2. This mode permits parallel loading of the register or, with external interconnection, shift-left operation. In this mode, shift left can be accomplished by connecting the output of each flip-flop to the parallel input of the previous flip-flop (Q_D to P_C, etc.), and serial data is entered at input P_D.

Clocking for the shift register is accomplished through the AND-OR gate E, which permits separate clock sources to be used for the shift-right and shift-left modes. If both modes can be clocked from the same source, the clock input may be applied commonly to clocks 1 and 2. Information must be present at the R-S inputs of the master-slave flip-flops prior to clocking. Transfer of information to the output pins occurs when the clock input goes from a logical 1 to a logical 0. Figure 9-12 shows the symbol for the 9395 left-right shift register.

9.6   *RECIRCULATING SHIFT REGISTERS*

Data stored in long shift registers must be shifted in order for the data to be read or transferred because it is not practical to provide parallel outputs when many flip-flops are used. As the data is shifted, it appears

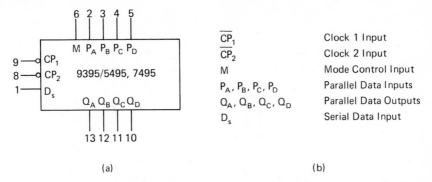

Figure 9-12. 9395/7495 4-bit left-right shift register: (a) symbol; (b) pin names. (Courtesy of Fairchild Semiconductor.)

at the output, one bit at a time. The output of the shift register is usually connected back to the input so that the data is recirculated, and, as a result, it is restored. If this were not done, the data would be lost.

Figure 9-13 shows the MSI 9328. It is a dual 8-bit shift register with data select inputs which are used to recirculate data. Provision is made for independent clock pulses or a single clock pulse for the two registers. Figure 9-14 shows the symbol for the 9328 shift register. It also shows the interconnection for recirculating the data.

If the data select input (D_S) is LOW, the input to the register is D_0 and the D_1 input is disabled. When D_S is high, D_0 becomes inactive and D_1 becomes active. This prevents the recirculation of the data and allows for the entry of new data.

9.7 MOS DYNAMIC SHIFT REGISTERS

One of the big advantages of MOS circuitry is in high capacity shift registers. You will recall from Chapter 6 that dynamic gates are most popular today because of their low power dissipation. This is particularly true in shift registers.

The shift registers previously discussed are implemented with flip-flops, which are bistable devices. Because of this, when a bit is clocked into the flip-flop, it remains there until the next clock pulse shifts it into the succeeding flip-flop. If the clock pulses are stopped, each flip-flop in the register will store the bits indefinitely.

This is not the case in dynamic shift registers. Because the bits of information are stored in capacitances, the circuit must be clocked continuously so that the charges which may leak off between clock pulses will be restored.

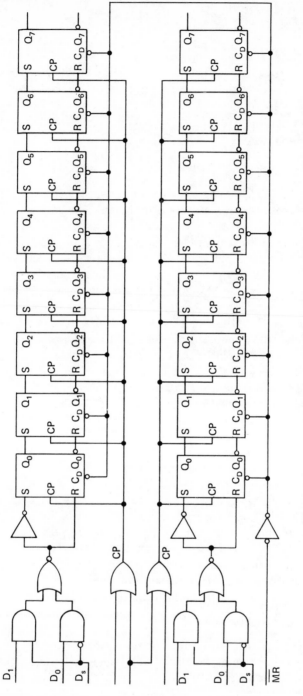

Figure 9-13. MSI 9328 dual 8-bit shift register. (Courtesy of Fairchild Semiconductor.)

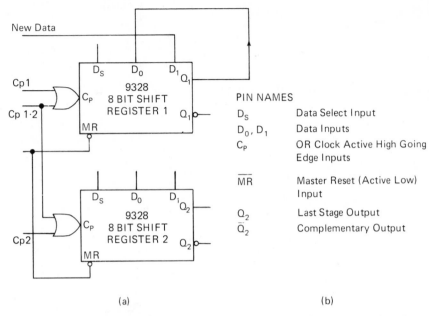

PIN NAMES

D_S	Data Select Input
D_0, D_1	Data Inputs
C_P	OR Clock Active High Going Edge Inputs
$\overline{MR}$	Master Reset (Active Low) Input
Q_2	Last Stage Output
$\overline{Q}_2$	Complementary Output

(a)   (b)

Figure 9-14.   MSI 9328 dual 8-bit shift register: (a) symbol; (b) pin names. (Courtesy of Fairchild Semiconductor.)

Figure 9-15 shows two ratioless inverters connected to form a single bit of storage in a dynamic shift register. Note that in the timing chart a 1 is represented by a minus voltage and a 0 is represented by 0 volts. This is true because PMOS transistors require a negative supply. This would of course be opposite if NMOS transistors were used.

This is a two-phase circuit, requiring a $\phi 1$ pulse followed by a $\phi 2$ pulse to store one bit of information. The chart shows an input bit appearing at t_0 (time zero). The register does not sense this until a $\phi 1$ pulse appears, turning on Q_1 and charging C_1. If a 1 is present at the input, Q_2 and Q_3 turn on and the inverted output charges C_2. At t_2, $\phi 2$ turns on Q_4 transferring the charge on C_2 to C_3. $\phi 2$ also turns on Q_6 and, since Q_5 is turned off by the charge on C_3, C_4 charges to a 1 level. Notice that after the two clock pulses, the input level is shifted to the output.

Figure 9-16 shows how 100 stages are connected to form a 100-bit dynamic shift register. Two of these registers appear in a single package.

The power dissipation may be reduced even further by using four-phase logic. Figure 9-17 shows two inverters connected to form a single bit position in a four-phase dynamic shift register. Notice that a DC power supply is not used.

The first inverter is energized by a $\phi 1$ pulse and Q_2 is turned on by

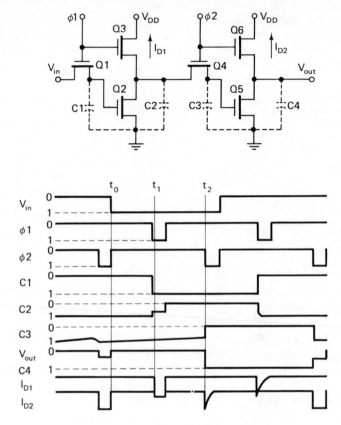

Figure 9-15. One-bit ratioless dynamic shift register and timing chart. (Courtesy of Motorola Semiconductor Products, Inc. *Motorola Monitor,* Vol. 9/No. 2.)

$\phi2$. This occurs at time t_1. The result is that if C_1 is not charged, it will charge regardless of the state of the input. This capacitance charges through Q_3 if the input is at a 0 level because Q_1 is turned off. However, if the input is at the 1 level, C_1 will charge to the potential effected by the ratio of Q_3 and Q_2.

At t_2, $\phi1$ has ended. If the input data is at the 0 level, C_1 will remain charged. If the input data is at the 1 level, C_1 will discharge through Q_2 and Q_1 because the $\phi2$ pulse has not yet ended. At the end of the $\phi2$ pulse, C_1 will hold the complement of the input level.

The second inverter operates in the same manner, using clock pulses $\phi3$ and $\phi4$. After a complete cycle of clock pulses, C_2 will be charged to a level equal to the input data.

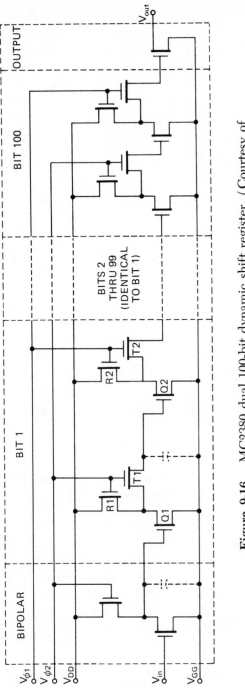

Figure 9-16. MC2380 dual 100-bit dynamic shift register. (Courtesy of Motorola Semiconductor Products, Inc. *Motorola Monitor*, Vol. 9/No. 2.)

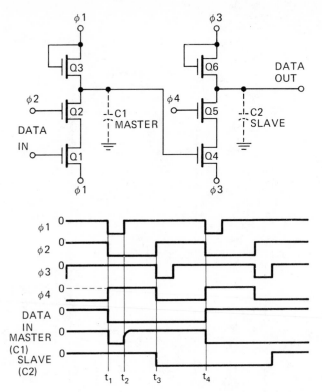

Figure 9-17. Four-phase dynamic shift register: 1-bit position. (Courtesy of Motorola Semiconductor Products, Inc. *Motorola Monitor*, Vol. 9/No. 2.)

The advantage of the four-phase dynamic shift register is that no DC paths exist, and therefore the power dissipated can only be reactive power. This means that the transistors act only as switches to charge or discharge the capacitances. Remember, these are stray capacitances that are inherent in MOS transistors. When the power dissipation is reduced to a minimum, the geometries of the transistors may also be reduced.

Data in dynamic shift registers must be recirculated or it will be lost. Another disadvantage of dynamic shift registers is that they require a more complex pulse generator. However, because of the low power dissipation, large amounts of circuitry can be placed on a single chip, allowing for placement of a clock pulse generator on the same chip as the shift register.

Figure 9-18 shows a dual 128-bit shift register that is compatible with TTL logic. It has its own clock generator and can be clocked with a single phase clock pulse at a 2 MHz rate.

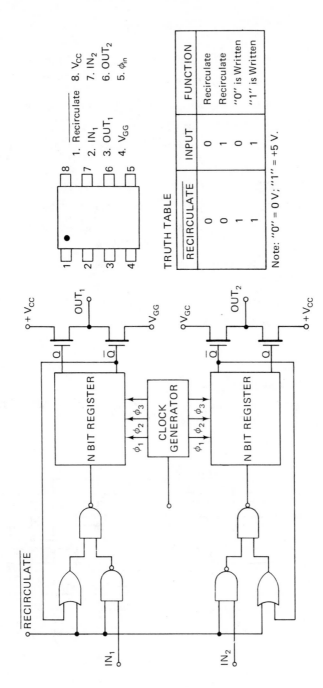

Figure 9-18. Dual 128-bit shift register with clock generator.

1. Recirculate
2. IN_1
3. OUT_1
4. V_{GG}

8. V_{CC}
7. IN_2
6. OUT_2
5. ϕ_{in}

TRUTH TABLE

RECIRCULATE	INPUT	FUNCTION
0	0	Recirculate
0	1	Recirculate
1	0	"0" is Written
1	1	"1" is Written

Note: "0" = 0 V; "1" = +5 V.

227

QUESTIONS AND PROBLEMS—CHAPTER 9

1. What is the difference between *parallel* and *serial* registers?
2. What does a register do?
3. How many flip-flops may be used in a shift register?
4. What are *buffer registers?*
5. Can data be shifted both directions through a shift register? Explain.
6. In Figure 9-11, through which gate is clocking accomplished?
7. In Figure 9-11, when is information transferred to the output pins?
8. If the clock pulses are stopped, how long will the flip-flops in a shift register store the bits of information?
9. Why must data in dynamic shift registers be continuously recirculated?
10. When jam entry is used, when may new data be entered?
11. How are the flip-flops in a jam entry register reset?
12. What are the two basic classifications of registers?
13. What kind of registers are used *between* circuits where timing differences exist?
14. For the registers discussed in this chapter, which type of data transfer requires the least amount of time?

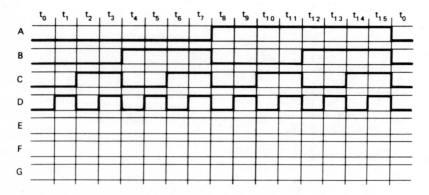

Figure 9-19. Waveforms for problems 18 through 20.

15. If a 9 (1001) is entered into the register in Figure 9-1, and then a 7 (0111) is immediately entered without resetting, what will be the number stored? Why?

16. If a 9 (1001) is entered into the register in Figure 9-2, and immediately followed by entering a 7 (0111), what will be the number stored? Why?

17. For what purpose are recirculating shift registers used?

18. Using the waveforms in Figure 9-19, draw the output waveforms (E, F, G, H) for the circuit in Figure 9-20. (Refer to Figure 9-9.)

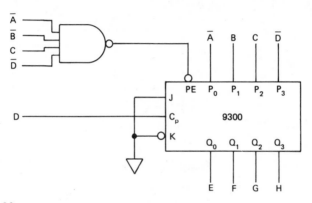

Figure 9-20.

19. Using the waveforms in Figure 9-19, draw the output waveforms (E, F, G, H) for the circuit in Figure 9-21. (Refer to Figure 9-9.)

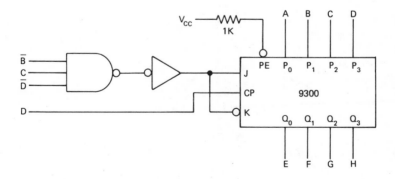

Figure 9-21.

20. Using the waveforms in Figure 9-19, draw the output waveforms (E, F, G, H) for the circuit in Figure 9-22. (Refer to Figures 9-11 and 9-12.)

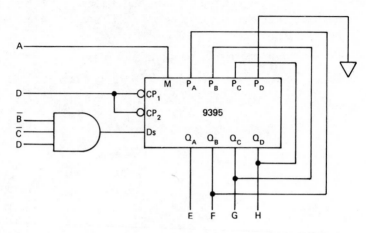

Figure 9-22.

Encoders, Decoders, and Code Converters

Data that is transmitted between peripherals and the CPU or over telephone lines is usually in the form of a character-oriented code such as ASCII, which was discussed in Chapter 2. In order to produce these codes from an alphanumeric source (such as a keyboard), an encoder is required. Most modern CPUs process numerical data in straight binary. Therefore, the numerical portion of the ASCII code (BCD) must be converted to binary at the input of the CPU.

In order to print out data, the process is reversed. The binary numbers must be converted back to BCD (ASCII) before they can be decoded. *Decoding* is the process wherein a code is translated to the character it represents. The block diagram in Figure 10-1 illustrates the process from input to output for a computer system.

Since the process of converting from BCD to binary or binary to BCD is an arithmetic process, these converters will be discussed in Chapter 13.

Because it is sometimes necessary to convert from one BCD code to another, such as from the 8421 code to the excess-three code, code converters as well as basic concepts related to data transfer and data multiplexing are also included in this chapter.

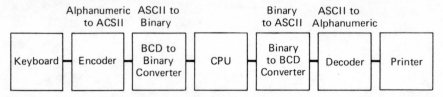

Figure 10-1.   Block diagram of data conversion steps from input to output.

The topics covered in this chapter are:

10.1 Encoders

10.2 Decoders

10.3 Code Converters

10.4 Multiplexers, Gating, and Data Steering

10.1 ENCODERS

Four binary bits (8421) are required for BCD. This means a BCD encoder will have four outputs. Looking at Figure 10-2 we see that to produce an 8 bit at the output, an 8 or 9 is required at the input; to produce

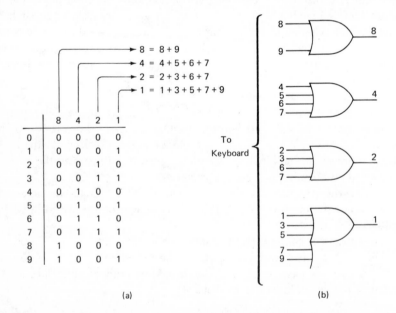

Figure 10-2.   BCD encoder: (a) truth table; (b) logic circuit.

a 4 bit at the output, a 4, 5, 6, or 7 is required at the input; and so on. Notice that the logic for each output line is the result of the OR function. Only one input line is active for any given character. For example, decimal 5 and decimal 7 cannot appear simultaneously. Each input line will activate one or more output lines (with the exception of 0, where the output is 0000). An active 7 at the input would activate the 4, 2, and 1 bits at the output. In order to construct an encoder for any base, one should consider the logic necessary to activate each bit of the binary code as required by the input character. Codes other than BCD, such as alphanumeric codes, are generated in the same manner; however, a greater number of bits may be required. Most terminals provide encoding at the keyboard so that the output connections are simplified.

10.2 DECODERS

A binary word, n bits in length, can represent 2^n different combinations of information. The 2-bit register in Figure 10-3(a) has four possible combinations: 00, 01, 10, and 11. Each one of these four combinations can be detected through a process known as *decoding*, shown in Figure 10-3(b). For the zero count (00), both flip-flops are in the reset state and the count is detected by ANDing the reset outputs $\overline{1}$ AND $\overline{2}$. In like manner, the 1 count is detected by ANDing 1 AND $\overline{2}$; the 2 count is detected by ANDing 2 AND $\overline{1}$; and the 3 count is detected by ANDing 2 AND 1.

10.2.1 Decoding hexadecimal. When hexadecimal characters are decoded from binary, the binary bits are broken into groups of four. Each group of four bits represents one hexademical character. There are 16 hexadecimal characters: 0 through 9 and A through F. Each hexadecimal character requires one 4-input AND gate so that its output is only one of the 16 possible input conditions. For example, in decoding the character 6 (binary 0110), all inputs of the AND gate must be true in order to provide an active output (6). Since the 8 and 1 bits are zero, $\overline{8}$ and $\overline{1}$ must be true. Then, the input to the AND gate must be $\overline{8}$ AND 4 AND 2 AND $\overline{1}$. In a similar manner, a 9 would be decoded as 8 AND $\overline{4}$ and $\overline{2}$ AND 1; "A" as 8 AND $\overline{4}$ AND 2 AND $\overline{1}$; and "B" as 8 AND $\overline{4}$ AND 2 AND 1, as shown in Figure 10-4.

The TTL/MSI 9311 is shown in Figure 10-5. It is a 4-bit input decoder with 16 outputs used to decode hexadecimal characters. The enable inputs are used to activate the circuit. If either enable input is HIGH, *all* outputs are HIGH. The outputs are active LOW.

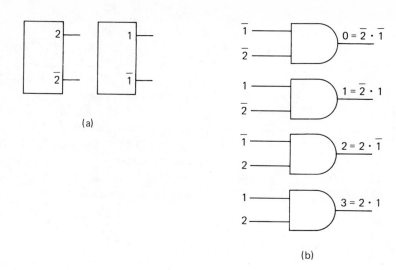

(a)

(b)

Figure 10-3.   Decoding a 2-bit register: (a) 2-bit register; (b) decoder.

10.2.2 Decoding BCD. Figure 10-6 shows ten AND gates used as a BCD decoder. The input is BCD and the output is ten lines, representing the decimal numbering system (0–9). Since there are six input conditions which are never experienced in BCD (invalid), the number of inputs required to some gates is reduced. For example, when observing the truth table in Figure 10-6(a), notice that the 8 bit is only required for the 8th

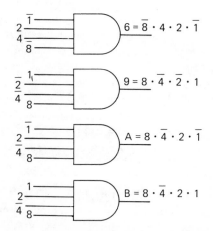

Figure 10-4.   First-level logic diagram for decoding the hexadecimal characters 6, 9, A, and B.

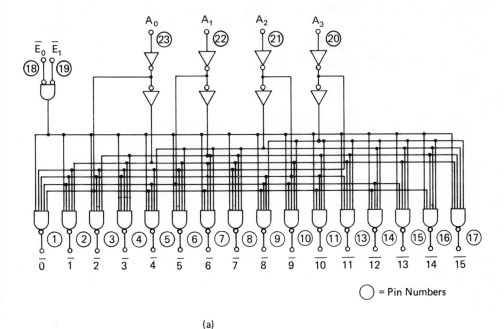

(a)

○ = Pin Numbers

V_{CC} = PIN 24
GND = PIN 12

(b)

Figure 10-5. TTL/MSI 9311 hexadecimal decoder: (a) logic diagram; (b) symbol.

and the 9th counts. It is not necesary to consider the value of the 4 and 2 bits because they never are true when the 8 bit is true. Therefore, 2-input gates are used to decode 8 and 9.

The gates decoding 2 through 7 are 3-input gates. Since the value

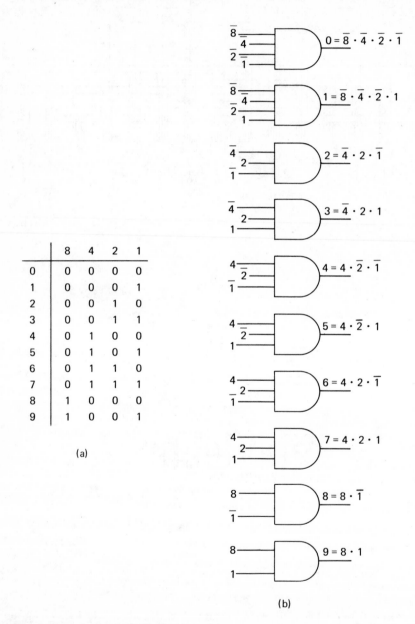

	8	4	2	1
0	0	0	0	0
1	0	0	0	1
2	0	0	1	0
3	0	0	1	1
4	0	1	0	0
5	0	1	0	1
6	0	1	1	0
7	0	1	1	1
8	1	0	0	0
9	1	0	0	1

(a)

$$0 = \overline{8} \cdot \overline{4} \cdot \overline{2} \cdot \overline{1}$$

$$1 = \overline{8} \cdot \overline{4} \cdot \overline{2} \cdot 1$$

$$2 = \overline{4} \cdot 2 \cdot \overline{1}$$

$$3 = \overline{4} \cdot 2 \cdot 1$$

$$4 = 4 \cdot \overline{2} \cdot \overline{1}$$

$$5 = 4 \cdot \overline{2} \cdot 1$$

$$6 = 4 \cdot 2 \cdot \overline{1}$$

$$7 = 4 \cdot 2 \cdot 1$$

$$8 = 8 \cdot \overline{1}$$

$$9 = 8 \cdot 1$$

(b)

Figure 10-6. BCD decoder: (a) truth table; (b) logic circuit.

of the 8 bits is always 0 through these counts, it is not considered. The gates decoding 0 and 1 are 4-input gates. All bits must be considered because the values of the 4, 2, and 1 bits reappear in other counts. Leaving out any bit will result in the gate's being activated on other counts.

The TTL/MSI 9301 is a BCD to decimal decoder; it is shown in Figure 10-7. It accepts four active HIGH BCD inputs and provides ten mutually exclusive outputs as shown by the logic symbol. The logic design of the 9301 insures that all outputs are HIGH when binary codes

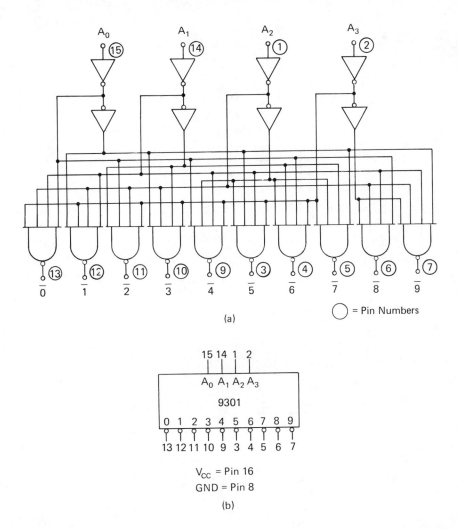

Figure 10-7. TTL/MSI 9301—BCD to decimal decoder: (a) logic diagram; (b) logic symbol. (Courtesy of Fairchild Semiconductor.)

greater than 9 are applied to the inputs. This prevents two output lines from becoming active when an invalid BCD input occurs.

Some decoders are used to drive specific types of readout devices. The 7441 shown in Figure 10-8 is a BCD to decimal decoder/driver designed specifically to drive cold cathode indicator tubes. High-perfor-

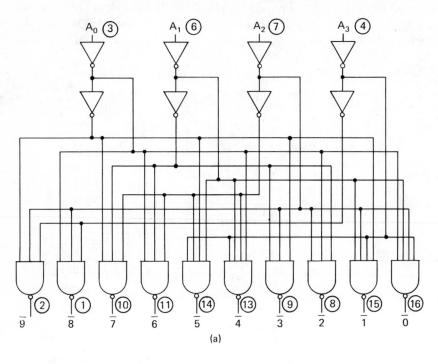

(a)

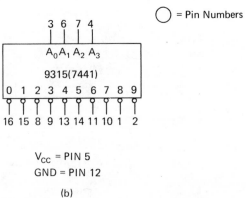

V_{CC} = PIN 5
GND = PIN 12

(b)

Figure 10-8.   TTL/MSI 7441 BCD decoder: (a) logic diagram; (b) logic symbol. (Courtesy of Fairchild Semiconductor.)

mance NPN transistors having high voltage breakdown, and low leakage characteristics are used to drive the cathodes directly. Full decoding is provided for all possible input states. For binary inputs greater than 9, all outputs are HIGH, thereby preventing invalid outputs.

10.2.3 Seven-segment decoders. Some numerical readout devices, such as light emitting diodes (LEDs), use a seven-segment configuration to produce the decimal characters 0–9. A decoder used to drive such a display must provide appropriate outputs for each of the seven segments. If the segments are labeled as in Figure 10-9(a), the following expressions can be used to define the decoding process:

$$a = 0 + 2 + 3 + 5 + 7 + 8 + 9$$
$$b = 0 + 2 + 3 + 4 + 7 + 8 + 9$$
$$c = 0 + 3 + 4 + 5 + 6 + 7 + 8 + 9$$
$$d = 0 + 2 + 3 + 5 + 6 + 8$$
$$e = 0 + 1 + 2 + 6 + 8$$
$$f = 0 + 1 + 4 + 5 + 6 + 8 + 9$$
$$g = 2 + 3 + 4 + 5 + 6 + 8 + 9$$

The resulting numerical display for each of the ten characters would appear as in Figure 10-9(b).

The 9317 is a TTL/MSI seven-segment decoder/driver designed to accept four inputs in the 8421 BCD code and to provide the appropriate output to drive a seven-segment numerical display. The decoder can be

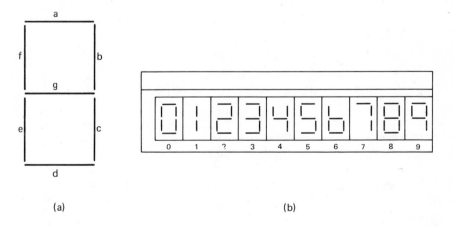

(a)   (b)

Figure 10-9.   Seven-segment numeric display: (a) segment designations; (b) ten digit display. (Courtesy of Fairchild Semiconductor.)

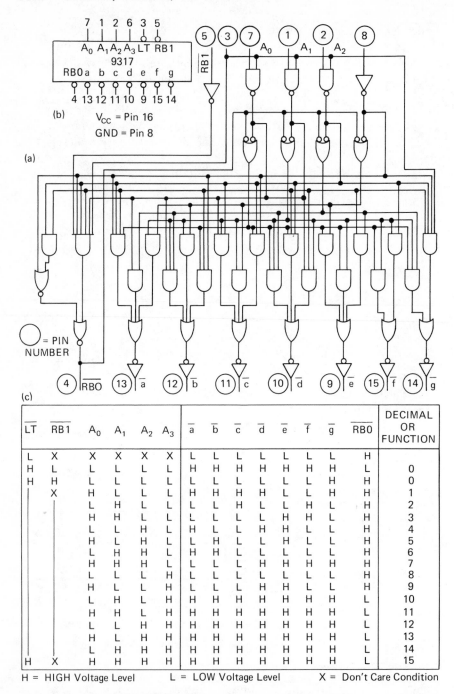

(c)

$\overline{LT}$	$\overline{RB1}$	A_0	A_1	A_2	A_3	$\overline{a}$	$\overline{b}$	$\overline{c}$	$\overline{d}$	$\overline{e}$	$\overline{f}$	$\overline{g}$	$\overline{RB0}$	DECIMAL OR FUNCTION
L	X	X	X	X	X	L	L	L	L	L	L	L	H	
H	L	L	L	L	L	H	H	H	H	H	H	H	L	0
H	H	L	L	L	L	L	L	L	L	L	L	H	H	0
	X	H	L	L	L	H	H	H	H	L	L	H	H	1
		L	H	L	L	L	L	H	L	L	H	L	H	2
		H	H	L	L	L	L	L	L	H	H	L	H	3
		L	L	H	L	H	L	L	H	H	L	L	H	4
		H	L	H	L	L	H	L	L	H	L	L	H	5
		L	H	H	L	H	H	L	L	L	L	L	H	6
		H	H	H	L	L	L	L	H	H	H	H	H	7
		L	L	L	H	L	L	L	L	L	L	L	H	8
		H	L	L	H	L	L	L	H	H	L	L	H	9
		L	H	L	H	H	H	H	H	H	H	H	L	10
		H	H	L	H	H	H	H	H	H	H	H	L	11
		L	L	H	H	H	H	H	H	H	H	H	L	12
		H	L	H	H	H	H	H	H	H	H	H	L	13
		L	H	H	H	H	H	H	H	H	H	H	L	14
H	X	H	H	H	H	H	H	H	H	H	H	H	L	15

H = HIGH Voltage Level   L = LOW Voltage Level   X = Don't Care Condition

Figure 10-10. TTL/MSI 9317 seven-segment decoder/driver: (a) logic diagram; (b) logic symbol; (c) truth table. (Courtesy of Fairchild Semiconductor.)

used to directly drive seven-segment incandescent lamp displays or light-emitting diode indicators. Figure 10-10 shows the symbol for the 9317, its truth table, and the logic diagram. The seven outputs (a, b, c, d, e, f, g) of the decoder select the corresponding segments of the display, as shown in Figure 10-11. Code configurations in excess of binary 9 disable the outputs.

The 9317 makes provision for automatic blanking of the leading and/or trailing edge zeros in a multidigit decimal number. This results in an easily readable decimal display conforming to normal writing practice. In an eight-digit integer fraction decimal representation using the automatic blanking capability, 060.0300 would be displayed as 60.03. Leading edge zero suppression is obtained by connecting the ripple blanking output (RBO) of a decoder to the ripple blanking input (RBI) of the next lower stage device. The most significant decoder stage should have the RBI input grounded; and since suppression of the least significant interger zero in a number is not usually desired, the RBI input of this decoder stage should be left open. A similar procedure for the fractional part of a display will provide automatic suppression of trailing edge zeros. The lamp test input overrides all other input combinations and allows checking for possible display malfunctions. A wide variety of MSI decoders are available as off-the-shelf items.

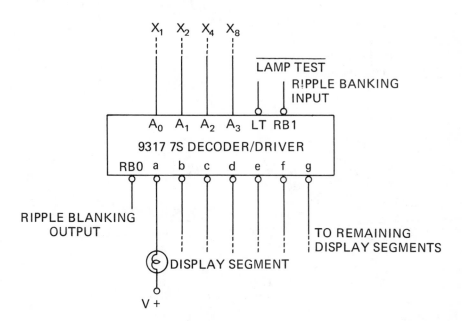

Figure 10-11.   Incandescent lamp display driven by seven-segment decoder/driver. (Courtesy of Fairchild Semiconductor.)

10.3   CODE CONVERTERS

Sometimes it is desirable to convert from one binary code to another. For instance, suppose the 2'421 code is used in one piece of equipment and the 8421 code is used in another. If data is to be transferred from one to the other, a code converter is required.

Let's design a logic circuit that will convert the 2'421 code to the 8421 code. The first step in our design would be to construct truth tables for the two codes as in Figure 10-12. (The student may need to review Chapter 2 for an explanation of the truth table for the 2'421 code.) Next we would write the logic for each of the four bits (8421). Finally, we would build the circuit based on our results.

An examination of the truth table shows that: 0 through 4 are identical in the two codes, but 5 through 9 differ; the 1 bit is identical in both codes. Therefore, we may use the 1 bit of the 2'421 code to produce the 1 bit of the 8421 code; the 2 bit in the 8421 code equals the 2 bit of the 2'421 code when 2' is NOT true. When 2' is true, the complement of the 2 bit in the 2'421 code is used. We can express this as

$$2 = (\overline{2'} \cdot 2) + (2' \cdot \overline{2}) = 2' \oplus 2$$

Further inspection of the truth table reveals that the 4 bit in the 8421 code is produced whenever the 4 bit is true and the 2 bit is NOT true in the 2'421 code. This produces the 4 bit for counts 4, 6, and 7. The 4 bit for the 5th count can be produced by 2' and $\overline{4}$ since this combination does not appear on any other count. This may be expressed as

$$4 = (4 \cdot \overline{2}) + (2' \cdot \overline{4})$$

Decimal	(2'	4	2	1)	(8	4	2	1)	
0	0	0	0	0	0	0	0	0	
1	0	0	0	1	0	0	0	1	
2	0	0	1	0	0	0	1	0	$2 = 2 \cdot \overline{2'} + \overline{2} \cdot 2'$
3	0	0	1	1	0	0	1	1	
4	0	1	0	0	0	1	0	0	
5	1	0	1	1	0	1	0	1	$4 = 4 \cdot \overline{2} + 2' \cdot \overline{4}$
6	1	1	0	0	0	1	1	0	
7	1	1	0	1	0	1	1	1	
8	1	1	1	0	1	0	0	0	$8 = 4 \cdot 2$
9	1	1	1	1	1	0	0	1	

Figure 10-12.   Truth tables for 2'421 BCD code and 8421 BCD code.

Finally, the 8 bit in the 8421 code can be produced with 4 and 2 in the 2′421 code since (4·2) are NOT true for any other counts. Eight can be expressed as

$$8 = 4 \cdot 2$$

After the logic is determined, the logic circuit is constructed as in Figure 10-13. Other codes can be converted in a similar manner. However, more complex codes such as alphanumeric codes are usually converted using ROMs (read-only memories). ROMs are discussed in Chapter 14.

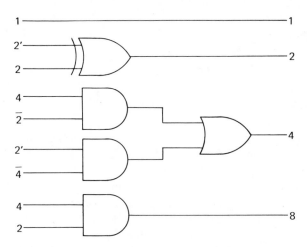

Figure 10-13.   Logic circuit for 2′421 to 8421 BCD code converter.

10.4 MULTIPLEXERS, GATING, AND DATA STEERING

In systems where data is handled in serial, there is a need to control the direction of data flow. For example, the register in Figure 10-14 has four possible sources of data. The method of selecting the appropriate data is called *gating* or *data steering*. Only one of the four AND gates is made active at a given time. This is usually done by the data to be selected AND a signal level called a *micro-instruction*, which has been produced specifically for this purpose (22, 24, 26, 28). A third input to the gates is usually required to select the time at which the gates are to be made active. These *timing pulses* are generated by the master clock. (Timing is discussed in Chapter 12.) For example, when micro-instruction 24 is generated, Data B will be transferred to the x Register when timing pulse t is present. Gating circuits such as these are commonly used in systems (such as calculators) where data is handled in serial.

MSI devices are also available for use in more complex data steering. The TTL/MSI 9312 in Figure 10-15 is an eight-input digital multiplex

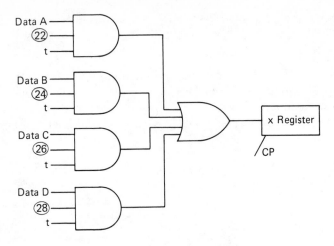

Figure 10-14.   Gating or data steering arrangement.

circuit. It provides in one package the ability to select one bit of data from up to eight sources. The data is selected by the three select inputs (S_0, S_1, S_2) and appears at the output in either the true or complement form. We can look at the 9312 as a logical implementation of a single-pole eight-position switch with the switch position controlled by the state of the three select inputs. The TTL/MSI 8263 is a 3-input 4-bit digital multiplexer. Its function is analogous to that of a four-pole, three-position switch. The 8263 is shown in Figure 10-16. Four bits of data are selected by a 2-bit channel selection code (S_0, S_1).

A third input (data complement) controls the conditional complement circuit at the multiplexer output to effect either inverting or noninverting data flow. This feature is desirable to provide the one's complement to the input of arithmetic units in order to facilitate the add-subtract functions.

The term *multiplex* usually refers to the handling of data over common lines. This is accomplished by time sharing the use of these lines. These lines are sometimes referred to as "buses." Data transfer in many computers is accomplished by multiplexing circuits and by using time-sharing techniques.

Tri-state logic circuits have been developed specifically for the purpose of driving bus lines. Figure 10-17 shows the circuit employed by the 8T09 quad bus driver. Note the additional input called *disable*. When this input is held LOW, the circuit is activated and operates as an inverter driver. When the output is active HIGH, Q_7 is turned on and Q_8 is turned off. When the output is active LOW, Q_8 is turned on and Q_7 is turned

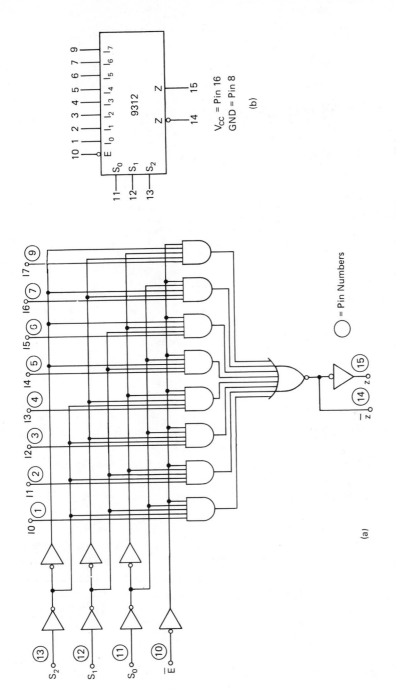

Figure 10-15. TTL/MSI 9312 multiplexer: (a) logic diagram; (b) logic symbol.

245

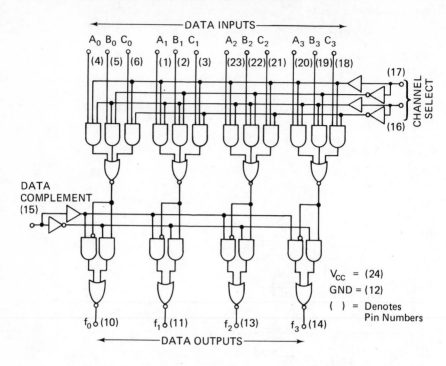

Figure 10-16.   TTL/MSI 8263 three-input 4-bit multiplexer. (Courtesy of Signetics Corporation.)

off. When the disable line is held HIGH, drive current is removed from both Q_7 and Q_8. The resulting high impedance output virtually disconnects the driver from the bus line. This third state of the driver allows many drivers to be connected to a common bus. Only the driver which is activated by the control input drives the bus.

Of special interest is the bidirectional data bus shown in Figure 10-18. It uses standard TTL gates such as the 7400 as receivers and the 8T09 as transmitters. As many as 25 receiver-transmitter pairs may be tied onto the same bus without exceeding the drive capability of the 8T09.

Standard MSI circuits may be adapted easily to bus-organized systems. Using the 8T09, one can design a minicomputer by employing a single high-speed bus as illustrated in Figure 10-19. The arithmetic logic unit, scratch-pad memory, and I/0 devices (multiplexers) may communicate directly. Moreover, micro-programmed instructions from a 1024-bit tri-state ROM may be put on the bus without need for an interface.

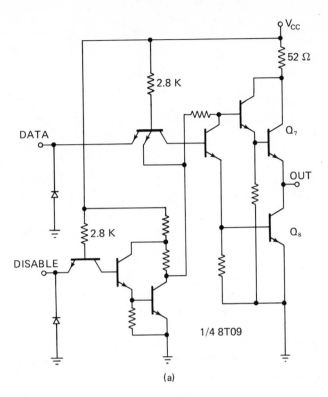

(a)

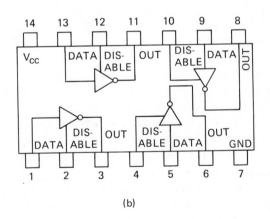

(b)

DATA	DISABLE	OUTPUT
0	0	1
1	0	0
0	1	HI-Z
1	1	HI-Z

(c)

Figure 10-17. 8T09 quad bus driver: (a) circuit schematic; (b) logic diagram; (c) truth table. (Courtesy of Signetics Corporation.)

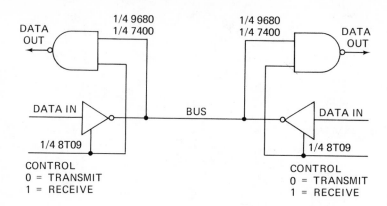

Figure 10-18.   Bidirectional data bus. (Courtesy of Signetics Corporation.)

Another example of a tri-state device is the 8T10 bus flip-flop. In one integrated circuit, a quad D-type flip-flop has been combined with tri-state output drivers for use in bus-organized systems. As shown in Figure 10-20, the outputs are disabled—i.e., switched to the high impedance state—when one or both of the inputs to the output disable NOR gate are HIGH. All four D-type flip-flops operate from a common clock and the data is transferred from the input to the output on the low-to-high transition of the clock pulse. With one or both of the input disable inputs HIGH, the flip-flops are in the hold mode and will store the information clocked in prior to the disable signal. The disable lines may change while the clock is HIGH or LOW without altering the data. A common reset input has also been provided.

The buffered tri-state outputs of the 8T10 allow the device to be used directly with other 8T10s in high-speed bus-organized systems without the need for interface gates. Multiplexing of data in bus-organized systems is rather simple, as illustrated in Figure 10-21. Each 8T10 provides data storage and is selected onto the bus by a LOW on both output disable lines. By means of a one-out-of-ten decoder, any one of the ten bus flip-flop devices can be selected.

Multiplexing of displays is greatly simplified and hardware is significantly reduced when using the 8T10 quad D-type bus flip-flop. Figure 10-22 shows that one decoder driver may be time-shared among several displays. If the display is large enough, the digit select drivers and decoding circuitry will cost much less than individual decoder drivers. In addition, strobing displays will result in a new power savings. The digit displays must be selected (strobed) in synchronization with the data on the bus. This is accomplished by the timing select circuit and the digit select drivers.

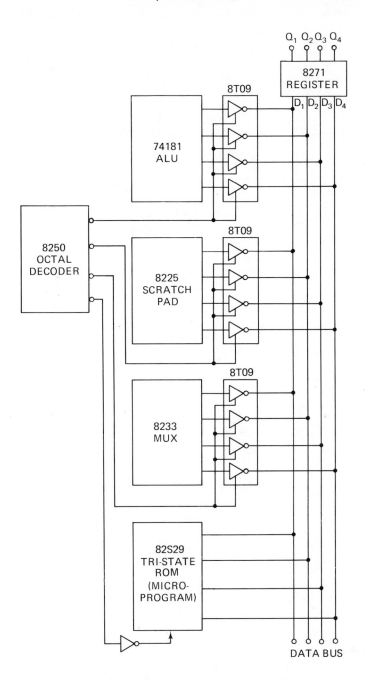

Figure 10-19. Bus-organized minicomputer. (Courtesy of Signetics Corporation.)

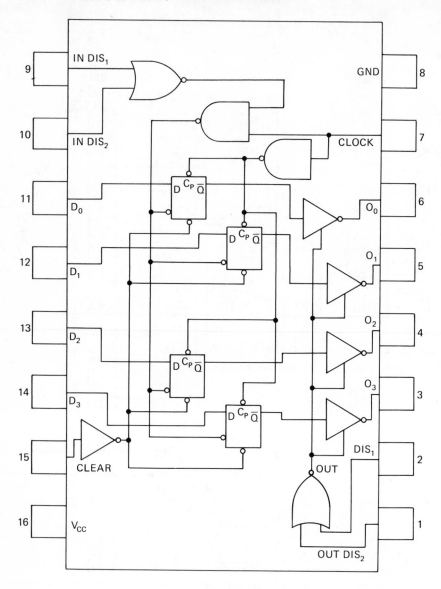

Figure 10-20.   Logic diagram of the 8T10 quad D-type bus flip-flop. (Courtesy of Signetics Corporation.)

Tri-state bus devices are easy to use, but caution should be exercised in systems design. Because of their high output current capability, they should be adequately decoupled just as other TTL drivers; that is, by placing a 0.01 to 0.1 μF high-frequency capacitor as close as possible to

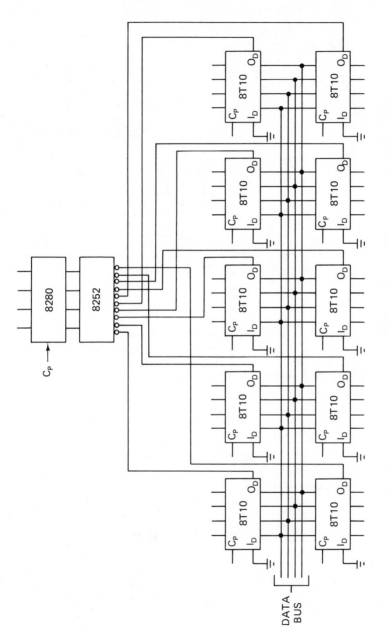

Figure 10-21. Multiplexing of data in bus-organized systems. (Courtesy of Signetics Corporation.)

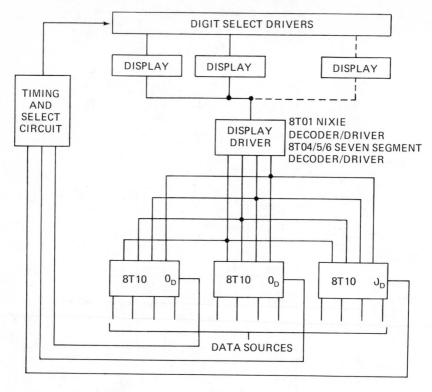

Figure 10-22.   Multiplexing of displays. (Courtesy of Signetics Corporation.)

the package. In this system, only one tri-state bus device per common bus is allowed to be enabled at a time. Realizing the high current handling capabilities, note that when two bus drivers are enabled simultaneously where one is active LOW and the other is active HIGH, a virtual short circuit will exist!

QUESTIONS AND PROBLEMS—CHAPTER 10

1. Design a logic circuit to convert from the excess-three code to the 8421 code. Show truth table, logic for each bit, and logic circuit.

2. When would code converters be used in the computer systems?

3. Explain the purpose of a *micro-instruction*.

4. How many outputs has the 9301? What conditions would cause all output to be HIGH?

5. Referring to Figure 10-2(b), determine the number of inputs that can be activated at the same time. How many gates can be activated at the same time?

6. What is meant by *activating* a gate?

7. What does the word *multiplex* mean when used in computer circuits? How is multiplexing accomplished in a computer system?

8. What is the difference between an *encoder* and a *decoder*?

9. Draw the first-level logic diagram for decoding the following hexadecimal characters: (a) 1, (b) 8, (c) D.

10. Referring to Figure 10-6, determine why the gate for decoding 2 must be a 4-input gate.

11. How many gates are required to encode the decimal digit 6?

12. How many and what type of gates are required to encode the hexadecimal digit E?

13. Why is it possible to use a 2-input AND gate to decode the decimal digit 9?

14. For the bi-directional data bus in Figure 10-18, what will occur if both control inputs are in the zero state?

15. For the 9312 Multiplexer in Figure 10-15, which input is selected when $S_0 = 0$, $S_1 = 1$, and $S_2 = 1$?

16. For the 8263 4-bit multiplexer in Figure 10-16, which input is selected when both channel select inputs are LOW?

17. Using the waveforms in Figure 10-23, draw the waveforms for the indicated outputs (E, F, G, H, I, J, K) in Figure 10-24. (Refer to Figure 10-5.)

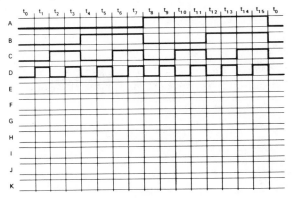

Figure 10-23.

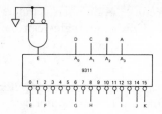

Figure 10-24.

18. Using the waveforms in Figure 10-23, draw the waveforms for the indicated outputs (E, F, G, H, I, J, K) in Figure 10-25. (Refer to Figures 10-10 and 10-11.)

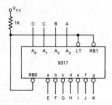

Figure 10-25.

19. Using the waveforms in Figure 10-23, draw the waveforms for the indicated outputs (E, F) in Figure 10-26. (Refer to Figure 10-15.)

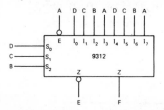

Figure 10-26.

20. Using the waveforms in Figure 10-23, draw the waveforms for the indicated outputs (G, H, I, J) in Figure 10-27. (Refer to Figure 10-16.)

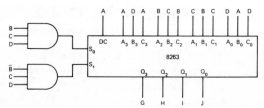

Figure 10-27.

11

Counters

Flip-flops may be programmed as counters, which are found in nearly all digital systems. Counters are used for frequency division, counting, sequencing digital operations, and mathematical operations.

Basically, digital counters are memory devices that store the number of input pulses. This chapter covers the more frequently used counting schemes. The present state of the art allows the designer to choose from a large variety of MSI devices which are available as off-the-shelf items.

The topics covered in this chapter are:

11.1 Binary Ripple-Carry Counter

11.2 BCD Ripple-Carry Counter

11.3 Synchronous Binary Counter

11.4 Synchronous BCD Counter

11.5 Presetting Counters

11.6 Down Counters

11.7 Up-Down Counters

11.8 Shift Counters

11.1   BINARY RIPPLE-CARRY COUNTER

The J-K or T flip-flops (as discussed in Chapter 7) have the ability to complement with each input pulse. This unique characteristic is required in binary counters. The J-K flip-flop will complement when both J and K inputs are active and when an active transient (CP) is applied to the T input. Figure 11-1(a) is a diagram demonstrating this condition. If a square wave is applied to the T input, as in Figure 11-1(b), the active transient (the change of state from LOW to HIGH) will cause the flip-flop to change state. Since only the active transient will cause the flip-flop to complement, the output frequency is one-half of the frequency at the T input.

Four flip-flops are connected in Figure 11-2 to form a binary ripple-carry counter. Notice that the T input of the second flip-flop is taken from the RESET output of the first since it is this output that will provide the necessary active transient to trigger the second flip-flop. When the first flip-flop changes from the 1 state to the 0 state, a HIGH going transient appears at the RESET output, causing the next flip-flop to change state. Because the output of each flip-flop is connected to the T input of the next, the input frequency is divided in half as many times as there are flip-flops. Therefore, the output frequency of the counter in Figure 11-2 is $\frac{1}{16}$ that of its input.

The output of each flip-flop in the counter may be considered to have a weighted binary value—that is, the output of the first flip-flop has a value of 1; the second, a value of 2; the third, a value of 4; the fourth,

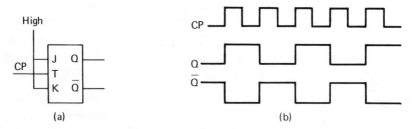

Figure 11-1.   J-K flip-flop: (a) flip-flop connected to complement; (b) input and output waveforms.

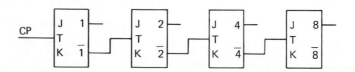

Figure 11-2.   First-level diagram of a binary ripple-carry counter.

a value of 8; and so on. By adding the value of all stages in the 1 (SET) state, we can obtain a total count. In this way, the simple frequency divider acts as a binary counter.

Observe that when any stage in a ripple-carry counter is to change, all preceding stages must also be changing from 1s to 0s. The situation is analogous to that encountered in adding 1 to an all-1 binary number. A 1 added to the least significant bit causes that bit to become a zero with a carry to the next bit. Each bit in turn becomes a zero with a carry to the next bit, and so on. The carry propagates bit by bit, until it finally encounters a zero.

If a stage changes from zero to one, this will terminate the carry because its output will not trigger the next stage. This is because the active transient to the next flip-flop will not be present.

A certain amount of time is used as each stage changes sequentially, and the carry seems to "ripple" through the counter. The outputs of the ripple-carry counter may be decoded. (Decoding was covered in Chapter 10.)

Figure 11-3 shows TTL 7476 J-K flip-flops connected to form a 0 to 15 (0000-1111) binary ripple-carry counter. Notice that the set output is used instead of the reset output because of the implied inversion at the T input of the next flip-flop. It should be apparent that any number of flip-flops may be added to increase the range of the counter.

Each flip-flop increases the range by a factor of two.

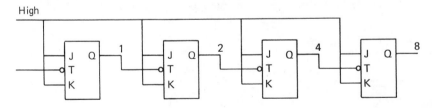

Figure 11-3. TTL 7476 J-K flip-flops connected to form a 0 to 15 (0000-1111) binary ripple-carry counter.

11.2 BCD RIPPLE-CARRY COUNTER

It is often desirable to construct binary ripple-carry counters so that they will count in decades; that is, their count is from 0 to 9 (0000-1001). The normal pattern of a four-state binary counter is a repetitive sequence of 16 discrete steps. In the base 10 ripple-carry counter, six counts must somehow be eliminated. If the six counts (1010-1111) are eliminated, the remaining ten maintain the same weighting as normal binary, namely

8-4-2-1. For this reason, the resulting BCD is sometimes called NBCD or the 8-4-2-1 code.

Figure 11-4 shows the truth table for the 8-4-2-1 BCD code. Note that the 8 bit appears only on the 8th and 9th counts. The 8-bit flip-flop cannot be triggered from the output of the 4-bit or 2-bit flip-flop because they do not change state after the 9th count. This means that the 8-bit flip-flop must be triggered by the input to the counter or the output of the 1-bit flip-flop. The counter in Figure 11-5 uses the latter. Therefore, the 8-bit flip-flop becomes synchronized with the 2-bit flip-flop. In order to prevent the 8-bit flip-flop from setting on counts 2, 4, or 6, the J input must be conditioned only when a 2 bit and a 4 bit are present. This is done by the AND gate at the J input of the 8-bit flip-flop.

The 10th input pulse causes the 1-bit and the 8-bit flip-flops to reset. However, unless it is prevented, the 2-bit flip-flop will set because of the carry from the 1-bit flip-flop. The 2-bit flip-flop is inhibited from setting by connecting the reset output of the 8-bit flip-flop to the J input of the 2-bit flip-flop. When the 8-bit flip-flop is set, its reset output is inactive, causing the J input of the 2-bit flip-flop to be inactive. Hence, the 2-bit flip-flop cannot set if an 8 bit is present. The 8-bit flip-flop will set on the eighth count because the J input is active (2 AND 4). On the tenth count,

	8	4	2	1
0	0	0	0	0
1	0	0	0	1
2	0	0	1	0
3	0	0	1	1
4	0	1	0	0
5	0	1	0	1
6	0	1	1	0
7	0	1	1	1
8	1	0	0	0
9	1	0	0	1
0	0	0	0	0

Figure 11-4.   Truth table for 8-4-2-1 BCD code.

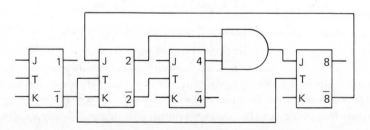

Figure 11-5.   First-level diagram of a BCD ripple-carry counter.

the 1-bit flip-flop as well as the 8-bit flip-flop will return to zero (complement). The 8-bit flip-flop's set output may be used as a carry to the next base 10 ripple-carry counter (decade). Figure 11-6 shows a BCD ripple-carry counter that uses RTL 9926 J-K flip-flops.

Decade counters are available in medium-scale integrated circuits (MSI), such as the Fairchild CμL 9958. The truth table and output waveforms are shown in Figure 11-7. Note that the output waveforms indicate

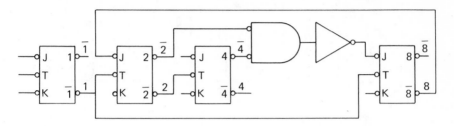

Figure 11-6.   BCD ripple-carry decade counter implemented with RTL 9926 J-K flip-flops.

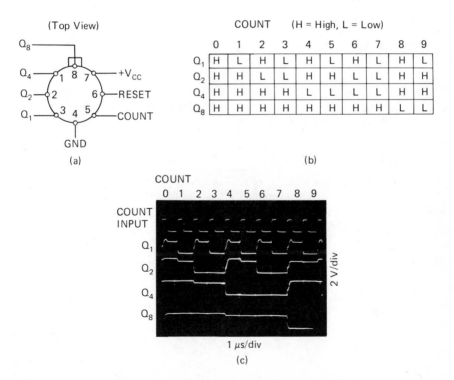

	COUNT					(H = High, L = Low)				
	0	1	2	3	4	5	6	7	8	9
Q_1	H	L	H	L	H	L	H	L	H	L
Q_2	H	H	L	L	H	H	L	L	H	H
Q_4	H	H	H	H	L	L	L	L	H	H
Q_8	H	H	H	H	H	H	H	H	L	L

Figure 11-7.   CμL/MSI 9958 decade counter: (a) connection diagram; (b) table of output states; (c) output waveforms. (Courtesy of Fairchild Semiconductor.)

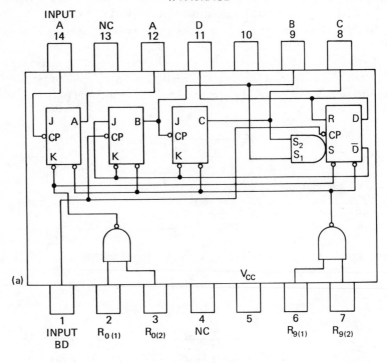

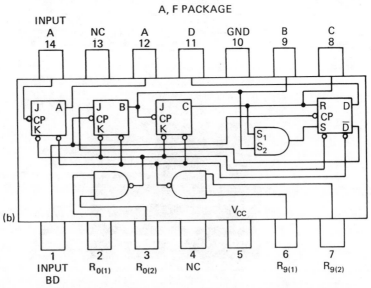

Figure 11-8. TTL/MSI 7490 counter: (a) W package; (b) A, F package. (Courtesy of Signetics Corporation.)

that low level is active. This circuit is compatible with the RTL family of logic circuits.

Figure 11-8 shows a highly versatile counter available in the TTL 7400 MSI series. It is a high-speed monolithic decade counter consisting of four master-slave flip-flops internally interconnected to provide a divide-by-two counter and a divide-by-five counter. Gated direct reset lines are provided to inihibit count inputs and return all outputs to a logical 0 or to a binary coded decimal (BCD) count of 9. As the output from flip-flop A is not internally connected to the succeeding stages, the count may be separated into three independent count modes:

1. When used as a BCD decade counter, the BD input must be externally connected to the A output. The A input receives the incoming count, and a count sequence is obtained in accordance with the BCD count sequence truth table shown in Figure 11-9. In addition to a conventional 0 reset, inputs are provided to reset a BCD 9 count for nine's complement decimal applications.

2. If a symmetrical divide-by-ten count is desired for frequency synthesizers or other applications requiring division of a binary count by a power of ten, the D output must be externally connected to the A input. The input count is then applied at the BD input, and a divide-by-ten square wave is obtained at output A.

BCD COUNT SEQUENCE (See Note 1)   RESET/COUNT (See Note 2)

COUNT	OUTPUT				RESET INPUTS				OUTPUT			
	D	C	B	A	$R_{0(2)}$	$R_{0(2)}$	$R_{9(1)}$	$R_{9(2)}$	D	C	B	A
0	0	0	0	0	1	1	0	X	0	0	0	0
1	0	0	0	1	1	1	X	0	0	0	0	0
2	0	0	1	0	X	X	1	1	1	0	0	1
3	0	0	1	1	X	0	X	0	COUNT			
4	0	1	0	0	0	X	0	X	COUNT			
5	0	1	0	1	0	X	X	0	COUNT			
6	0	1	1	0	X	0	0	X	COUNT			
7	0	1	1	1								
8	1	0	0	0								
9	1	0	0	1								

NOTES:

1. Output A connected to input BD for BCD count.

2. X indicates that either a logical 1 of a logical 0 may be present.

3. Fanout from output A to input BD and to 10 additional Series 54/74 loads is permitted.

Figure 11-9. Truth table for the 7490 BCD counter. (Courtesy of Signetics Corporation.)

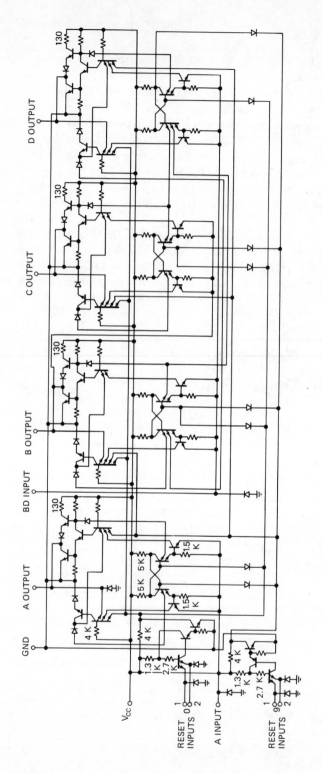

Figure 11-10. Schematic diagram for 7490 BCD counter. (Courtesy of Signetics Corporation.)

3. For operation as a divide-by-two counter and divide-by-five counter, no external interconnections are required. Flip-flop A is used as a binary element for the divide-by-five operation at B, C, and D outputs. In this mode, the two counters operate independently; however, all four flip-flops are reset simultaneously. (The schematic diagram for the 7490 appears in Figure 11-10.)

11.3 SYNCHRONOUS BINARY COUNTER

The ripple-carry counter requires a finite amount of time for each flip-flop to change state. This propagation delay causes the last flip-flop in the counter to change state at a later time than the first, resulting in a serious limitation in the operating frequency. In the synchronous binary counter, all flip-flops are triggered from the input. Therefore, all flip-flops are triggered together, and all outputs which are scheduled to change do so simultaneously.

Figure 11-11 shows a 4-bit synchronous counter. The first flip-flop changes state with each input clock pulse. The input to the second flip-flop is conditional; that is, the first flip-flop must be in the 1 state in order for the second to change with the next clock pulse. This is accomplished by wiring the J and K inputs to the SET outputs of the first flip-flop. In order to condition the third flip-flop to change state, both the first and second flip-flops must be in the 1 state. This is achieved by ANDing the SET outputs of the first and second flip-flops and connecting to the J and K inputs of the third flip-flop.

Conditioning the fourth flip-flop requires all the preceding flip-flops to be in the 1 state. The output of the first, second, and third flip-flops are ANDed and connected to the J and K inputs of the fourth. Therefore, for the synchronous binary counter, each flip-flop can change state only if all the preceding flip-flops are in the 1 state.

The advantage of the synchronous binary counter is that it allows for decoding specific counts without concern for unwanted gating spikes or shortened pulses due to the cumulative delay of the ripple-carry counter.

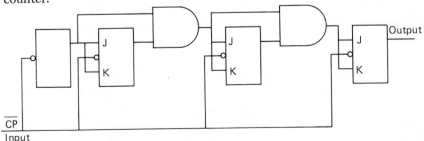

Figure 11-11. Four-bit synchronous counter.

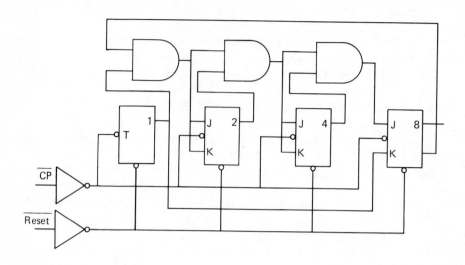

Figure 11-12.   Synchronous BCD counter.

11.4   *SYNCHRONOUS BCD COUNTER*

The 8-4-2-1 decade counter may be wired in the synchronous mode, as shown in Figure 11-12. All trigger inputs are connected together and to the input clock pulse so that all flip-flops that are scheduled to change state will do so together.

The truth table for the synchronous BCD counter appears in Figure 11-13. The J and K input gates of the 2-, 4-, and 8-bit flip-flops must be conditioned as follows:

1. To set the 2-bit flip-flop, a 1 bit and an $\overline{8}$ bit must be present.

2. To set the 4-bit flip-flop, a 1 bit, an $\overline{8}$ bit, and a 2 bit must first be present.

3. To set the 8-bit flip-flop, a 1 bit, a 2 bit, and a 4 bit must first be present.

On the tenth count, the 1-bit flip-flop will reset by complementing. The 2-bit and 4-bit flip-flops are not conditioned to SET because of the presence of the 8 bit. The 8-bit flip-flop will RESET because the 2 bit and the 4 bit are not present to condition its J input.

	8	4	2	1
0	0	0	0	0
1	0	0	0	1
2	0	0	1	0
3	0	0	1	1
4	0	1	0	0
5	0	1	0	1
6	0	1	1	0
7	0	1	1	1
8	1	0	0	0
9	1	0	0	1
0	0	0	0	0

Figure 11-13.   Truth table for synchronous BCD counter.

11.5   PRESETTING COUNTERS

Counters may be preset to a desired count, or number, in the same manner as registers, which were discussed in Chapter 9. The most common method is jam entry, as shown in Figure 11-14. The desired count is applied to the inputs A, B, C, and D. The count is transferred in parallel to the counter when the clock pulse (CP_2) appears. The counter will increment from the preset count with additional input pulses (CP_1).

Figure 11-15 shows a presettable synchronous binary counter. Figure 11-16 shows a presettable synchronous decade counter. These counters

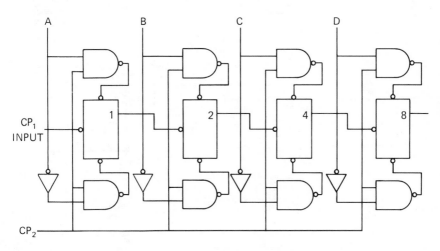

Figure 11-14.   Presetting counter using jam entry.

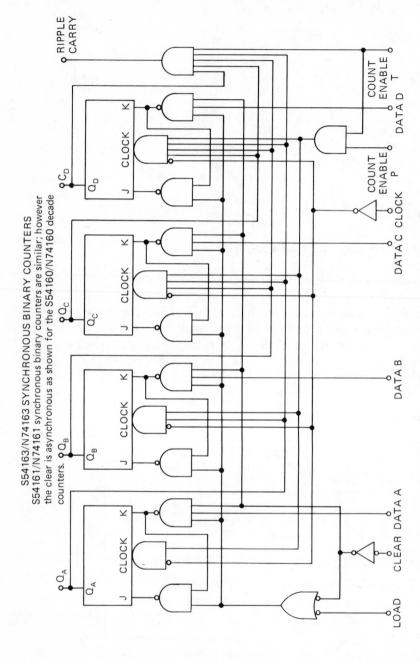

Figure 11-15. Synchronous binary counter. (Courtesy of Signetics Corporation.)

266

are desirable for application in high-speed counting schemes. Figure 11-17 shows the simplified logic symbol, including pin configurations of the counters in Figures 11-15 and 11-16. Figure 11-18 depicts the various waveforms of the counters discussed above.

11.6   DOWN COUNTERS

Counters may be made to count down by reversing the output connections of each flip-flop. Figure 11-19 shows how the trigger pulse is taken from the SET output instead of the RESET output of each flip-flop. In

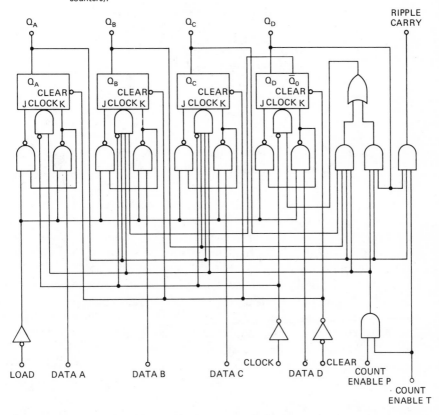

S54160/N74160 SYNCHRONOUS DECADE COUNTERS

(S54162/N74162 synchronous decade counters are similar; however the clear is synchronous as shown for the S54163/N74163 binary counters).

Figure 11-16.   Synchronous decade counter. (Courtesy of Signetics Corporation.)

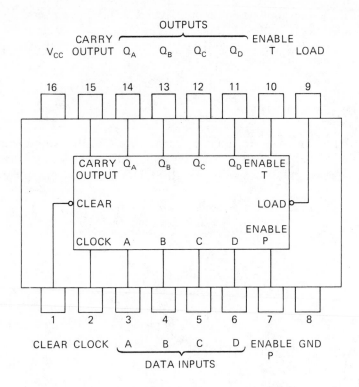

Figure 11-17. Pin configuration for S54160/N74160. (Courtesy of Signetics Corporation.)

order to count down, each flip-flop should change state when its preceding flip-flop changes from a 0 to a 1.

Figure 11-20 shows an implemented logic diagram of a binary down counter using TTL flip-flops. Remember, because of the implied inversion (low-going transient active) at the T input, the opposite connection is required. All of the direct set inputs (S_D) are tied together, providing a method of presetting the counter to its maximum count (1111). The counter will *decrement* (count down) with each input clock pulse. After it reaches 0000, the next input pulse will cause it to return to 1111.

11.7   UP-DOWN COUNTERS

Figure 11-21 shows how logic gates may be connected between the flip-flops so that the counter can be made either to count up or to count down.

If the level at the count mode input is HIGH, gate B is active so

TYPICAL CLEAR, PRESET, COUNT, AND INHIBIT SEQUENCES

Illustrated below is the following sequence:

1. Clear outputs to zero.
2. Preset to binary twelve.
3. Count to thirteen, fourteen, fifteen, zero, one, and two.
4. Inhibit.

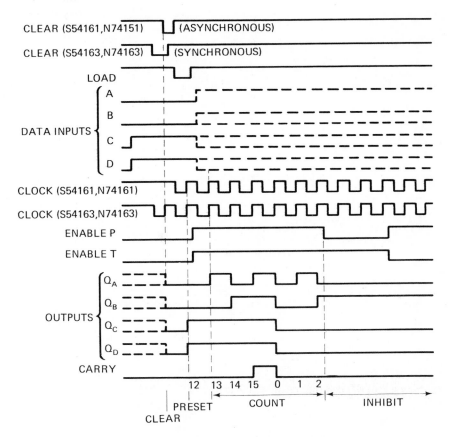

Figure 11-18. Waveforms for S54163/N74163 synchronous binary counter, and S54160/N74160 synchronous decade counter. (Courtesy of Signetics Corporation.)

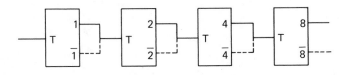

Figure 11-19. First-level diagram of a ripple-carry down counter.

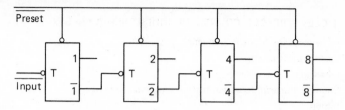

Figure 11-20. Implemented logic diagram of a binary down counter using TTL flip-flops.

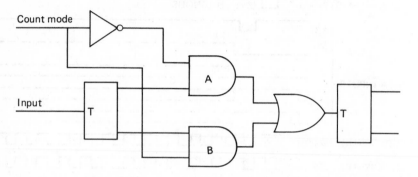

Figure 11-21. Logic gates may be connected between flip-flops to make the counter count up or count down.

that the counter will count up. However, if a LOW level is present at the mode input, gate A becomes active and the counter will count down.

Figure 11-22 shows a hexadecimal (4-bit binary) synchronous up-down counter. The "up/$\overline{\text{down}}$" input indicates that if this level is HIGH, the counter will count up, and if it is LOW, it will count down.

Figure 11-23 shows a BCD (8-4-2-1) up-down counter. Internal logic is provided to modify the count for BCD as well as to provide the mode of the count.

11.8 SHIFT COUNTERS

Shift registers may be used in a number of ways to produce counters. A simple ring counter is shown in Figure 11-24. The number of counts possible equals the number of flip-flops in the counter (N). Since there are six flip-flops, the count is 6, or 0 to 5, as indicated in the truth table in Figure 11-25.

This counter is not self-starting; that is, it must be preset so that a 1 appears only in the first flip-flop. If this is not done, an erroneous count

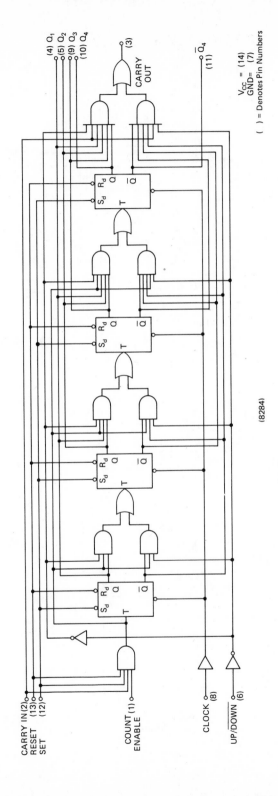

Figure 11-22. MSI 8284 synchronous up-down binary counter. (Courtesy of Signetics Corporation.)

271

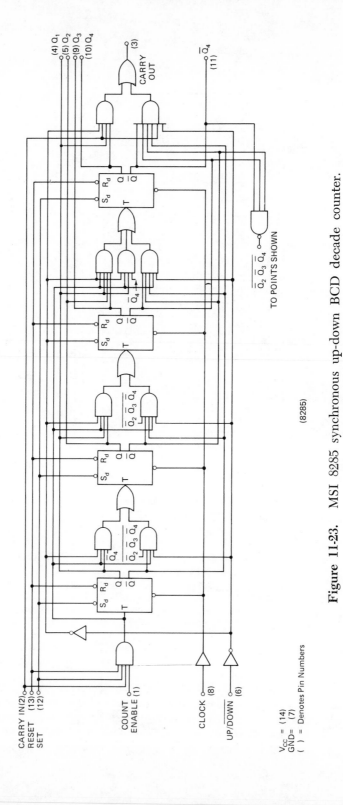

Figure 11-23. MSI 8285 synchronous up-down BCD decade counter. (Courtesy of Signetics Corporation.)

will be obtained. The 1 is shifted through the register and back again for each count cycle.

The advantage of this counter is that it does not require decoding. The outputs are available directly from the flip-flops, as indicated in the diagram.

The counter in Figure 11-26 is called a *feedback shift counter*. The number of counts in a complete cycle equals twice the number of flip-flops used (2N). All of the flip-flops in the counter are triggered simultaneously from a common input pulse.

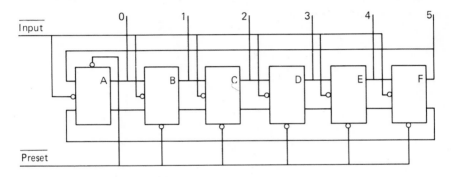

Figure 11-24.   Simple ring counter.

	A	B	C	D	E	F
0	1	0	0	0	0	0
1	0	1	0	0	0	0
2	0	0	1	0	0	0
3	0	0	0	1	0	0
4	0	0	0	0	1	0
5	0	0	0	0	0	1

Figure 11-25.   Truth table for simple ring counter.

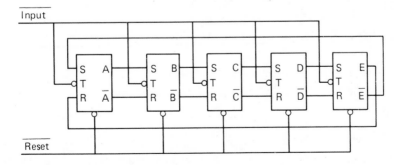

Figure 11-26.   Feedback shift counter.

The feedback from the output is cross-connected back to the input so that the SET output is connected to the RESET input and the RESET output is connected to the SET input. The output flip-flop (E) conditions the input flip-flop so that A sets to the opposite state of E; that is, if E is SET, the next clock pulse will cause A to RESET. All of the other flip-flops are conditioned to set the state of the preceding flip-flop.

The result is that each flip-flop will SET in sequence and then RESET in sequence, as indicated by the truth table in Figure 11-27.

One advantage of the feedback shift counter is that pulses of both time and duration can be decoded. This characteristic is desirable in timing networks. An example of this counter is discussed in Chapter 12.

	A	B	C	D	E
0	0	0	0	0	0
1	1	0	0	0	0
2	1	1	0	0	0
3	1	1	1	0	0
4	1	1	1	1	0
5	1	1	1	1	1
6	0	1	1	1	1
7	0	0	1	1	1
8	0	0	0	1	1
9	0	0	0	0	1
	0	0	0	0	0

Figure 11-27.   Truth table for the feedback shift counter.

QUESTIONS AND PROBLEMS—CHAPTER 11

1. Why are J-K flip-flops used in counters?

2. How many flip-flops would be required to form an octal (0–7) binary counter?

3. How are the last six counts eliminated in the base 10 ripple-carry counter?

4. What is the purpose of the NOR gate in Figure 11-6?

5. Why is the inverter necessary in Figure 11-6?

6. How are ripple-carry counters made to return to zero?

7. Some counters have a "race" problem. How is it solved?

8. What is the advantage of a synchronous counter?

9. What is the purpose of the two AND gates in Figure 11-13?

10. What is *jam entry?*

11. What is the purpose of CP_2 in Figure 11-14?

12. How can counters be made to count down?

13. What is the purpose of the OR gate in Figure 11-21?

14. What is an advantage of a shift counter?

15. What is an advantage of a feedback shift counter?

16. For the circuit in Figure 11-11, draw the four output waveforms. Start at time zero and continue through the next time zero.

17. Draw the output waveforms for the circuit in Figure 11-12. Start at time zero and continue through the next time zero.

18. Draw the output waveforms for the circuit in Figure 11-24. Start at time zero and continue through the next time zero.

19. Draw the output waveforms for the circuit in Figure 11-26. Start at time zero and continue through the next time zero.

20. Draw the output waveforms for the circuit in Figure 11-20. Start at time zero and continue through the next time zero.

Timing

A computer is programmed to perform *operations,* such as ADD, SUB-TRACT, MULTIPLY, and STORE, etc. The computer is under the control of a *program,* which is a series of *instructions,* one instruction for each operation. In order to perform one of these operations, many small steps are necessary. These small steps are called *micro-operations.* Each micro-operation is under the control of an instruction called a *micro-instruction,* which is usually stored permanently. The purpose of micro-instructions is to produce logic levels which control the various circuits in order to perform the instruction. The correct time and sequence of these logic levels are controlled by timing circuits. The "heart" of the computer, or of any digital system for that matter, is the timing circuitry—often called the *clock.*

A grandfather's clock is analogous to the clock in a computer. The pendulum is the master oscillator, and its frequency is 12 cycles per minute. This frequency is divided by a series of gears. The first set of gears divides the master oscillator frequency by 12. Therefore, a gear goes through one revolution per minute. The next set of gears again divides the frequency by 60, causing a gear to go through one revolution per hour. The minute hand is attached to this gear. The purpose of this hand is to provide a means of decoding the minutes. Minutes are decoded by interpreting the position of the hand relative to the face of the clock. One final set of gears divides the frequency by 12, causing the hour hand to revolve once every 12 hours. The clock may therefore be considered to consist of an oscillator, a series of frequency dividers (counters), and a decoder.

The alarm clock is more sophisticated in that it can decode a specific time in order to sound an alarm. Other clocks or "timers" may decode periods of time (duration) as well. For example, a timer may be used to start a lawn sprinkler at a predetermined time and stop it after a specified duration.

Figure 12-1 is the logic diagram of an electronic clock. Notice that it consists of an oscillator, frequency dividers, and decoders, just like a grandfather's clock. The master oscillator operates at a frequency of 60 Hz. (The 60-CPS power line frequency is substituted for the oscillator in many clocks.) A base 60 counter is used to divide the frequency down to one pulse per second. The frequency is further divided by a base 10 counter. The outputs of these two counters, representing seconds, are decoded and displayed. Minutes are counted with base 10 and base 6 counters; from these, minutes are decoded and displayed. Finally, a base 12 counter, decoder, and display are used for the hours.

Figure 12-2 shows an electronic watch. Its circuitry is similar to that described above; however, it uses LSI monolithic circuits and a crystal oscillator.

One advantage of an electronic clock is that BCD outputs from the

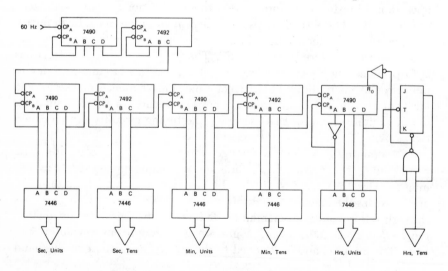

Figure 12-1.   Logic diagram of an electronic clock.

Figure 12-2.   Electronic watch.

counters may be provided so that times and durations may be decoded to control electrical appliances such as lights, radios, and ovens.

The topics covered in this chapter are:

12.1 Clock Cycles

12.2 Multiphase Clock Pulses

12.3 Timing in Electronic Calculators

12.4 Modifying the Clock Cycle

12.5 Sequencing Operations

12.1   CLOCK CYCLES

The clock discussed in Figure 12-1 completes one clock cycle in a 12-hour period. Other clock cycles, such as a 24-hour cycle (24-hour clocks), exist. Clock cycles for computers may range from nanoseconds to milliseconds, as determined by the master oscillator and frequency dividers.

Figure 12-3 is a diagram of an oscillator and frequency divider. The oscillator may be a multivibrator or a crystal-controlled oscillator, depending on the accuracy required. The frequency divider is a 4-bit binary counter. Since there are four bits, there are 16 counts. Because the oscillator frequency is 200 kHz, the basic time period is 5 microseconds

$$\frac{1}{200\,\text{kHz}}$$

The base of the counter is 16; hence, the total time for a clock cycle is 80 μs (5 μs $\times$ 16), which consists of sixteen 5 μs periods.

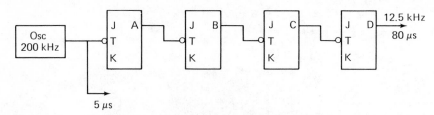

Figure 12-3.   Oscillator and frequency divider.

Figure 12-4 illustrates how each of these time periods may be decoded. A 9311 is used to decode the outputs of the frequency divider. These outputs are labeled to designate each of the 16 time periods ($\overline{t_0}$ through $\overline{t_{15}}$). It is important to note that each of these outputs represents a period of time.

Figure 12-5 is a timing chart for the above described clock. The output of each of the flip-flops in the counter is shown. Although all of the outputs from the decoder are not shown, $\overline{t_0}$, $\overline{t_1}$, $\overline{t_2}$, and $\overline{t_{15}}$ are representative outputs. Usually, a clock pulse is required during a time period. This pulse should not overlap the leading or trailing edge of the time period, but it should appear within the time period. Figure 12-6 shows how a short pulse can be produced with a monostable flip-flop (single shot) from the oscillator output. This pulse is delayed from the start of the period to the center of the period by triggering on the HIGH going transient of the oscillator pulse (CP1). The duration of the pulse is determined by the time constant R_x and C_x. This pulse is used to trigger a flip-flop or register after it has been enabled.

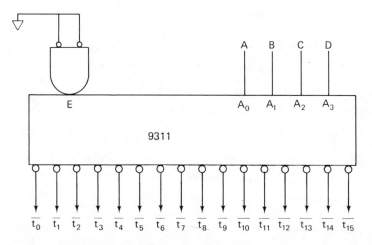

Figure 12-4.   Decoding 16 periods (t_0 through t_{15}).

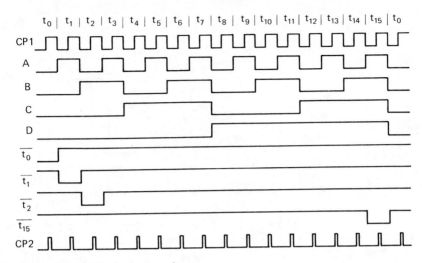

Figure 12-5. Clock timing chart.

In order to better understand the function of a clock, let's use the following example. Imagine that it is necessary to determine whether a sign bit is present in the x register at time t_1. This information is to be used between times t_1 and t_{15}. Therefore, it is stored in a flip-flop, as shown in Figure 12-7. Notice than the flip-flop is enabled only when a sign bit is present and during time t_1. The flip-flop is set with CP2. The advantage of using CP2 is that it removes ambiguity in the time when the flip-flop is set. First the flip-flop is enabled; then it is set with a clock pulse. The flip-flop is reset directly from the clock at time t_{15}.

A J-K flip-flop is commonly used, as shown in Figure 12-8. The J

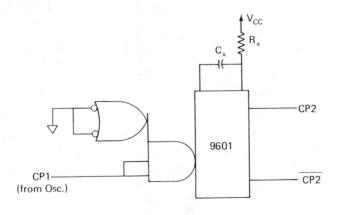

Figure 12-6. Monostable flip-flop used to generate delayed clock pulse.

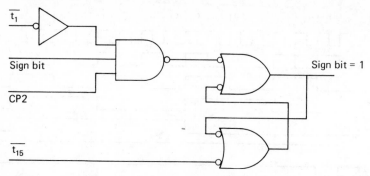

Figure 12-7.   Circuit to determine state of sign bit.

input is enabled with the sign bit and t_1. The flip-flop is triggered with $\overline{CP2}$. The reset enable input (K) is enabled at t_{15} and is reset with the $\overline{CP2}$ pulse. Notice that at other time periods neither J nor K is enabled. Therefore, $\overline{CP2}$ will have no effect on the state of the flip-flop during these times. It is possible to reset the J-K flip-flop at t_{15} directly from the clock by using the direct reset input (S_D). This is an example of one operation in a given clock cycle. Many operations can be performed in sequence between t_0 and t_{15}.

When data is transferred serially through shift registers, the appropriate shift pulses must be supplied by the clock. Let's assume that we have a 12-bit shift register. In such a case, 12 shift pulses are required to shift data completely through the register. If we use the clock described in Figures 12-3 and 12-4, times t_3 through t_{14} could be used for the 12 pulses required. This allows t_1 and t_2 to be used for some preliminary operations, such as determining the status of some circuit. Normally, the last time period (t_{15}) in a clock cycle is employed to reset the various flip-flops used to store such data. Time t_0 is used as a reference time. Operations are usually not performed at t_0. The circuit is shown in Figure 12-9. Notice that the clock pulses are inhibited during t_0, t_1, t_2, and t_{15}.

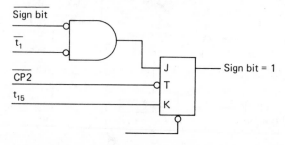

Figure 12-8.   J-K flip-flop used to store state of sign bit.

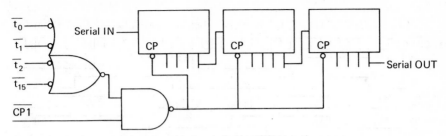

Figure 12-9.   Twelve-bit shift register with clock pulse decoder circuit.

The clock pulse $\overline{CP1}$ is used because it will cause the shift to occur at the beginning of each time period. This allows CP2 to be used *during* the period to effect an operation on the data after each shift pulse.

A counter often used in clocks is the feedback shift counter described in Chapter 11. A clock using this counter is shown in Figure 12-10. Its primary advantage is that it is easy to decode. Not only can times be decoded, but durations can also be easily decoded. Since the oscillator runs at 1 MHz and because the frequency division is 20, there are twenty 1-μs time periods. This may be thought of as two 10-μs cycles. Pulse designations may indicate both time and duration. For example, t2d4 indicates that a pulse starts at t_2 and has a duration of four periods. This pulse is indicated in the timing chart in Figure 12-11. Careful inspection of the timing chart reveals that t2d4 can be decoded by ANDing C and B and $\overline{F}$, as shown in Figure 12-12. Notice that $\overline{F}$ designates the first decade (t_0–t_9) and F the second decade (t_{10}–t_{19}). The pulse starts with C and ends with B.

How would t6d2 be decoded? Since at the beginning of t_6, A is going LOW, $\overline{A}$ is ANDed with C and $\overline{F}$. $\overline{A}$, C, AND $\overline{F}$ are all HIGH only during t6d2!

Figure 12-13 shows how the various pulses on the timing chart are

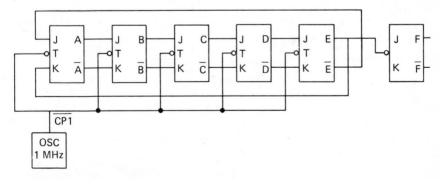

Figure 12-10.   Feedback shift counter.

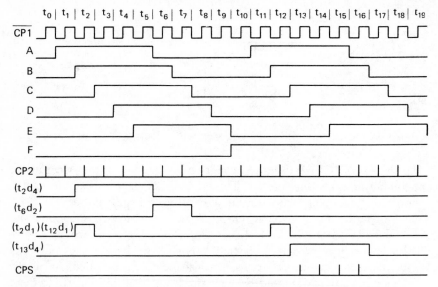

Figure 12-11. Timing chart for feedback shift counter.

decoded. Pulses often reoccur in the second cycle as is the case with t2d1 and t12d1. This is accomplished by ignoring the state of F.

Suppose four shift pulses (CPS) were required at t_{13} through t_{17}. The circuit in Figure 12-14 shows how they may be decoded. CP2 is generated as in Figure 12-6. First, t13d4 is the result of C and B and F. These

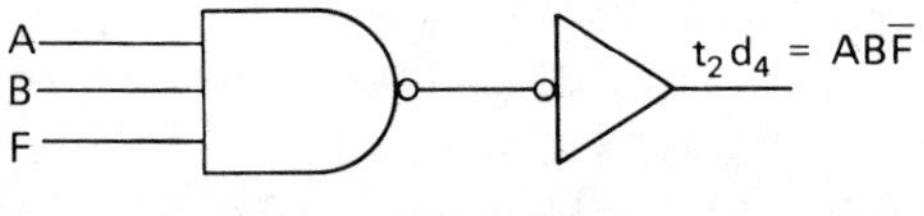

$$t_2 d_4 = AB\overline{F}$$

Figure 12-12. Decoding *t2d4*.

$$t_6 d_2$$

$$(t_2 d_1)(t_{12} d_1)$$

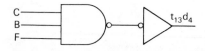

$$t_{13} d_4$$

Figure 12-13. Decoding various times as shown in Figure 12-11.

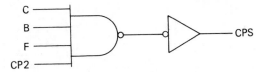

Figure 12-14. Circuit to decode CPS.

variables are ANDed, with CP2 producing CPS, as was shown in Figure 12-11.

12.2 MULTIPHASE CLOCK PULSES

In Chapter 6, dynamic gates were discussed; Chapter 9 dealt with dynamic shift registers. Dynamic circuits require multiphase clock pulses. The circuit in Figure 12-15 is a two-phase clock pulse generator. The associated waveforms appear in Figure 12-16. The oscillator frequency is divided in half by the J-K flip-flop. The outputs Q and $\overline{Q}$ are ANDed with the oscillator output to produce ϕ_1 and ϕ_2. The basic time period is taken from the output of the J-K flip-flop and applied to the counter. The time periods may be decoded from the output of the counter. The pulse widths of ϕ_1 and ϕ_2 are fixed because they are equal to one-half of the oscillator period.

The circuit in Figure 12-17 uses two monostable flip-flops (9602) to generate ϕ_1 and ϕ_2. This allows the pulse width to be independent of the oscillator. The pulse width is determined by the R and C of each one-shot. This means that ϕ_1 and ϕ_2 do not have to have the same pulse width. The complements of ϕ_1 and ϕ_2 are also available at the output of the one-shots.

The four-phase shift register discussed in Chapter 9 (Figure 9-17) requires four clock pulses. These pulses may be generated by the circuit

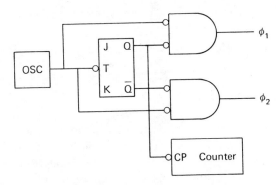

Figure 12-15. Multiphase clock pulse generator.

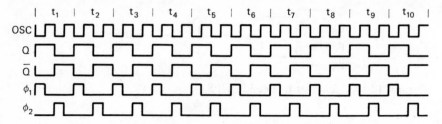

Figure 12-16.   Waveforms for two-phase clock pulse generator.

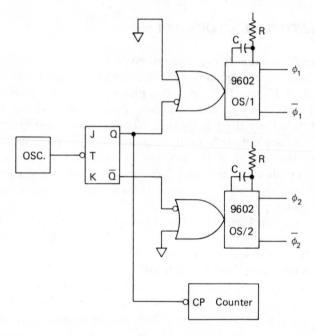

Figure 12-17.   Two-phase clock pulse generator using one-shots.

shown in Figure 12-18. Phases 2 and 4 (ϕ_2 and ϕ_4) are taken from a J-K flip-flop which is driven by the oscillator. Clock pulse ϕ_2 drives a one-shot to produce ϕ_3. The waveforms are shown in Figure 12-19. The times t_1, t_2, t_3, and t_4 correspond to the times discussed in Chapter 9. Notice that the four phases are produced during one cycle of the J-K flip-flop. The output of the flip-flop also drives a counter. The outputs of the counter may be decoded to produce the basic time periods of the clock. These are indicated in the waveforms as T_1, T_2, T_3, etc. It is common practice to indicate short times with a lower-case t and longer, inclusive times with an upper-case T.

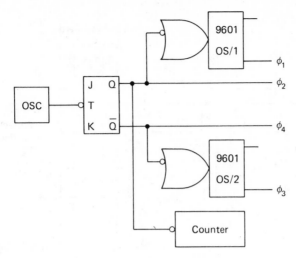

Figure 12-18. Four-phase clock pulse generator.

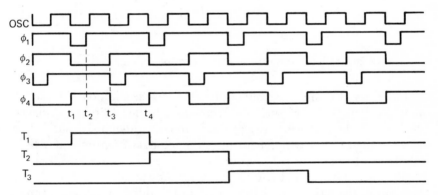

Figure 12-19. Waveforms for four-phase clock pulse generator.

12.3 TIMING IN ELECTRONIC CALCULATORS

An example of a system requiring precise timing is an electronic calculator. In order to simplify the hardware, all data is handled in serial. All operations are performed sequentially. Because of this, the timing cycle is relatively slow. However, fast cycle times are not required because the inputs and outputs are determined by the speed of the operator, which is *very* slow compared to the capability of the electronic circuitry.

Figure 12-20 is a block diagram of a hypothetical electronic calculator. It is capable of handling numbers eight digits in length. It has two

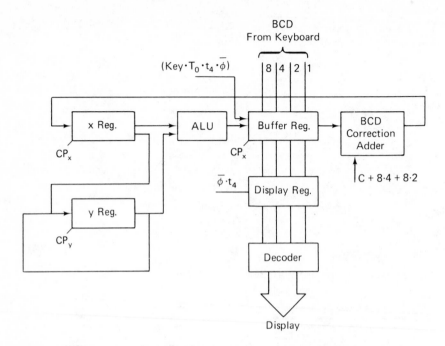

Figure 12-20.   Block diagram of an electronic calculator.

registers, x and y, and is capable of four functions: addition, subtraction, multiplication, and division. The data stored in the x Reg is displayed. Data enters the x Reg by way of the keyboard through the buffer Reg. The first digit of a new number causes the old data from the x Reg to be transferred to the y Reg. For example, if we added 123 to 456, the first number would be entered into the x Reg by pressing the equals key ($=$). This causes the data in the x Reg (123) to be added to the data in the y Reg, which contains 0. Next, keys 4, 5, and 6 are pressed in sequence. Again, the contents of the x Reg and y Reg are added. Notice that the sum (or difference or product or quotient) always returns to the x Reg so that it may be displayed.

During one machine cycle, the data in the x Reg is shifted through the arithmetic and logic unit (ALU), then through the buffer and correction adder, and then back to its original position in the x Reg. The addition or subtraction operation requires one machine cycle. The process of multiplication or division requires many machine cycles.

The machine cycle determines the clock cycle. Because the x Reg stores eight digits and because the BCD code is used, it must store 32 bits (4 bits $\times$ 8 digits $=$ 32). The clock must provide 36 shift pulses (32 for the x Reg and 4 for the buffer Reg) to complete one machine cycle. The clock pulse generator is shown in Figure 12-21. Two phases of the oscillator are provided—ϕ and $\bar{\phi}$. Note that $\bar{\phi}$ drives a bit counter. The bit counter uses two J-K flip-flops (A and B) connected as a feedback shift counter. The output of this counter is decoded to produce the bit times t_1, t_2, t_3, and t_4. The digit counter (J-K flip-flops C, D, E, and F) is a base 10 counter. The output of the digit counter is decoded to produce the digit times T_0 through T_9. The waveforms for the clock appear in Figure 12-22. Notice that four bit times are required to produce one digit time.

A high-going transient of ϕ initiates a bit-time period and effects the shift for both the x and y Regs (CP$_x$ and CP$_y$). This permits the low-going transient of ϕ to be used as a trigger source *during* a bit-time period.

The decoder and display are multiplexed; that is, digits are displayed in sequence, starting with the least significant digit. At T_0, the reference time, the data occupies the reference position in the x Reg. The least significant bit (LSB) of the least significant digit (LSD) occupies the LSB position of the x Reg. Therefore, at the end of T_1 time (after four shift pulses), the LSD is stored in the buffer Reg. This digit is transferred in parallel to the *display* Reg. The purpose of the display Reg is to store a digit for a complete digit-time period. Notice that the transfer

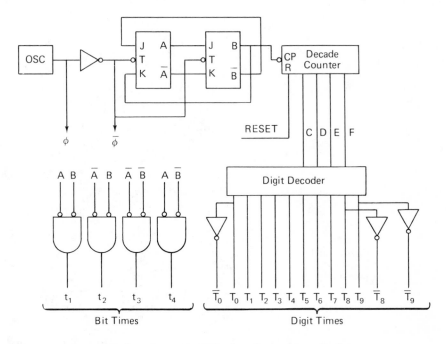

Figure 12-21.   Clock pulse generator for electronic calculator.

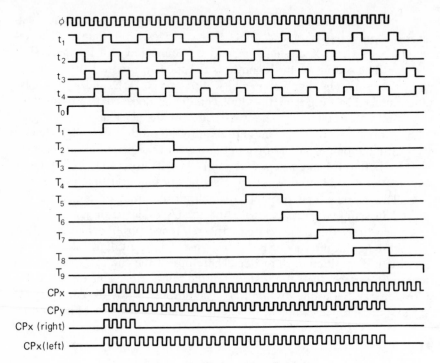

Figure 12-22.   Timing chart for electronic calculator.

takes place with the low-going transient at t_4 time, *after* the fourth shift pulse, by $\overline{\phi}$ and t_4.

The LSD is stored in the display Reg during T_2 time. At this same time, it is decoded by the seven-segment decoder and displayed in the position of the LED display, as shown in Figure 12-23. Notice that only the LSD position of the display is activated at T_2 time. This process is continued. At $T_2 \cdot \overline{\phi} \cdot t_4$, the next significant digit is transferred to the display Reg where it is decoded and displayed during T_3 time. Although the LED displays are pulsed, we see the digits without flicker due to the persistence of vision.

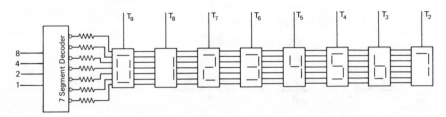

Figure 12-23.   Eight-digit seven-segment display.

As mentioned previously, 36 shift pulses are required by the x Reg. How many shift pulses are required for the y Reg? Because data in the y Reg is not shifted through a buffer register, only 32 pulses are required for a normal shift. When multiplying and dividing, it is often necessary to shift digits either left or right. Remember, a digit consists of four bits. Therefore, four shift pulses are required to shift a digit one position.

A right digit shift consists of only four shift pulses in one machine cycle. A left digit shift is accomplished by providing four shift pulses less than normal. For example, if the x Reg is pulsed 32 times (four less than normal), the data will be four bit positions shy of its original position. Therefore, a left digit shift has been accomplished.

Figure 12-24(a) shows how the shift pulses CP_x and CP_y are decoded. For a normal x Reg shift cycle, the micro-instruction "x NORM" is activated. This causes CP_x to be produced by $\overline{T_0}$ and ϕ. For a normal shift cycle:

$$CP_x = (x \text{ NORM}) \, (\overline{T_0}) \, (\phi)$$

For a right digit shift in the x Reg, the micro-instruction "x RIGHT" is activated, causing CP_x to occur during T_1. For a right digit shift cycle:

$$CP_x = (x \text{ RIGHT}) \, (T_1) \, (\phi)$$

For a left digit shift in the x Reg, the micro-instruction "x LEFT" is activated, causing CP_x to occur during the period T_1 through T_8. For the left digit shift cycle:

$$CP_x = (x \text{ LEFT}) \, (\overline{T_0})(\overline{T_9}) \, (\phi)$$

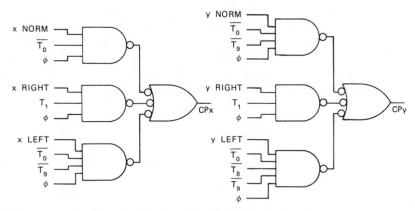

Figure 12-24. Shift pulse decoder: (a) CP_x; (b) CP_y.

CP_y is generated in a similar manner, as shown in Figure 12-24(b). For a normal cycle:

$$CP_y = (y \text{ NORM}) (T_0) (T_9) (\phi)$$

For a right digit shift:

$$CP_y = (y \text{ RIGHT}) (T_1) (\phi)$$

For a left digit shift:

$$CP_y = (y \text{ LEFT}) (\overline{T_0}) (\overline{T_8}) (\overline{T_9}) (\phi)$$

Data is encoded to BCD at the keyboard and entered into the x Reg through the buffer Reg. Each digit is entered at time $(T_0) (t_4) (\overline{\phi})$. With each entry, an "$x$ LEFT" micro-instruction is provided. This causes a left digit shift with each entry such that the last digit entered will occupy the LSD position of the x Reg.

Although many aspects of the various operations of our hypothetical calculator were not covered, the intent of this discussion was to provide an insight into the timing concepts as they relate to electronic calculators or other systems handling data in the serial mode.

12.4 MODIFYING THE CLOCK CYCLE

It is important that the machine clock cycle is efficiently designed; that is, the clock cycle should be long enough to perform the required operations in a given cycle but not so long as to waste time. Efficiency of a clock cycle can be improved by modification of the clock cycle relative to the operation to be performed. For example, when a right shift operation is to be performed in the electronic calculator, only four shift pulses were required (1 digit time). This means that the 10 digit times could be reduced considerably, perhaps to 2 or 3. (Refer to the timing chart in Figure 12-22). The right shift operation is complete after T_1. The clock could therefore be terminated at T_2. This can be accomplished by resetting the decade (digit) counter in Figure 12-21 at $T_2 \cdot t_4 \cdot \overline{\phi}$. The circuit in Figure 12-25 decodes the reset pulse when the micro-instruction

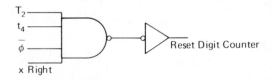

Figure 12-25.   A method for terminating the electronic calculator clock cycle.

"x RIGHT" occurs. The output of this circuit is connected to the direct reset input of the decade counter. Although this is an important concept, it probably would not be considered in the calculator because time is not as important as the amount of circuitry required.

The circuit in Figure 12-26 shows a method of modifying the normal sequence of a clock cycle. The 9316 is a presettable decade counter. The output of the counter is decoded by the 9311 decoder producing outputs $\overline{T_0}$ through $\overline{T_{15}}$. A priority encoder (9318) provides an input for eight possible conditions that can alter the sequence of the counter. The ouput of the encoder is a 3-bit binary code representing the active input condition. If two or more inputs are active simultaneously, the input with the highest priority is represented at the output, with input line 7 having the highest priority. The outputs of the encoder are connected through inverters (to provide proper active levels) to the parallel inputs (P_1, P_2, P_3) of the counter. When the parallel enable input (PE) is active, the

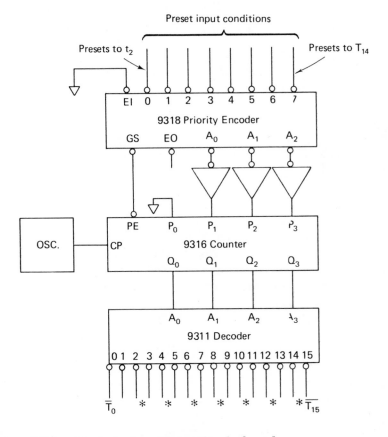

Figure 12-26. Method of modifying the clock cycle.

next clock pulse causes the counter to be set to the state of the levels at the parallel inputs. The group select (GS) of the encoder is active whenever one or more of the inputs are active.

Notice that the state of the counter can change, causing the skipping of certain time periods; alternatively, it can change by returning to a previous count. The latter allows for the repetition of operations requiring a number of cycles.

You will notice that this circuit can only be preset to 8 of the 16 possible time periods and that these are the even numbered times because the 1-bit parallel input (P_0) is never active. However, variations of this scheme can be used to solve many timing problems.

12.5   SEQUENCING OPERATIONS

Many operations require a number of clock cycles to complete. A program counter, sometimes referred to as a *micro-instruction address counter*, may be used to keep track of these cycles. The micro-instruction address counter remains at a given address until an operation is complete, regardless of the number of clock cycles required. When the operation is complete, it is allowed to increment to the next address.

Figure 12-27 shows a 3-bit micro-instruction counter which can control eight operations. The counter starts in the zero address (000). This is the address that enables Function A, which is the first operation in the sequence. However, the "Enable Function A" line becomes active only if the "Condition A" input line is active. Assuming Condition A is active, the output of the multiplexer will also be active; this inhibits the counter because the J and K inputs are held LOW. The output of the multiplexer also enables the decoder so that the "Enable Function A" line is active. When the "Condition A" line becomes inactive, indicating that operation A is complete, the output of the multiplexer becomes inactive; this disables the decoder, causing the "Enable Function A" line to be inactive. The address counter is now enabled, and with the next input pulse, the counter will increment to address 001. If the "Condition B" line is active, the counter will be inhibited and the "Enable Function B" line will be activated. However, if the "Condition B" line is not active, the counter remains enabled and will increment with each input pulse until it reaches an address corresponding to an active input line. The counter is then inhibited until the indicated operation is complete. This process continues until all operations are complete. It is important to note that this scheme allows for a selection of operations. This kind of system is sometimes called a *function sequencer*.

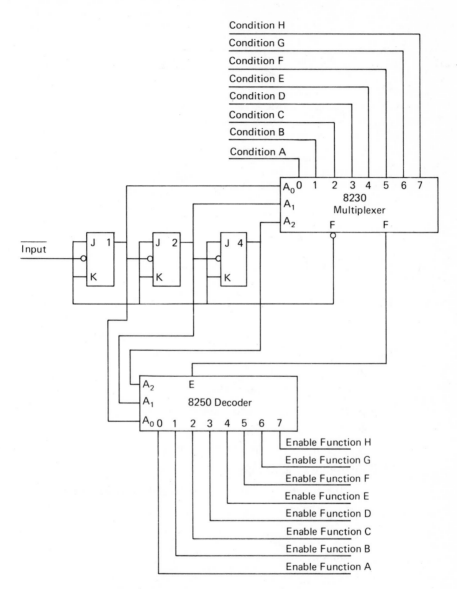

Figure 12-27. Function sequencer.

Figure 12-28 is a flow diagram of the process. Notice that if an operation is indicated, the function is enabled until the operation is complete. If an operation is not indicated, the process is continued to the next operation.

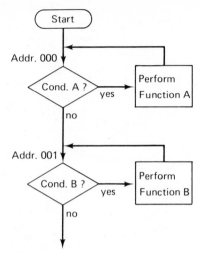

Figure 12-28.   Flow diagram for function sequencer.

QUESTIONS AND PROBLEMS—CHAPTER 12

1. When data is transferred serially through shift registers, what supplies the shift pulses?

2. In a 16-bit shift register, how many shift pulses are required to shift data completely through the register?

3. What is the advantage of using the feedback shift counter in some clocks?

4. What is the purpose of the inverters in Figures 12-12 and 12-13?

5. What determines the pulse width of ϕ_1 and ϕ_2 in Figure 12-16?

6. What are the functions of the x and y registers in Figure 12-20?

7. What is the purpose of ϕ and $\overline{\phi}$ in Figure 12-21?

8. Why are the decoder and display multiplexed?

9. Describe the operation of a function sequencer.

10. Normally a given clock cycle has a fixed duration. For some operations, the duration of the clock cycle may be shortened in order to achieve greater efficiency. Under what conditions is this possible?

11. How can the duration of the clock cycle in Problem 10 be shortened?

12. For the electronic calculator discussed in Figure 12-20, how is a right digit shift accomplished?

13. For the electronic calculator discussed in Figure 12-20, how is a left digit shift accomplished?

14. Referring to the timing chart in Figure 12-5, draw the waveform for $(t_2 + t_5)$.

15. Referring to the timing chart in Figure 12-5, draw the waveform for $(A \cdot \overline{B} \cdot \overline{C} \cdot D \cdot CP2)$.

16. Referring to the timing chart in Figure 12-11, draw the waveform for $(\overline{C} \cdot \overline{A})$.

17. Referring to the timing chart in Figure 12-11, draw the waveform for $\overline{F}(E + C)$.

18. Referring to the timing chart in Figure 12-11, decode $t_4 d_3$. Include the logic circuit and waveforms.

19. Referring to the timing chart in Figure 12-11, decode $t_5 d_5$. Include the logic circuit and waveforms.

20. Referring to the timing chart in Figure 12-11, decode $(t_3 d_2) + (t_{13} d_2)$. Include the logic circuit and waveforms.

13

Arithmetic Logic Unit (ALU)

The basic computer consists of a central processing unit (CPU) and various input and output units. The most essential section of the CPU is that which performs the arithmetic operations and/or makes decisions. This section is called the *arithmetic logic unit* (ALU).

In many CPUs, the ALU reduces all arithmetic operations to a series of additions and/or subtractions. This includes multiplication, division, and the more complex mathematical operations. Since this is true, all the arithmetic operations are performed using adders and subtracters.

Decision making also takes place in the ALU. Decisions such as comparison, determination of which of two numbers is larger or smaller, whether a number equals zero, and whether a number is positive or negative, and others are usually necessary steps in a program. The decision-making functions are performed by logic circuits such as AND, OR, EXCLUSIVE OR, comparators, etc.

The ALU works in conjunction with a number of operating registers, including an accumulator. These registers temporarily store data on which mathematical operations are performed by the ALU. Data may be transferred to these registers from memory or vice versa, depending upon what is required.

In this chapter, we will discuss arithmetic and logic circuits. Various applications of both arithmetic and logic circuits will be examined, as will some of the MSI devices currently available.

The topics covered are:

13.1   Binary Adder

13.2   Full Adder

13.3   Serial Adder

13.4   Parallel Adder

13.5   Binary Subtracter

13.6   Comparators

13.7   High-Speed Adders

13.8   Two's Complement Method

13.9   Logic Decisions

13.10 BCD Addition

13.11 BCD Subtracter

13.12 BCD Nine's Complementer

13.13 BCD-to-Binary Conversion

13.14 Binary-to-BCD Conversion

13.1   BINARY ADDER

Most operations in the ALU are reduced to simple addition, which can be performed sequentially for more complex operations such as multiplication or division. Therefore, let's discuss the simple binary adder. The following table shows all combinations for a 1-bit addend and augend:

0	0	1	1	addend	(A)
+ 0	+ 1	+ 0	+ 1	augend	(B)
0	1	1	10	sum	(S)
			↖— carry		(C)

Careful inspection reveals that the sum always equals A EXCLUSIVE OR B. A carry is generated only when A and B equal 1. Therefore the following statements can be written:

$$S = A \oplus B = (A + B)\,(\overline{AB}) = A\overline{B} + \overline{A}B \qquad (13.1)$$

$$C = AB \qquad\qquad\qquad\qquad\qquad\qquad (13.2)$$

The logic diagram in Figure 13-1 produdces both the sum and the carry outputs. Such a logic circuit is called a *half adder*.

Figure 13-2 shows a half adder implemented using TTL gates. Its truth table is the same as that in Figure 13-1(b), except that the complement of the sum and carry are provided.

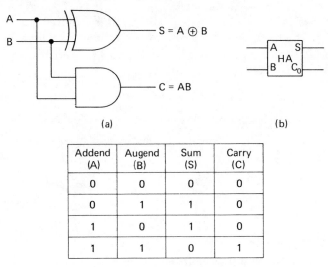

(a)   (b)

Addend (A)	Augend (B)	Sum (S)	Carry (C)
0	0	0	0
0	1	1	0
1	0	1	0
1	1	0	1

(c)

Figure 13-1. Half adder: (a) first-level logic diagram; (b) symbol for half adder; (c) truth table.

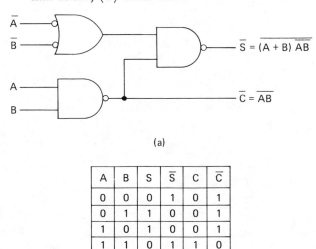

(a)

A	B	S	$\overline{S}$	C	$\overline{C}$
0	0	0	1	0	1
0	1	1	0	0	1
1	0	1	0	0	1
1	1	0	1	1	0

(b)

Figure 13-2. Half adder implemented using TTL gates: (a) logic diagram; (b) truth table.

13.2   FULL ADDER

In the more general case of the addition of multibit numbers, when a carry is generated, it must be added to the next significant bit position. Therefore, each adder should have an additional input for the carry bit. To provide the full addition operation, two half adders and an OR gate are used, as illustrated in Figure 13-3(a). This logic circuit is called a *full adder*. The truth table for a full adder is shown in Figure 13-3(b).

The logic diagram in Figure 13-4 depicts two TTL implemented half adders connected to form a full adder.

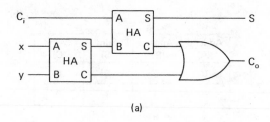

(a)

x	y	C_i	S	C_o
0	0	0	0	0
0	0	1	1	0
0	1	0	1	0
0	1	1	0	1
1	0	0	1	0
1	0	1	0	1
1	1	0	0	1
1	1	1	1	1

(b)

Figure 13-3.   Full adder: (a) first-level logic diagram; (b) truth table.

13.3   SERIAL ADDER

One full adder may be used to add multi-bit numbers if addition is performed one bit position at a time. A serial adder is shown in Figure 13-5. The input to the adder is provided by two shift registers. The output sum

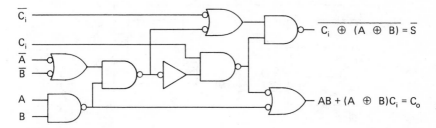

Figure 13-4. Logic diagram of a full adder using TTL implemented half adders.

is stored in a third shift register. The carry is stored in a flip-flop. The addition is performed on the least significant bit first. When a clock pulse appears, the sum is stored in the adder output register, and the carry (if present) is stored in the carry flip-flop. Simultaneously, the A and B registers (A Reg and B Reg) shift all bits one position to the right so that the next significant bits appear at the input of the adder. This process is continued for as many bit positions as are in the registers.

It is sometimes desirable to shift the sum back into the A Reg, as is done in Figure 13-6. This makes the A Reg an accumulator. New data entered into the B Reg are added to the sum accumulated in the A Reg and recycled to the A Reg. This procedure simplifies the arithmetic process when complex functions are performed. Serial addition utilizes a minimum of hardware. However, the savings in hardware may be offset by the considerable amount of time required to complete the process.

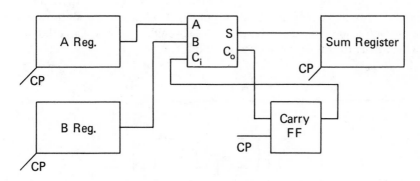

Figure 13-5. Block diagram of a serial adder.

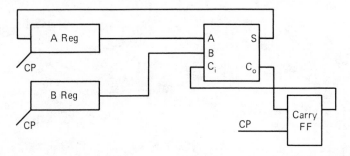

Figure 13-6. Clock diagram of a serial adder that uses the A Reg as an accumulator.

13.4 PARALLEL ADDER

In order to save time, many high-speed computers use parallel addition. The time required is dependent on the propagation delay of the carry bit from the least significant bit position to the most significant bit position. A block diagram for a 4-bit parallel adder appears in Figure 13-7. The data must be available in parallel at the input of the adder. After sufficient time to allow the output to settle due to the propagation delay of the carry bit, it is entered in parallel to the adder output Reg or accumulator.

Figure 13-8 shows implementation of a 4-bit full adder using TTL gates. Each adder is identical to the full adder of Figure 13-4 except that here inverters have been provided to realize true inputs and outputs. Note the connections to propagate the carry bit from position to position. All four sum outputs (Σ_1, Σ_2, Σ_3, and Σ_4) are available in parallel. The circuit provides only one carry output (C_o), which is the output of the most significant bit adder. One input is provided for the carry bit (C_i), which goes to the least significant bit adder.

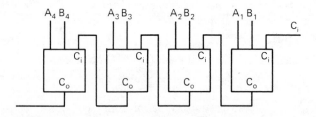

Figure 13-7. Block diagram of a 4-bit parallel adder.

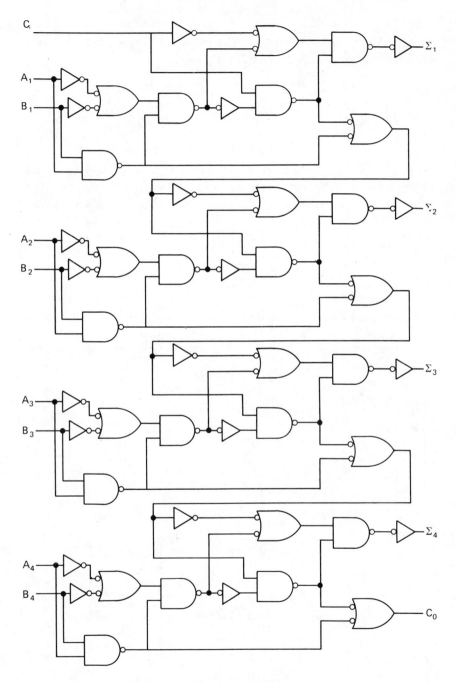

Figure 13-8.   Four-bit full adder using TTL gates.

13.5 BINARY SUBTRACTER

Some ALUs require a subtracter. The following table shows all possible combinations for minuend and subtrahend:

0	0	1	1	minuend (x)
0	1	0	1	subtrahend (y)
0	11	1	0	difference (D)
				borrow (B)

Careful inspection reveals that the difference always equals A EX-CLUSIVE OR B. Note that when subtracting a 1 from a 0, a borrow is required. This is the second condition in the table. The borrow is generated only when $x = 0$ and $y = 1$. Therefore, the following statements can be written:

$$D = x \oplus y$$

$$B = \overline{x} \cdot y$$

Figure 13-9(a) is the logic diagram for a half subtracter. The truth table for the half subtracter is shown in Figure 13-9(b). Figure 13-10 shows a half subtracter implemented using TTL gates. The truth table is the same as that in Figure 13-9(b). Both the borrow and its complement are provided.

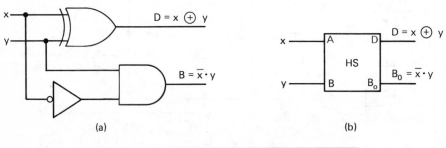

(a)

(b)

Minuend (x)	Subtrahend (y)	Difference (d)	Borrow (B)
0	0	0	0
0	1	1	1
1	0	1	0
1	1	0	0

(c)

Figure 13-9.   Half subtracter: (a) first-level logic diagram; (b) symbol for half subtracter; (c) truth table.

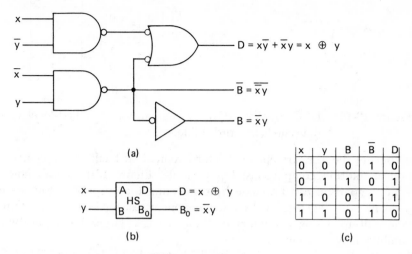

$$D = x\bar{y} + \bar{x}y = x \oplus y$$

$$\bar{B} = \overline{\overline{xy}}$$

$$B = \bar{x}y$$

(a)

x —[A   D]— $D = x \oplus y$
   [HS]
y —[B   B_0]— $B_0 = \bar{x}y$

(b)

x	y	B	$\bar{B}$	D
0	0	0	1	0
0	1	1	0	1
1	0	0	1	1
1	1	0	1	0

(c)

Figure 13-10. Half subtracter implemented using TTL logic: (a) first-level logic diagram; (b) symbol; (c) truth table.

13.6   COMPARATORS

Other examples of logic decisions that the ALU is required to make are inclusive OR, EXCLUSIVE OR, AND, and comparison. The concept of a *comparator* may require discussion at this point. A comparator produces a true output if all inputs are true OR if all inputs are false.

Consider the EXCLUSIVE OR and its truth table in Figure 13-11. Note that when the two inputs are NOT the same, a 1 appears in its output. If the ouput of the EXCLUSIVE OR is inverted as shown in Figure 13-12, a 1 appears at the output when BOTH inputs are 1s or when BOTH inputs are 0s. This EXCLUSIVE NOR may also be called a comparator because it produces a true output whenever both inputs are the same.

The TTL/MSI-8242 contains four independent EXCLUSIVE-NOR gates which may be used to implement digital comparison functions. The

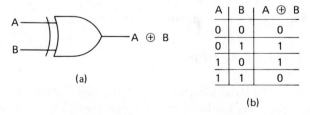

A ——⟩⟩—
         ⟩— $A \oplus B$
B ——⟩⟩—

(a)

A	B	$A \oplus B$
0	0	0
0	1	1
1	0	1
1	1	0

(b)

Figure 13-11. EXCLUSIVE OR: (a) first-level logic diagram; (b) truth table.

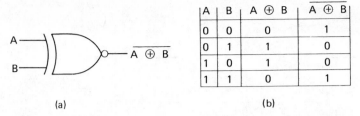

A	B	A $\oplus$ B	$\overline{A \oplus B}$
0	0	0	1
0	1	1	0
1	0	1	0
1	1	0	1

(a)   (b)

Figure 13-12.   EXCLUSIVE NOR, or comparator: (a) first-level logic diagram; (b) truth table.

outputs of the 8242 are open collector circuits that facilitate implementation of multibit comparisons. For example, Figure 13-13 shows how a 4-bit comparison can be made by connecting the outputs of four independent gates together. A discrete component resistor is required to provide pull-up. Using a number of these devices, larger binary or BCD numbers may be compared.

Another comparator is shown in Figure 13-14. It is the TTL/MSI 9324 5-bit comparator, which provides comparison between two 5-bit words and gives three outputs. Besides the "equal to" output, an output is provided for "A is greater than B" and another for "A is less than B." The enable input forces all outputs LOW when it is HIGH.

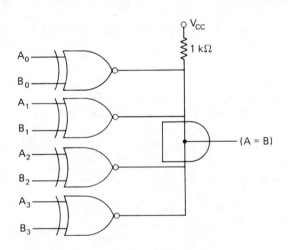

Figure 13-13.   Four-bit comparator using 8242s.

13.7   HIGH-SPEED ADDERS

When we deal with large binary numbers, parallel addition is performed by the arithmetic units to obtain the greatest speed. The limitation to the speed of the adder is primarily due to the propagation of the carry bit.

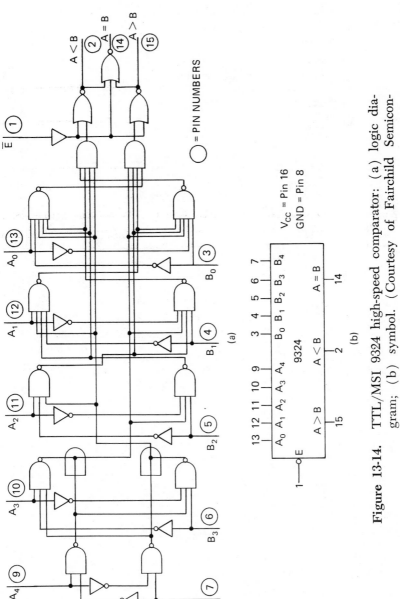

Figure 13-14. TTL/MSI 9324 high-speed comparator: (a) logic diagram; (b) symbol. (Courtesy of Fairchild Semiconductor.)

309

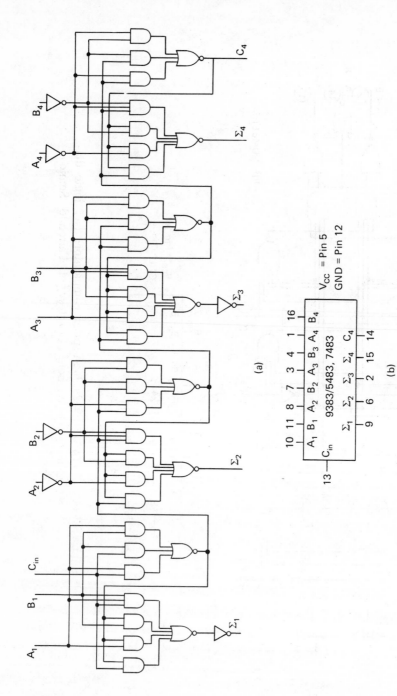

Figure 13-15. TTL/MSI 7483 4-bit binary full adder: (a) logic diagram; (b) symbol. (Courtesy of Fairchild Semiconductor.)

The sums produced from the most significant bits are delayed because of the propagation delay of the carry bits from the lesser significant bit positions. This problem is overcome by using *carry look ahead* circuitry. A high-speed 4-bit binary full adder logic diagram is shown in Figure 13-15.

The carry bit is propagated internally from the carry input (C_i) through each full adder until it reaches the output of the most significant bit (Σ_4). The time required for this propagation is typically 30 nanoseconds, but is guaranteed by the manufacturer not to exceed 50 nanoseconds. If this adder is to be used for binary digits greater than four bits, each four bits requires another 7483. For instance, a 16-bit binary sum would require four 7483s. If the carry bit were allowed to propagate as described, it would cause a possible total propagation delay of 4×50 or 200 nanoseconds. This is overcome in the 9340 through the use of the carry look ahead circuitry, as shown in Figure 13-15. The carry for each four bits is produced separately by gating the inputs of the adder. The guaranteed maximum time that will produce this carry bit is 20 nanoseconds.

One may consider the four bits as a single hexadecimal character and the carry produced by the *look ahead circuit* as a hexadecimal character carry. An example of a hexadecimal addition is shown in Figure 13-16.

The outputs from the x and y registers (x Reg and y Reg) are entered into the three-position adder. The least significant hexadecimal characters (7 and 9) are entered into the position on the right: 7 (0111) from the x Reg and 9 (1001) from the y Reg. The sum of x_1 and y_1 is 0 with a carry. The sum of x_2, y_2, and a carry is 0 and a carry. The addition continues as the carry propagates from right to left. A carry out of the least significant hexadecimal character position becomes a carry into the

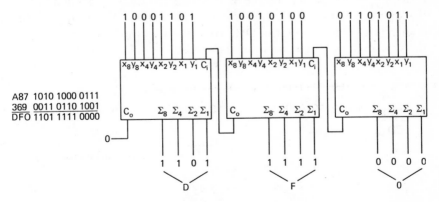

Figure 13-16. Twelve-bit/3-digit hexadecimal position parallel adder.

next position. Into this position are entered the characters 8 and 6 from the x Reg and y Reg. The carry out of this position (if any) becomes a carry into the most significant hexadecimal character position. The output of the adder is 1101 1111 0000, or $DF0_{16}$.

13.8 TWO'S COMPLEMENT METHOD

Most computers use either the one's complement or the two's complement method to perform subtraction because of the simplicity of handling the data. Figure 13-17 shows a 12-bit position adder which uses the two's complement method for subtraction. Two previously discussed MSI devices are used—the 8263 multiplexer discussed in Chapter 10 and the 7483 4-bit adder. You will recall that the 8263 could select data from one of three sources as determined by the channel select inputs (S_0 and S_1). This device also has the ability to provide true output data or data in complement form. When the data complement input (DC) is active, the output data appears in the one's complement form.

Let's add the hexadecimal numbers 473 and 2A6:

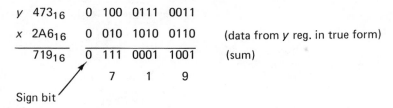

y 473_{16} 0 100 0111 0011

x $2A6_{16}$ 0 010 1010 0110   (data from y reg. in true form)

719_{16} 0 111 0001 1001   (sum)

            7   1   9

Sign bit

Remember, the most significant bit is the sign bit. Because the numbers are positive, the sign bit equals 0. The data is selected from the y Reg by activating S_0 and S_1 (11) on the 8263s. Since we are performing addition, the subtract input (SUB) is *not* active. The output of the multiplexers is therefore in the true form. Data from the x and y Regs is added, and the sum appears at the output of the 7483s.

Now let's subtract 2A6 from 473:

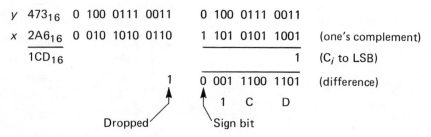

y 473_{16} 0 100 0111 0011   0 100 0111 0011

x $2A6_{16}$ 0 010 1010 0110   1 101 0101 1001   (one's complement)

$1CD_{16}$ 1   (C_i to LSB)

                  1   0 001 1100 1101   (difference)

                              1   C   D

Dropped   Sign bit

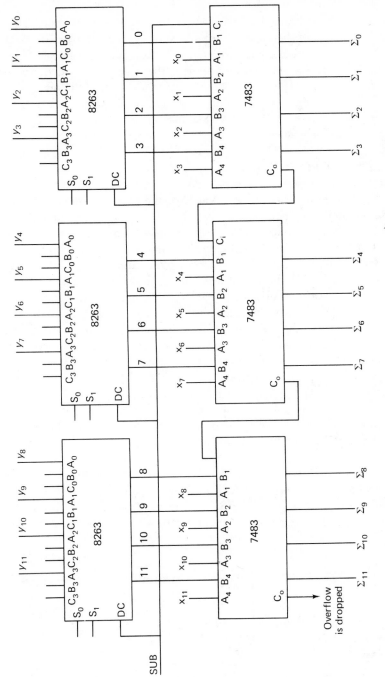

Figure 13-17. Twelve-bit high-speed parallel adder using the two's complement method for subtraction.

The data from the y Reg is selected as before by activation of the appropriate channel select inputs. Because subtraction is to be performed, the subtract input (SUB) is activated. The data from the y Reg is therefore inverted and appears at the output of the multiplexers in the one's complement form. You will recall that 1 must be added to the one's complement to produce the two's complement. This is accomplished by activating the carry input (C_i) to the LSB position by the SUB input.

13.9 LOGIC DECISIONS

The adder is many times called upon to make logic decisions such as: "Are the contents of the x register equal to the contents of the y register?"; or "Are the contents of the x register greater than or less than the contents of the y register?" Such decisions may be made by subtracting one register from the other and then testing the output of the adder. For example, in order to determine if x equals y, the circuit in Figure 13-17 may be used. First, y is subtracted from x, and then all of the bits of the difference are ORed, as shown in Figure 13-18. If the output of the OR gate is true (1), it indicates that a difference between x and y exists and that x and y are not equal. If a 0 appears, it indicates that x does equal y.

Let's suppose we wish to determine if y is greater than x. First, we would subtract y from x as before. Then, we would test the state of the sign bit at the output of the adder. If the sign bit equals 1, it indicates that y is greater than x because the difference is negative. In order to determine if x is greater than y (and not equal to zero), x would be subtracted from y. This can be accomplished if multiplexers are used at both adder inputs.

Circuits used for making logic decisions as well as for performing addition and subtraction are called *arithmetic logic units* (ALUs). The 9340 in Figure 13-19 is a TTL/MSI high-speed 4-bit ALU with full on-chip carry look ahead circuitry. It can perform the arithmetic operations,

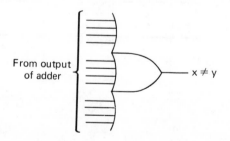

Figure 13-18.   OR gate used to test for equality of x and y.

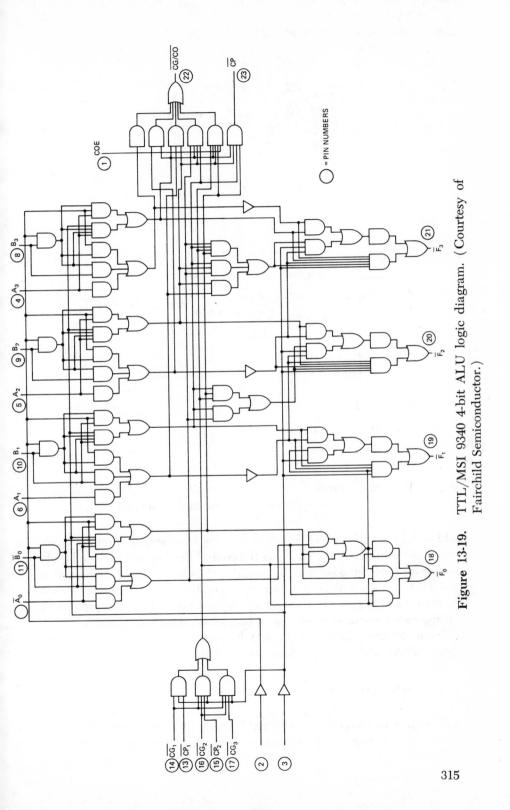

Figure 13-19. TTL/MSI 9340 4-bit ALU logic diagram. (Courtesy of Fairchild Semiconductor.)

315

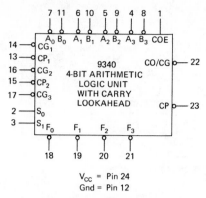

Figure 13-20(a).   Logic equivalents of the 9340 with active low operands. (Courtesy of Fairchild Semiconductor.)

add or subtract in parallel, or any of six logic functions. The output function is determined by the states on the control lines S_0 and S_1. The inputs and outputs of the 9340 may be considered to be active LOW or active HIGH. Logic equivalents for four representations of the 9340 are shown in Figure 13-20(a)–(d). The add and subtract operations are performed on the entire word, with carries or borrows propagated between bits of different weight. The arithmetic may be performed in one's complement, two's complement, or sign-magnitude notation.

13.10 BCD ADDITION

Although most computers use straight binary for their arithmetic operations, some earlier computers, as well as present-day calculators, use

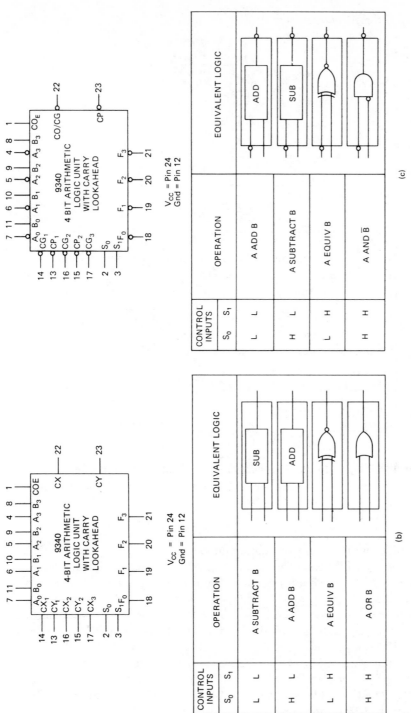

Figure 13-20(b). Logic equivalents of the 9340 with active high operands; (**c**) logic equivalents with active low operands and inverted B input. (Courtesy of Fairchild Semiconductor.)

317

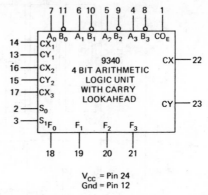

V_{CC} = Pin 24
Gnd = Pin 12

CONTROL INPUTS		OPERATION	EQUIVALENT LOGIC
S_0	S_1		
L	L	A ADD B	ADD
H	L	A SUBTRACT B	SUB
L	H	A EX OR B	
H	H	A OR $\overline{B}$	

(d)

Figure 13-20(d).   Logic equivalents of the 9340 with active high operands and inverted B input. Courtesy of Fairchild Semiconductor.)

BCD. BCD addition and/or subtraction is complicated by the fact that some sums or differences are invalid, as was discussed in Chapter 3. When two BCD digits are added, it is possible to produce 16 different sums, 6 of which are invalid. Furthermore, when a carry-in is considered, an additional four more conditions are invalid. All of these invalid sums are corrected by adding BCD 6 (0110). Note the table of BCD sums in Figure 13-21. Column 1 shows uncorrected sums. Note that all sums greater than 9 are invalid. Column 2 shows the sums after correction.

The circuit in Figure 13-22 is a BCD adder including the correction circuitry. The first adder is a straight 4-bit parallel binary adder (7483). The output of this adder may be a valid or an invalid sum. Invalidity is determined by the logic expression $C_o + 8\cdot4 + 8\cdot2$, where C_o = carry out.

Decimal	Uncorrected BCD sum $C_4\Sigma_4\Sigma_3\Sigma_2\Sigma_1$	Corrected BCD sum $C_4\Sigma_4\Sigma_3\Sigma_2\Sigma_1$	
0	0000	0000	
1	0001	0001	
2	0010	0010	
3	0011	0011	
4	0100	0100	No correction required
5	0101	0101	
6	0110	0110	
7	0111	0111	
8	1000	1000	
9	1001	1001	
10	1010	10000	
11	1011	10001	
12	1100	10010	
13	1101	10011	
14	1110	10100	Correction required
15	1111	10101	
16	10000	10110	
17	10001	10111	
18	10010	11000	
19	10011	11001	

Figure 13-21. Table of BCD sums.

This expression represents all of the conditions that need correction. The need for correction is detected by the gates at the output of the first binary adder. This output is used to produce a binary 6 at the input of a second 7483 and is added to the invalid sum. You will observe that all sums that require correction must also produce a carry. Therefore, the output of the correction detection circuit is used as the carry. This adder may be used with other identical adders to perform addition in parallel. As many adders are required as there are BCD digits. This adder may also be used for serial addition. Remember, the data must be shifted 4-bit positions for each BCD digit.

13.11 BCD SUBTRACTER

You will recall from Chapter 3 that adding the ten's complement of the subtrahend to the minuend produces the difference. The nine's complement is easier to produce because no borrows are involved in the process.

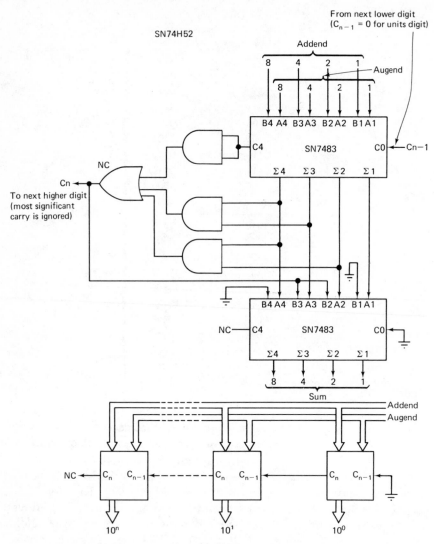

Figure 13-22.   BCD adder including correction circuitry.

However, an end-around carry (EAC) is required. For example, let's subtract 216 from 921.

minuend	921	921	
subtrahend	(−)216	783	(nine's complement)

$$\begin{array}{r} 921 \\ 783 \\ \hline 1\ \ 704 \end{array}$$

EAC ⟶ 1

705   (difference)

The same procedure may be followed when using BCD:

| minuend | 1001 0010 0001 | 1001 0010 0001 | |
| subtrahend | (-) 0010 0001 0110 | 0111 1000 0011 | (nine's complement) |

```
                            1 0000 1010 0100
                              0110 0110
                              0111 0000 0100
                    EAC                      1
                              0111 0000 0101    (difference)
```

Keep in mind that all invalid sums must be corrected by adding 0110.

13.12   BCD NINE'S COMPLEMENTER

Figure 13-23 shows a table of valid BCD digits with the BCD nine's complement. You will notice that the 1 bit is inverted to produce the 1 bit for the nine's complement. The 2 bit is not modified. The 3 bit is produced by "EXCLUSIVE-ORing" the 4 and 2 bits. The 8-bit is produced when the 2, 4, and 8 bits are all equal to 0. A BCD nine's complementer circuit appears in Figure 13-24. Figure 13-25 shows how a 4-bit full adder may be used to produce the nine's complement. This circuit adds the one's complement of the BCD digit to the two's complement of 6 (1010) to produce the nine's complement.

A nine's complementer is combined with a BCD adder in Figure 13-26 to form a BCD adder/subtracter. The EXCLUSIVE-OR gates at the "Addend-Subtrahend" input are used instead of the inverters in Figure 13-25. If the "SUB" input is active, the EXCLUSIVE OR's invert the

DECIMAL	BCD DIGIT DCBA	NINES COMPLEMENT PQRS
0	0000	1001
1	0001	1000
2	0010	0111
3	0011	0110
4	0100	0101
5	0101	0100
6	0110	0011
7	0111	0010
8	1000	0001
9	1001	0000

Figure 13-23.   Table of valid digits and BCD nine's complement.

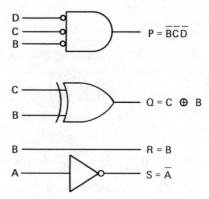

Figure 13-24.   Nine's complementer circuit.

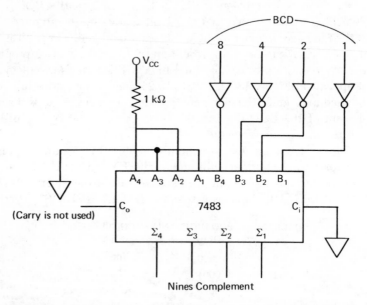

Figure 13-25.   BCD nine's complementer using 4-bit full adder.

subtrahend and add the two's complement of 6 (1010), producing the nine's complement. When the "SUB" input is not active, the EXCLUSIVE OR's do not invert and 0 is added to the addend. This applies the addend in its true form to the input of the BCD adder. Therefore, if the "SUB" input is active, the *difference* appears at the output; if the "SUB" input is not active, the sum appears.

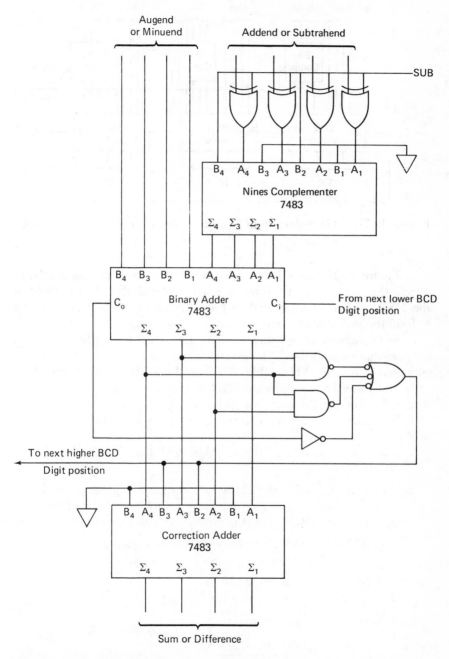

Figure 13-26. One-decade BCD adder/subtracter using nine's complement.

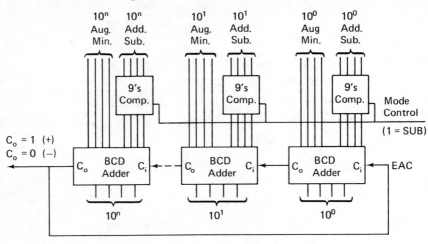

Figure 13-27.   Three-decade BCD adder/subtracter using nine's complement.

Figure 13-27 shows how three adder/subtracters are connected to form a three-decade BCD adder/subtracter. Notice that the carry-out of the MSD position is connected to the carry input of the LSD position to form the end-around carry.

Self-complementing codes are sometimes used to simplify the nine's complementing procedure. The excess-three code is an example of a self-complementing code. The table in Figure 13-28 shows the excess-three code and the nine's complement. Notice that the nine's complement is achieved by taking the one's complement.

When adding with the excess-3 code, one must follow two simple rules:

1. If a carry is produced after one digit addition, a 3 (0011) must be added; and

2. If no carry is produced, 3 must be subtracted. The latter can easily be accomplished by adding the two's complement of 3 (1101). Ignore the carry (if any) that is produced by the correction addition.

Let's add 537 to 389 using the excess-three code.

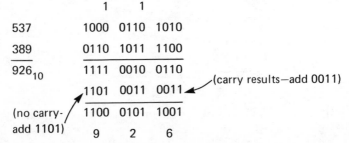

DECIMAL	EXCESS – 3	9's COMPLEMENT
0	0011	1100
1	0100	1011
2	0101	1010
3	0110	1001
4	0111	1000
5	1000	0111
6	1001	0110
7	1010	0101
8	1011	0100
9	1100	0011

Figure 13-28.   Excess-three code and nine's complement.

Now let's subtract 389 from 537 using the nine's complement method and the excess-three code.

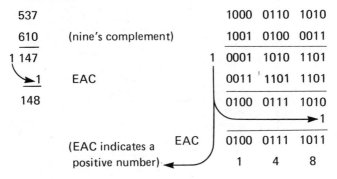

A negative difference is produced if 537 is subtracted from 389:

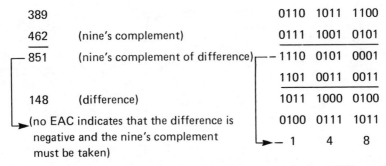

A negative result is indicated when no carry-out of the MSD position occurs. Figure 13-29 illustrates how a one-decade excess-3 BCD adder/

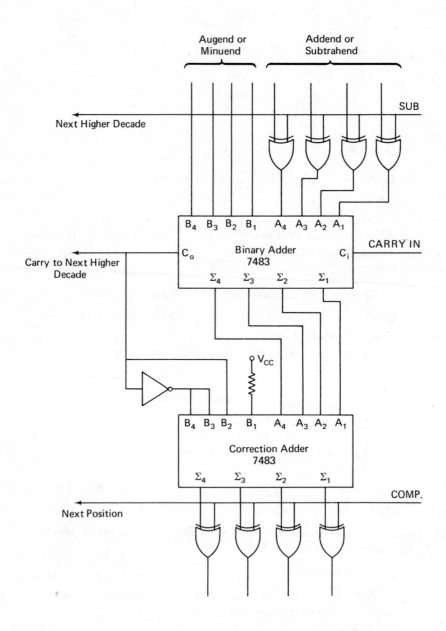

Figure 13-29. One-decade excess-three BCD adder/subtracter.

subtracter is implemented. The EXCLUSIVE-OR's at the addend-subtra-hend inputs are used to produce the one's complement (nine's complement) when the "SUB" input is active. Whenever an end-around carry is not produced, the "COMP" input is activated and the EXCLUSIVE ORs at the input complement the output to provide the difference in the true form of the excess-three code. Notice that if a carry results from the binary adder, a 3 (0011) is entered into the correction adder. However, if a carry does not result, 1101 (the two's complement of 3) is entered.

Figure 13-30 shows how a three-decade excess-3 adder/subtracter is implemented using three of the circuits in Figure 13-29. Notice that an end-around carry is allowed only if the "SUB" input is active ($EAC = SUB \cdot C_o$). The output must be complemented if no end-around carry exists when the "SUB" input is active ($COMP = \overline{C_o} \cdot SUB$).

13.13 BCD-TO-BINARY CONVERSION

Data entering the CPU is usually in the form of an interchange code, such as EBCDIC. You will recall that the numerical portion of that code is BCD. Before numerical data is processed, it is usually converted to binary. This is accomplished with a BCD-to-binary converter.

Shift registers are well suited for BCD-to-binary conversion. Figure 13-31 illustrates the process. Each time the data is shifted to the right, a division by 2 is accomplished. For example, if a bit is shifted from the 8-bit position to the 4-bit position, its weight is divided by 2. Suppose, that a bit is shifted from the 1-bit position of a BCD digit to the 8-bit

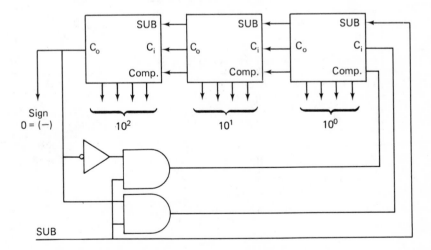

Figure 13-30. Three-decade excess-three BCD adder/subtracter.

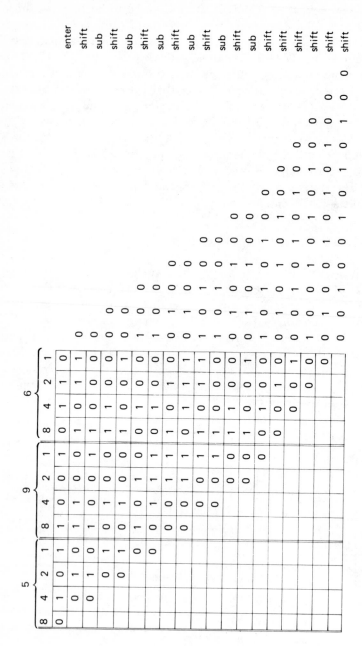

Figure 13-31. Example of BCD-to-binary conversion.

BCD (0101 1001 0110) = 1001010100₂

position of the next BCD digit. Its weight reduces from 10 to 8 instead of from 10 to 5. This means that the resulting BCD number will have a value of 3 greater than it should be. It is possible to subtract 3 (0011) after the shift to make the appropriate correction. You will notice in the example that if an 8-bit is detected in any decade, 3 (0011) is subtracted from that decade.

The process continues with as many shifts as there are bit positions in the register. In practice, the data is shifted from one register to another so that the binary number appears in the second register after the shifting process has been completed.

The BCD-to-binary converter in Figure 13-32 uses the MSI 7495. It uses the two's complement method to perform the subtraction of 3. The two's complement of 0011 is 1101. Notice that if an 8-bit is present, 1101 is added by the 4-bit full adder (7483). When this is done, ignore the carry-out of the adder. A right shift is accomplished by a clock pulse ($\overline{CP2}$) when the parallel input is not active (PE = LOW). After the right shift, the output of the adder is parallel transferred back into the shift register. This is accomplished by activating the parallel entry (PE = HIGH) and the clock pulse $\overline{CP1}$.

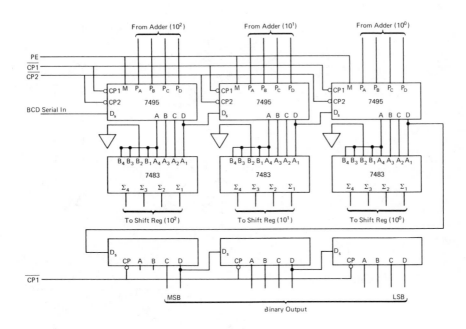

Figure 13-32. BCD-to-binary converter.

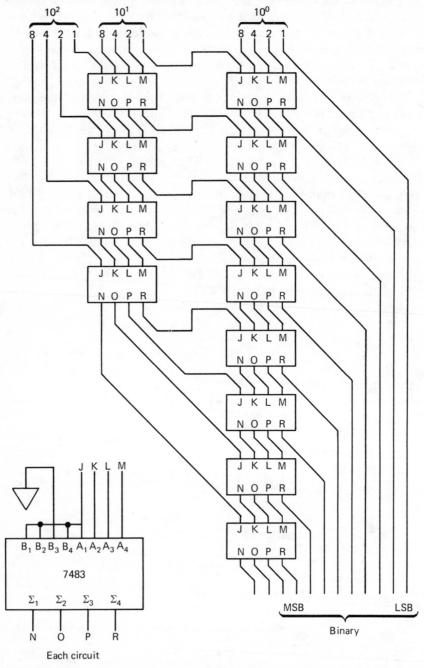

Figure 13-33.   Three-decade BCD-to-binary converter.

BCD-to-binary conversion can also be achieved using a nonclocked approach. Figure 13-33 shows how a number of 4-bit adders can be connected to perform the conversion in parallel mode. The shifting is accomplished by hard-wiring the LSB position of one adder to the MSB position of an adder in the next lower decade. Notice that if there is an 8-bit at the input of an adder, 1101 is added. Although 12 outputs are indicated, 12 bits of BCD data reduce to 10 bits of binary data. The number of inputs can be increased by increasing the number of adders.

13.14 BINARY-TO-BCD CONVERSION

Shift registers are also used to convert binary numbers to BCD. The process is similar to BCD-to-binary conversion, except that the shifting process is from right to left and the correction is somewhat different. A left shift results in a multiplication by 2. If a logical 1 is shifted from the 8-bit position to the 1-bit position of the next higher digit, its value should increase from 8 to 16 instead of from 8 to 10, as it would for BCD. Therefore, to correct for binary, a 6 must be added after the shift. (One can make this correction before the shift by adding 3 whenever an 8-bit exists.) If, after shifting, a number greater than 9 exists in any decade, 10 must be subtracted and a value of 1 must be added to the next decade.

All corrections can be made, if *before* shifting, all decades having a value greater than 4 (0100) are detected and 3 (0011) is added. Figure 13-34 illustrates the process. The process is complete when the last bit has been shifted in.

Figure 13-35 shows a table of corrections for each decade. Notice that only values greater than 4 are corrected. Since values greater than 9 (1001) will not exist at the input, they are not included. The expressions

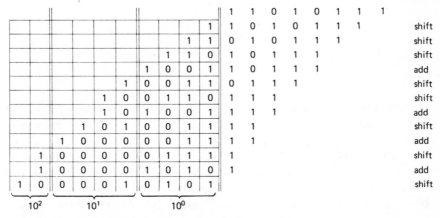

Figure 13-34. Example of binary-to-BCD conversion.

INPUT				OUTPUT			
A	B	C	D	E	F	G	H
0	0	0	0	0	0	0	0
0	0	0	1	0	0	0	1
0	0	1	0	0	0	1	0
0	0	1	1	0	0	1	1
0	1	0	0	0	1	0	0
0	1	0	1	1	0	0	0
0	1	1	0	1	0	0	1
0	1	1	1	1	0	1	0
1	0	0	0	1	0	1	1
1	0	0	1	1	1	0	0

Figure 13-35.   Binary-to-BCD correction table.

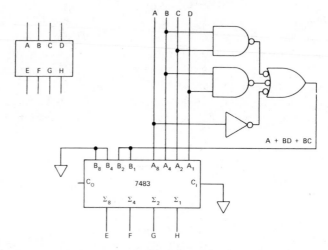

Figure 13-36.   Basic binary-to-BCD converter circuit.

for the output are included. Figure 13-36 shows a logic circuit used to detect a number greater than 4 (0100) and make correction by adding 3 (0011).

This circuit is employed in the binary-to-BCD converter shown in Figure 13-37. The 8-bit shift register in the circuit can shift either left or right, depending on the status of the mode control inputs (S_0 and S_1). If S_0 is HIGH and S_1 is LOW, the register will shift left with a clock pulse

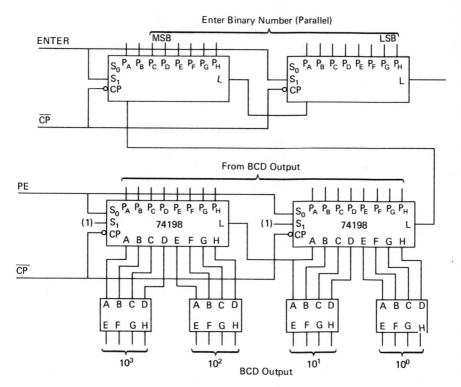

Figure 13-37. Binary-to-BCD converter.

($\overline{CP}$). If both S_0 and S_1 are HIGH, data will enter in parallel with a clock pulse. Binary data is entered into the upper two registers in parallel. Bits are left-shifted into the bottom two registers where the conversion occurs. Each time a number greater than 0100 is detected, 0011 is added and then reentered into the appropriate digit position of the register. Correction is made before each left shift. When the process is complete (after 16 shift pulses), the data is available in parallel at the output of the registers.

Binary-to-BCD conversion can also be accomplished with non-clocked circuitry. Figure 13-38 shows a 12-bit (four-decade) binary-to-BCD parallel converter. It uses the correction circuit in Figure 13-36. Eighteen of these circuits are hard-wired to perform the correction process. The circuit can be expanded to accommodate larger binary numbers.

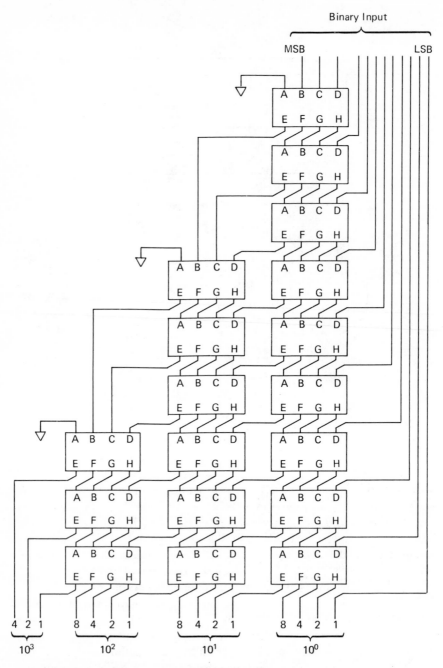

Figure 13-38. Twelve-bit binary-to-BCD converter.

QUESTIONS AND PROBLEMS—CHAPTER 13

1. Write the logic expression for each of the following.

    (a) half adder

    (b) half subtracter

2. What are the advantages and disadvantages of serial and parallel adders?

3. What is the purpose of the inverters in the full adder in Figure 13-4?

4. What is the difference between the INCLUSIVE OR and the EXCLUSIVE OR?

5. Must the half adder be changed in order to add binary numbers having more than one bit position? If so, why?

6. How would you implement a half adder?

7. How would you implement a half subtracter?

8. How would you implement a full adder?

9. Write the logic expression for a comparator.

10. Implement an adder/subtracter using DTL gates.

11. Implement a 2-bit comparator using 7400 NAND gates.

12. Implement a 2-bit comparator using 7402 NOR gates.

13. What is the purpose of "carry look ahead" circuitry?

14. The output of the OR gate in Figure 13-18 is 0. What does this mean?

15. Subtract using the nine's complement:

    (a) 862 − 345

    (b) 670 − 913

16. Subtract using the nine's complement:

    (a) 793 − 176

    (b) 342 − 674

17. Add using excess-three code:

    (a) 360 + 741

    (b) 709 + 119

18. Add using excess-three code:

(a) $435 + 816$

(b) $675 + 173$

19. Perform the following subtractions using the nine's complement and excess-three code.

(a) $834 - 573$

(b) $336 - 941$

20. Perform the following subtractions using the nine's complement and excess-three code.

(a) $615 - 273$

(b) $428 - 870$

21. What is the purpose of the inverter in Figure 13-30?

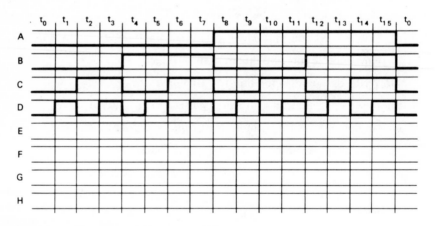

Figure 13-39.   Waveforms for problems 22 and 23.

22. For the logic diagram in Figure 13-4, draw the output waveforms. Use the timing chart in Figure 13-39. Use variables A, B, and C for inputs A, B, and C_i.

23. Using the timing chart in Figure 13-39, draw the output waveforms for the logic circuit in Figure 13-29. Substitute A for the 8-bit, B for the 4-bit, C for the 2-bit, and D for the 1-bit.

14

Memory

Memory systems may be placed in two separate categories: random access memories (RAMs) and read-only memories (ROMs). A better name for RAM is "read/write" memory because data is accessed in the same manner for both functions. The term RAM connotes that data may either be written into the memory or read from the memory repeatedly. However, data may be written into the ROM only once. Some ROMs are "programmed" during the manufacturing process.

RAMs are used for main memories, smaller memories associated with arithmetic operations called *scratch pad* memories, and *buffer* memories used for input-output applications. ROMs are used for fixed instructions, such as subroutines or micro-instructions. They are also used as look-up tables for data, such as trigonometric tables.

The topics covered in this chapter are:

14.1   Magnetic Core Memories

14.2   Three-Dimensional Array

14.3   Bipolar Memories

14.4   MOS Static Memories

14.1   MAGNETIC CORE MEMORIES

Although LSI technology is now providing the industry with semiconductor memory devices, most computers still use core memory. Magnetic cores are small doughnut-shaped pieces of ferromagnetic material. They are designed to exhibit a square-loop hysteresis characteristic, as shown in Figure 14-1(a).

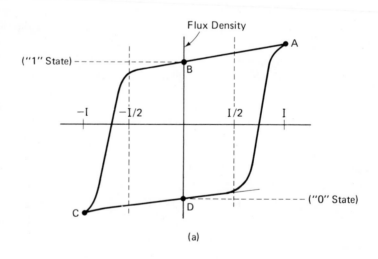

(a)

(b)   (c)

Figure 14-1.   Memory cores: (a) square hysteresis loop characteristics; (b) core in "one state"; (c) core in "zero state."

These cores are magnetized in one of two possible conditions (bi-stable). These two conditions are indicated by points B or D on the curve. If current is passed through the core, as indicated in Figure 14-1(b), the core becomes magnetized in the indicated direction. The resulting field is indicated by point A on the curve. When the current ceases, the strength of the field will decrease slightly to point B and remain at that strength due to its residual magnetism. For purposes of discussion, let's assume that this is the "1" state. A current in the opposite direction, as indicated in Figure 14-1(c), will cause the flux field to reverse, bringing us to point C on the curve. Upon removal of this current, the flux field will reduce slightly to point D and remain there. This is the "0" state. Be-cause of the characteristics of the material, a discrete amount of current is required to switch the core from one state to the other. This is indicated on the graph as I or $-I$. Notice that one-half of this current ($I/2$ or $-I/2$) is not sufficient to cause the core to switch states.

Cores are usually assembled to form a core plane, as shown in Figure 14-2. Notice that two wires are shown passing through each core. These are identified as x and y lines. The purpose of these lines is to address one core in the plane. For example, if the x_3 line and the y_3 line are both active, the core labeled 1010 will be selected and switched to the "1" state. No other core will be switched because there will either be no cur-rent through that core or only half of the current needed to cause it to

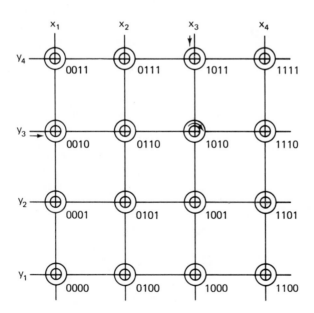

Figure 14-2.   Core memory plane.

switch. The y_3 line and the x_3 line each carry only half of the current necessary to cause a core to switch. Only where these two lines pass through the same core will the sum of the currents be sufficient. The core at address 1010 can be set to the "0" state by providing a current in the opposite direction in the x_3 and y_3 address lines.

How is data retrieved (read) from a core memory plane? Figure 14-3 shows a core memory plane with an additional wire passing through each core. This wire is called the *sense line*. It senses a change of state of any core in the plane. When data is to be read, the appropriate x and y lines are activated to select a given core. The current is such that it causes the core to switch to the "0" state. Of course, if the core was already in the "0" state, it will not switch. However, if it were in the "1" state, it would switch, causing a small amount of current to be induced in the sense line. This current activates the sense amplifier, which in turn causes a flip-flop to be SET only if the core switched. The strobe input

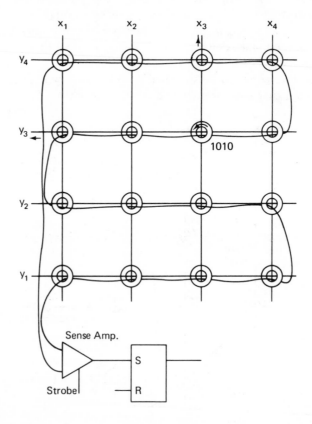

Figure 14-3.   Core plane showing sense line.

to the sense amplifier is provided to allow the amplifier to become active only during the read time. After a core is "read," its state will always be "0," and the data at that address is destroyed. This is called *destructive readout*. It should be emphasized, however, that the data is not lost but is stored in the flip-flop, as shown in Figure 14-3. At the end of a clock cycle, data is written back into core.

Figure 14-4 shows various waveforms that represent the process of reading a core. Notice that during read time, both x_3 and y_3 lines are activated with a $-I/2$ current. This causes the total current through core 1010 to equal $-I$. If that core is in the "1" state, it will switch to the "0" state, causing a current to be induced in the sense line. This signal is amplified by the sense amplifier, and its output is fed into the flip-flop. This flip-flop will store a "1" if the core switched, or a "0" if the core did not switch. During write time, the core may switch again. However, because the sense amplifier is not activated, there will be no output pulse.

After a core is "read," we may "write" data into it. Since the cores are in the "0" state (because of the "read" operation), the x and y lines must each be activated with a $+I/2$ current. Activating both the x and y lines in a selected core will cause the core to switch to the "1" state.

A fourth line, called the *inhibit line*, is passed through the cores, as illustrated in Figure 14-5. The inhibit line prevents a core from switching

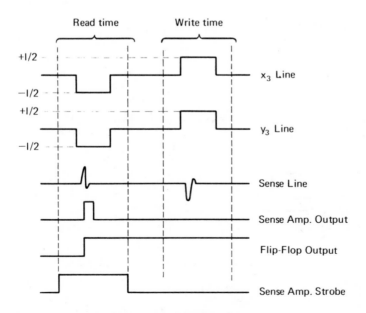

Figure 14-4. Waveforms for the read operation.

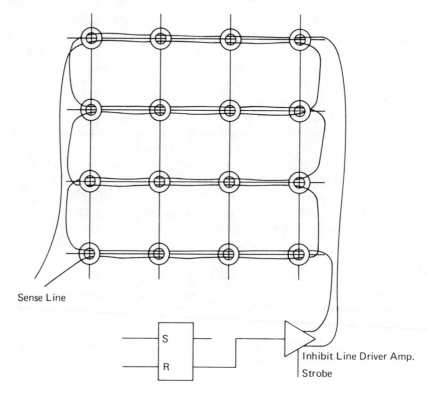

Figure 14-5.   Core plane showing inhibit line.

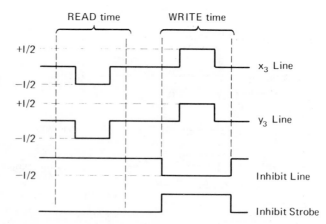

Figure 14-6.   Waveforms for writing.

to the "1" state whenever it is necessary to write a "0." This is accomplished by providing a current of $-I/2$ in the inhibit line. The inhibit current opposes the current in the x and y lines. The resulting current is not sufficient to cause the core to switch; therefore, a "0" is stored.

Figure 14-6 shows various waveforms of the "writing" process. During write time, both x_3 and y_3 lines are activated with a current of $+I/2$, causing the total current through the selected core to equal $+I$. However, if the inhibit line is active, the resulting current will be insufficient to cause the core to switch to the "1" state. Thus, a "0" is written. Note that the inhibit line driver amplifier is activated by the reset output ("0" state) of the flip-flop.

14.2   THREE-DIMENSIONAL ARRAY

Core memories may be arranged to accommodate either word- or character-organized data. This is accomplished by stacking a number of core planes to form a three-dimensional array. Figure 14-7 shows a three-dimensional array consisting of four planes. Each plane stores a bit according to its place value. A memory organized to store EBCDIC characters would require eight planes.

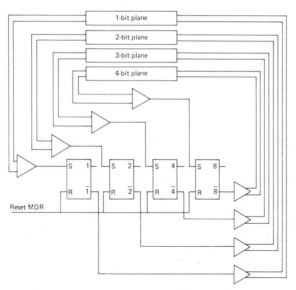

Figure 14-7.   Three-dimensional core memory array with sense amplifiers, inhibit drivers, and memory data register.

In a three-dimensional array, the x lines of each plane are connected in series. For example, x_1 of the 1-bit plane is connected in series with x_1 of the 2-bit plane. The same is true for the y lines. This means that when a given core is selected on the 1-bit plane, cores occupying the same position on all other planes are also selected. Each plane has a sense line of its own, independent of other planes. A sense amplifier and a flip-flop are required for each plane. These flip-flops form the memory data register (MDR).

Each plane has an inhibit line and an inhibit line driver, which are independent of the other planes. During the "write" cycle, these line drivers are activated by the "0" output of the flip-flops in the MDR. The "read" operation occurs near the beginning of the machine clock cycle, and the "write" operation occurs near the end of the machine clock

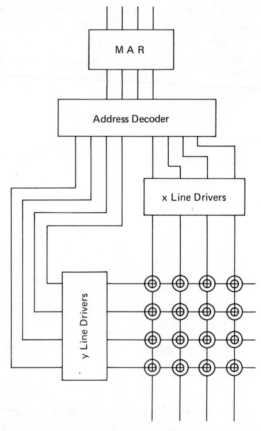

Figure 14-8.   Address decoder and line drivers.

cycle. Between the "read" and "write" times, data may be transferred to or from the MDR. For example, if data is to be read, it is transferred from the MDR to the appropriate register between "read" and "write" times. The same data would then be written back into the core at the "write" time. If new data is to be entered into core, it would be transferred into the MDR between the "read" and "write" times.

How is memory addressed? You may recall that the second part of an instruction word is the operand. The *operand* is used to designate the address of data. The operand can be used to preset the memory address register (MAR). Figure 14-8 shows how a core plane is addressed. A decoder appears at the output of the MAR. The outputs of the decoder are amplified by the x and y line drivers, which provide the current required to switch the cores. The MAR is a presettable register which may also increment (count) so that addresses may be selected in sequence.

The primary advantage of core memory over semiconductor memory is that the former is *nonvolatile*. This means that if the power should fail, data stored in core will remain. However, semiconductor memories require continuous power to retain data. On the other hand, semiconductor memories are smaller in size, consume less power, and are progressively decreasing in cost. In addition, data in semiconductor memories is not destroyed during the "read" operation.

14.3   BIPOLAR MEMORIES

Bipolar memories are made of cells similar to the one shown in Figure 14-9(a). Such a cell is a simple flip-flop that uses multi-emitters, which is typical in TTL circuits. The function of the bipolar cell is similar to that of a ferrite core. A cell is selected when both the x and y lines are activated (HIGH). You will note that when both the x and y lines are HIGH, the respective emitter-base junctions are reverse-biased; therefore the state of the flip-flop is determined by the 0 and 1 sense lines. These lines are either used to preset (write) or sense (read) the state of the flip-flop.

Figure 14-9(b) shows how 16 bipolar memory cells are connected to form an array. A cell is addressed by activating one x and one y line. Addressing a cell does not affect its state. The state of the flip-flop is determined by the sense lines. If the 0 sense line is LOW, a 0 is stored; conversely, if the 1 sense line is LOW, a 1 is stored. The states of sense lines are amplified by buffer amplifiers and provided at the "read" outputs as S_0 and S_1. The unaddressed cells do not affect the sense lines because their associated emitter-base junctions for both transistors are reverse-biased.

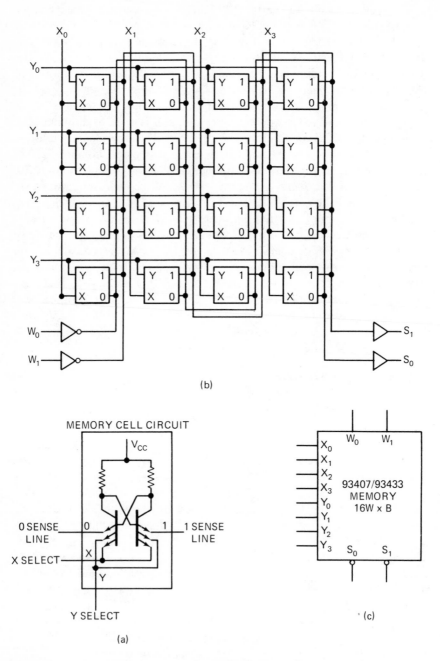

(b)

MEMORY CELL CIRCUIT

(a)

(c)

Figure 14-9. A 16W × 1B read/write memory: (a) circuit for memory cell; (b) logic symbol. (Courtesy of Fairchild Semiconductor.)

Note that both sense lines are active LOW, as shown in the logic symbol. Figure 14-9(c) shows the logic symbol for a 16-bit memory.

Data can be written by activating the 0 write (W_0) or the 1 write (W_1) lines while a cell is addressed. Activating the W_0 line causes the flip-flop to attain the 0 state. The 1 state is obtained by activating the W_1 line. The write inputs (W_0 and W_1) are buffered to prevent interference with the sense lines.

A 16-address bipolar memory is illustrated in Figure 14-10. Each address stores four bits. This circuit is analogous to a four-stack, three-dimensional array of cores. With this arrangement, 4-bit words may be stored ($16W \times 4B$). Each address is selected by activating one of the 16 lines (X_0 through X_{15}). If many of these devices are used to form a larger memory, a given device can be addressed by activating its chip-select (CS) input.

The four output data lines (O_0, O_1, O_2, and O_3) provide an output for the addressed data. Data to be stored (written) is applied to the inputs (I_0, I_1, I_2, and I_3) and stored when the write enable (WE) input is activated.

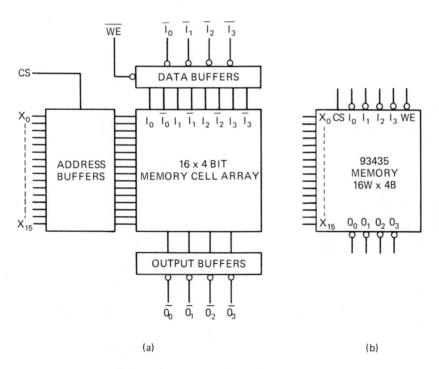

(a)   (b)

Figure 14-10.   93435, $16W \times 4B$ read/write memory: (a) block diagram; (b) logic symbol. (Courtesy of Fairchild Semiconductor.)

Uncommitted collector outputs (collectors do not have pull-up resistors) are provided to allow maximum flexibility in output connections. The outputs of the bipolar memories can be tied to outputs of other devices in order to expand the size of the memory. An example of an expanded memory constructed with these devices is shown in Figure 14-11. The diagram shows two 16W × 4B arrays of a bank of 10. The 9311

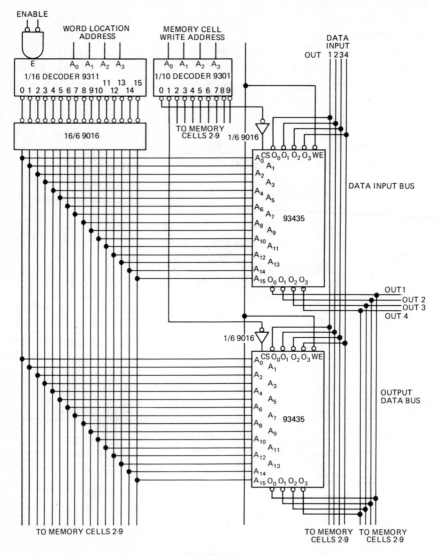

Figure 14-11. A 160W × 4B expanded memory. (Courtesy of Fairchild Semiconductor.)

decoder (discussed in Chapter 10) operates as a word address decoder while the 9301 decodes the chip-select inputs of the arrays. The 9301 decoder selects the various arrays, and since it has 10 outputs, it can select any one of 10 arrays. Thus the memory has been expanded to 160 words by four bits (160W $\times$ 4B).

As the number of addresses increases, it becomes more practical to provide on-chip decoders. Figure 14-12 illustrates a 16W $\times$ 4B bipolar memory with on-chip decoding. Note that only 4 input lines (binary code) are used to address 16 words.

An expanded memory using sixteen arrays of 16W $\times$ 4B are shown in Figure 14-13. Note that each of the arrays is organized as a cell in an

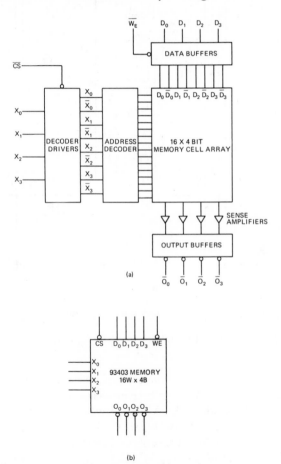

Figure 14-12. 93403, 16W $\times$ 4B bipolar memory with on-chip decoder: (a) block diagram; (b) logic diagram. (Courtesy of Fairchild Semiconductor.)

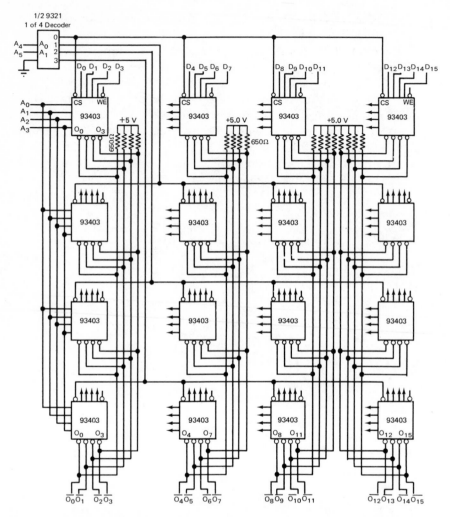

Figure 14-13.   Expanded 64W × 16B bipolar memory. (Courtesy of Fairchild Semiconductor.)

expanded array and may be selected by row and column. All address inputs (A_0, A_1, A_2, and A_3) are tied in parallel and are used to select columns. A 1 of 4 decoder (A_4, A_5) is provided to select rows. This results in six input lines and 64 addresses. Each address stores 16 bits. Thus, a memory of 64 words by 16 bits is produced. Because of the uncommitted collectors at the outputs of each chip, external pull-up resistors are required.

Figure 14-14(a) shows the logic symbol for a 93400, 256-bit read/write memory with decoder. It is addressed with a partially decoded x-y

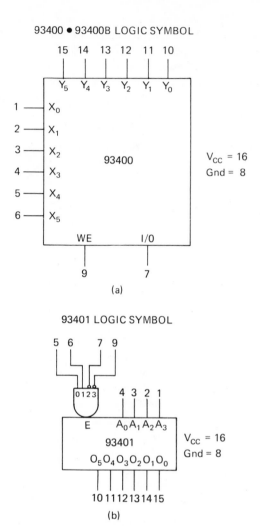

Figure 14-14. A 256-bit read/write memory: (a) symbol; (b) symbol for decoder. (Courtesy of Fairchild Semiconductor.)

coincident selection scheme. The six x-inputs require three binary bits and their complements; and the same is true for the y inputs. This results in a 16-bit × 16-bit array, or 256 1-bit addresses. When it is used in an expanded memory, a companion unit (93401), a decoder-driver, is required, as shown in Figure 14-14(b).

The 93401 is a partial decoder and driver for the 93400. It accepts a 4-bit binary code (A_0, A_1, A_2, and A_3) and produces a 3 of 6 code, which is required by the 93400. A four-input AND gate activates the enable (E) input. Since two of these inputs are HIGH and two are active LOW, it is possible to route two binary-coded lines to four different 93400s to get two additional bits of decoding with no extra packages. This is illustrated in the memory addressing scheme in Figure 14-15. The 93400s are arranged in an array of 16 rows and 8 columns. Each row requires one decoder/driver. Four 93401s are required to decode and drive the columns. The columns are selected by address lines A_0 through A_5. The rows are selected by address lines A_6 through A_{11}. Thus, 4096 8-bit words are stored.

Eight data input lines (D_0 through D_7) are provided, and data is entered when the $\overline{\text{STROBE I}}$ input is activated and the read-write mode is LOW. Data appears at the output when the $\overline{\text{STROBE W}}$ line is activated and the read-write mode input is HIGH.

14.4   MOS STATIC MEMORIES

You will recall from Chapter 6 that MOS circuits can be either static or dynamic. A basic static MOS memory cell is shown in Figure 14-16. This cell is made up from eight P channel MOS transistors. Note the x and y address lines and the 0 and 1 sense lines. The methods of addressing and reading or writing are similar to those in bipolar memories. Data in a static memory cell will remain until it is addressed and new data is written in. Of course, when power fails, data is lost.

A six-transistor static MOS cell is shown in Figure 14-17. Because of its relative simplicity, this cell is currently popular with static read/write memories.

It has been mentioned previously that complementary MOS (CMOS) circuits are rapidly becoming very popular. Figure 14-18 illustrates a six- and an eight-transistor cell. Although the six-transistor cell is less complex, it requires more complex decoding. Decoders, however, are usually constructed on the same chip. Complete words can be addressed with the word-select line. The eight-transistor cell allows for the x-y coincident decoding scheme, where one bit may be selected at a time.

An example of a fully decoded 256-bit static read/write memory (2501) is shown in Figure 14-19. Its organization is 256 1-bit words (256W $\times$ 1B). The first four address lines, A_1 through A_4, are decoded to produce the x coincident lines, and address lines A_5 through A_8 are decoded to produce the y coincident lines. Its package symbol with pin designation is shown in Figure 14-20.

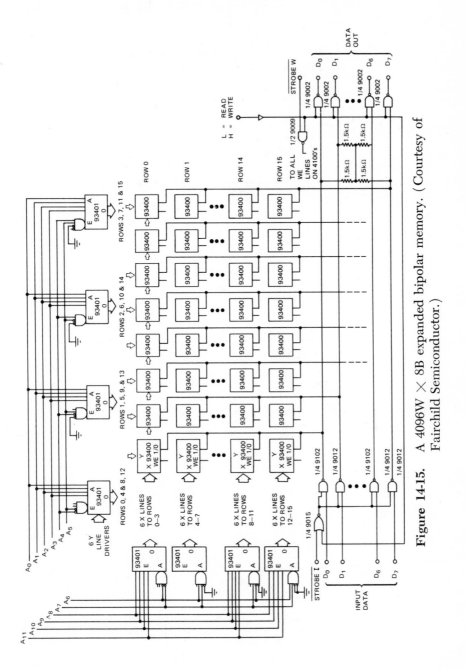

Figure 14-15. A 4096W × 8B expanded bipolar memory. (Courtesy of Fairchild Semiconductor.)

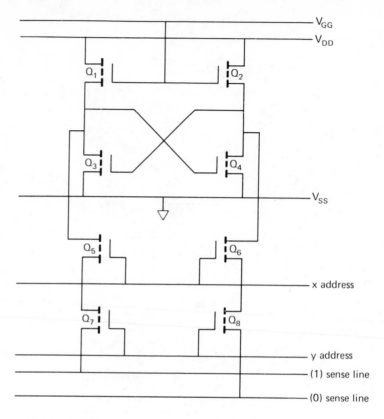

Figure 14-16.   Eight-transistor static P channel MOS cell.

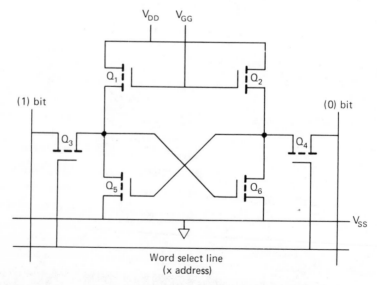

Figure 14-17.   Six-transistor static P channel MOS cell.

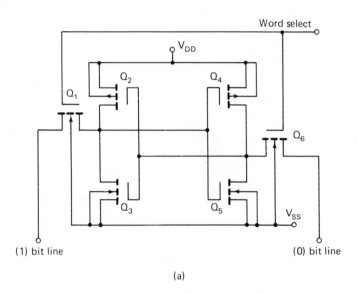

(a)

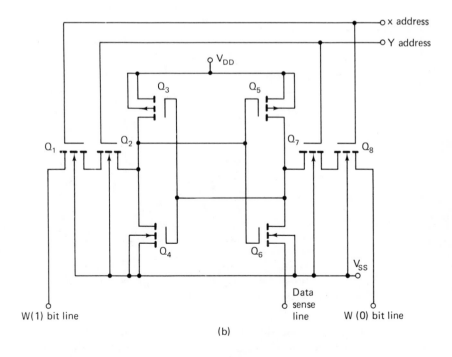

(b)

Figure 14-18. CMOS static read/write cells: (a) six-device cell; (b) eight-device cell.

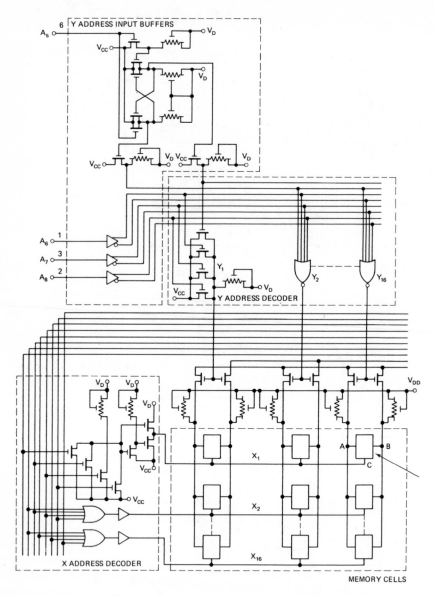

Figure 14-19.   MOS fully decoded read/write 256W $\times$ 1B memory. (Courtesy of Signetics Corporation.)

Figure 14-21(a) illustrates how a number of 2501s are organized to form a larger memory. An additional decoder is required for the column select (C_1, C_2, . . . C_n). Figure 14-21(b) shows the organization of a 4096 word by 12-bit memory.

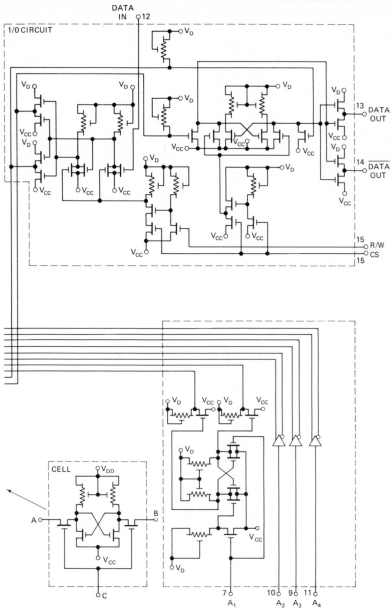

Figure 14-19. (continued).

Figure 14-22 is a larger (1024W × 1B) static read/write memory. Note that the memory cells are organized in 32 rows and 32 columns and are addressed by ten lines ($2^{10} = 1024$). Many of these devices may easily be used to form a very large expanded memory.

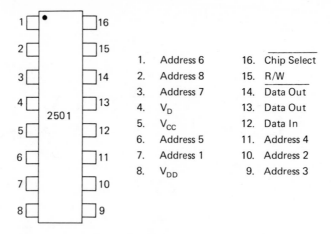

Figure 14-20.   2501 package symbol with pin designations. (Courtesy of Signetics Corporation.)

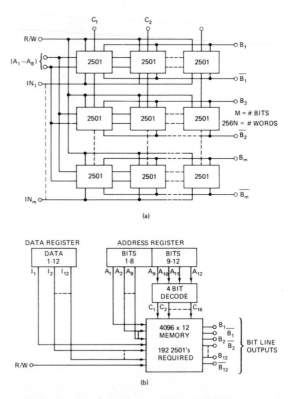

Figure 14-21.   Method of expanding 2501s: (a) basic scheme; (b) 4096W $\times$ 12B memory. (Courtesy of Signetics Corporation.)

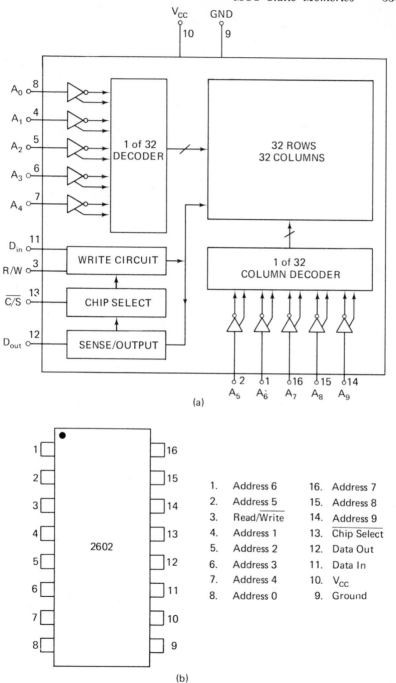

Figure 14-22. A 1024W × 1B static MOS read/write memory: (a) block diagram; (b) package symbol with pin designations. (Courtesy of Signetics Corporation.)

14.5   MOS DYNAMIC MEMORIES

A major disadvantage of the static cell is that its power dissipation limits the total number of bits per package. A method by which power dissipation may be reduced was discussed in Chapter 6. This method relies on clocking the supply lines. A four-transistor dynamic MOS memory cell is shown in Figure 14-23(a). Transistors Q_3 and Q_4 are the load devices for the flip-flop formed by Q_1 and Q_2. Transistors Q_3 and Q_4 are also used as the word-enable transistors. In order to read data from the cell, the word-select line is driven to the negative condition, and the current in the 0 and 1 lines is sensed. For example, if Q_1 is conducting, the sense line associated with Q_3 will carry current and the other sense line will not. To write, the sense lines are forced to the desired state and the word line is enabled. Because of the dynamic characteristics, the cell must be refreshed. This is accomplished when the sense lines and the word lines are brought to a negative potential. Figure 14-23(b) shows the waveforms for a write operation followed by a read operation. For this cell, a 0 is defined as the most negative potential.

The circuit diagram in Figure 14-24(a) shows a three-transistor cell which relies on the stored charge of the capacitance, C. It has four control lines: read-select, write-select, write-data line, and read-data line.

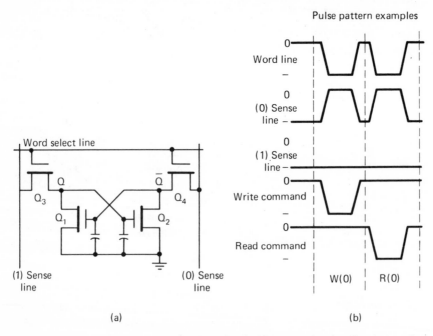

Figure 14-23.   Four-transistor dynamic MOS memory cell: (a) schematic; (b) waveforms.

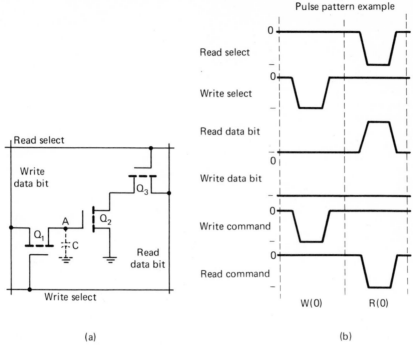

(a)   (b)

Figure 14-24.   Three-transistor dynamic MOS memory cell: (a) schematic; (b) waveforms.

The waveforms are shown in Figure 14-24(b). The read and write lines are normally held negative. In order to write a 0, the write-select line is brought to a negative potential and the read-select line is held at 0 volts. The write-data line is held negative, which turns on Q_1 and charges C to a negative potential. In order to read, the read-select line is driven negative and the write-select line is held at 0 volts. The write-data line is now isolated from C. The read-data line is discharged to 0 volts through Q_2 and Q_3. The logic state can now be interrogated by sensing the read-data line. Note that the inversion takes place through the cell. Therefore, the output must be complemented. This cell requires periodic refreshing, accomplished by a read cycle followed by a write cycle. A logic 0 for this cell is defined as the most negative voltage.

The three-transistor MOS cell is used in the 1103-1 fully decoded 1024-bit dynamic memory. A schematic diagram appears in Figure 14-25. The block diagram and pin configuration are shown in Figure 14-26(a) and (b). The 1103-1 is designed for main memory applications where high performance, low cost, and large storage are important design objectives. The dynamic circuitry dissipates power only during precharge. Information stored in the memory is nondestructively read. Refreshing of all 1024 bits is accomplished in 32 read cycles and is required every

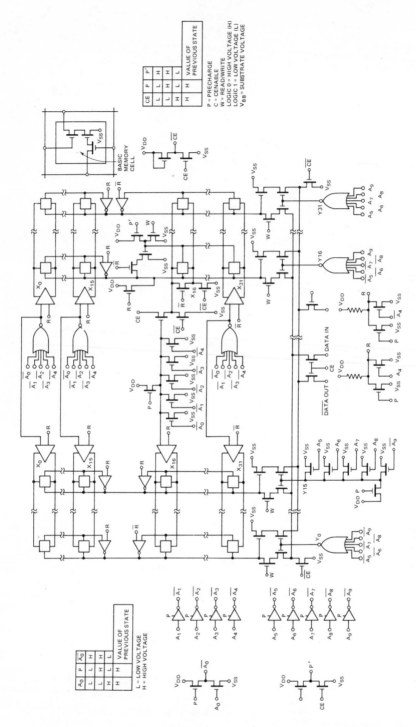

Figure 14-25. Schematic diagram of the 1103-1 dynamic read/write memory. (Courtesy of Signetics Corporation.)

two milliseconds. A separate chip-enable input allows selection of an individual package when the outputs are OR-tied. The waveforms for the 1103-1 appear in the timing chart in Figure 14-27.

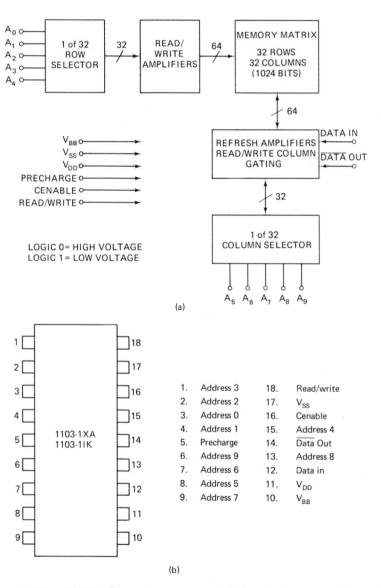

(a)

(b)

Figure 14-26. 1103-1 dynamic memory: (a) block diagram; (b) package symbol with pin designation. (Courtesy of Signetics Corporation.)

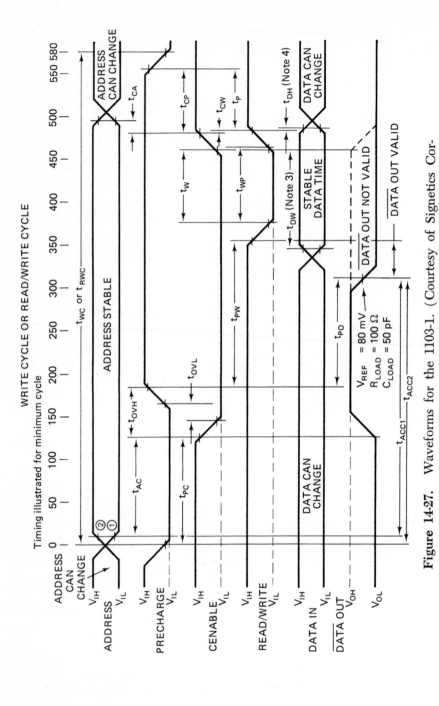

Figure 14-27. Waveforms for the 1103-1. (Courtesy of Signetics Corporation.)

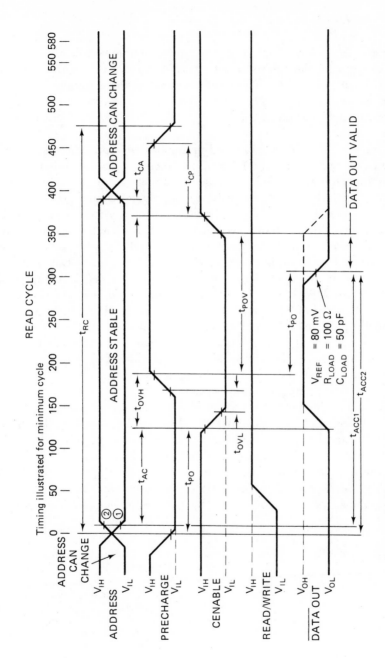

Figure 14-27. (continued).

14.6   BIPOLAR ROMs

The basic bipolar read-only memory cell is diagrammed in Figure 14-28. It is simply a bipolar transistor with its collector tied to V_{CC} and a resistor in series with its emitter. The base of the transistor is driven by the address decoder. If the resistor in the emitter has a low value of resistance (approximately 150 ohms), a 0 is stored in the cell. However, if the resistor is open, a 1 is stored.

The cell may be programmed in one of two ways. The first method is accomplished during the fabrication process. Interconnections on the monolithic chip are made by a process of aluminum deposition and etching. If a 1 is to be stored, the resistor is simply not connected.

The second method allows the user to program the device as he desires. Special devices that have emitter resistors made from deposited Nichrome are required. A 1 is stored in the cell by addressing the cell and pulsing it with a high current (25 to 35 mA), which causes the resistor to be blown open. These devices are called *field programmable ROMs* or *PROMs*. Figure 14-29 illustrates how 256 cells are connected to form an array of 32 words by eight bits. An on-chip address decoder is usually provided. The output of the decoder drives the word-select lines. This selects the eight bits of a single word (B_0 through B_7). If a 0 is stored, the sense line will be pulled to HIGH level by the selected cell. This causes the output buffer transistor to be turned on and an active LOW will appear at the output. If a 1 is stored, the sense line will not be affected by the activated cell because the resistor is open.

An example of a bipolar ROM is shown in Figure 14-30. The 93412 and 93434 are 256-bit read-only memories organized in 32 words having eight bits each. Eight outputs are provided with uncommitted collectors to allow wire-OR memory expansion. A chip-enable input is provided to simplify address decoding. The 93412 is field-programmable, which means the user may program the device by using special equipment. This is

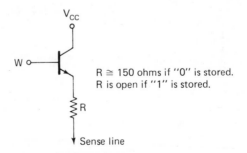

Figure 14-28.   Basic bipolar ROM cell.

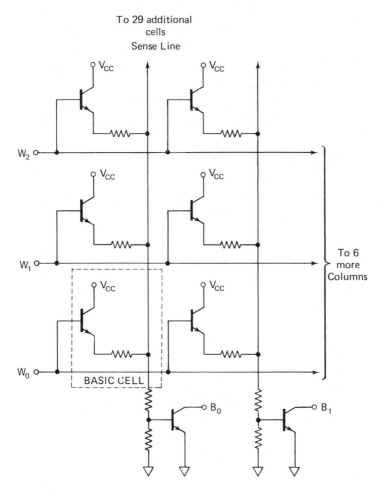

Figure 14-29. Six cells of a 256-bit ROM.

desirable in research and development applications. The 93434 is programmed during the manufacturing process, making it more economical in high volume use.

The method of addressing these devices is the same as with read/write memories. Note however that no data input lines are provided.

A similar device is found in the TTL 7400 series—the 7488. This is not field-programmable. The customer must provide the program information so that it may be programmed to his specifications. The programming information is provided by the customer by his completion of the truth

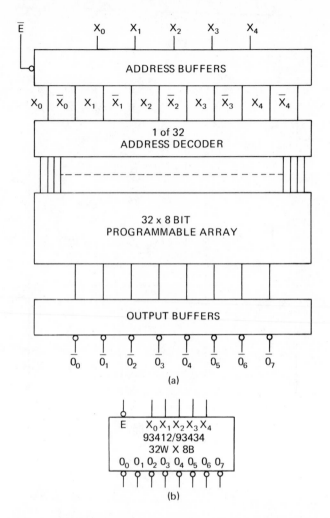

Figure 14-30.   93412/93434 bipolar ROM: (a) block diagram; (b) symbol. (Courtesy of Fairchild Semiconductor.)

table order blank as shown in Figure 14-31. For example, let's provide the information required to program the seven-segment encoder described in Chapter 10. Although it may not be a practical example, it shows the concepts involved. The Xs in the table indicate "don't care states." The BCD input is applied to addresses A_0 through A_3. The seven-segment outputs (a through f) are taken from the outputs B_0 through B_6. For convenience, the code sequence was repeated three times to fill the table.

A 7488 which has been especially programmed is the 74184 shown in Figure 14-32. It is an implementation of a 6-bit BCD-to-binary converter, which was discussed in Chapter 13. The outputs are labeled Y_1

256-BIT READ ONLY MEMORIES TRUTH TABLE/ORDER BLANK

CUSTOMER:_____
P.O. No.: _____
YOUR PART NO.:_____
DATE: _____

THIS PORTION TO BE COMPLETED BY SIGNETICS
PART NO.:_____
S.D. No.:_____
DATE RECEIVED:_____

			INPUTS							OUTPUTS				
WORD	A_4	A_3	A_2	A_1	A_0	ENABLE	B_7	B_6	B_5	B_4	B_3	B_2	B_1	B_0
0	0	0	0	0	0	0	X	0	1	1	1	1	1	1
1	0	0	0	0	1	0	X	0	1	1	0	0	0	0
2	0	0	0	1	0	0	X	1	0	1	1	0	1	1
3	0	0	0	1	1	0	X	1	0	0	1	1	1	1
4	0	0	1	0	0	0	X	1	1	0	0	1	1	0
5	0	0	1	0	1	0	X	1	1	0	1	1	0	1
6	0	0	1	1	0	0	X	1	1	1	1	1	0	0
7	0	0	1	1	1	0	X	0	0	0	0	1	1	1
8	0	1	0	0	0	0	X	1	1	1	1	1	1	1
9	0	1	0	0	1	0	X	1	1	0	0	1	1	1
10	0	1	0	1	0	0	X	0	1	1	1	1	1	1
11	0	1	0	1	1	0	X	0	1	1	0	1	0	0
12	0	1	1	0	0	0	X	1	0	1	1	0	1	1
13	0	1	1	0	1	0	X	1	0	0	1	0	1	1
14	0	1	1	1	0	0	X	1	1	0	0	1	1	0
15	0	1	1	1	1	0	X	1	1	0	1	1	0	1
16	1	0	0	0	0	0	X	1	1	1	1	1	0	0
17	1	0	0	0	1	0	X	0	0	0	0	1	1	1
18	1	0	0	1	0	0	X	1	1	1	1	1	1	1
19	1	0	0	1	1	0	X	1	1	0	0	1	1	1
20	1	0	1	0	0	0	X	0	1	1	1	1	1	1
21	1	0	1	0	1	0	X	0	1	1	0	1	0	0
22	1	0	1	1	0	0	X	1	0	1	1	0	1	1
23	1	0	1	1	1	0	X	1	0	0	1	0	1	1
24	1	1	0	0	0	0	X	1	1	0	0	1	1	0
25	1	1	0	0	1	0	X	1	1	0	1	1	0	1
26	1	1	0	1	0	0	X	1	1	1	1	1	0	0
27	1	1	0	1	1	0	X	0	0	0	0	1	1	1
28	1	1	1	0	0	0	X	1	1	1	1	1	1	1
29	1	1	1	0	1	0	X	1	1	0	0	1	1	1
30	1	1	1	1	0	0	X	0	1	1	1	1	1	1
31	1	1	1	1	1	0	X	0	1	1	0	1	0	0
ALL	X	X	X	X	X	1	1	1	1	1	1	1	1	1

Figure 14-31. Truth table order blank. (Courtesy of Signetics Corporation.)

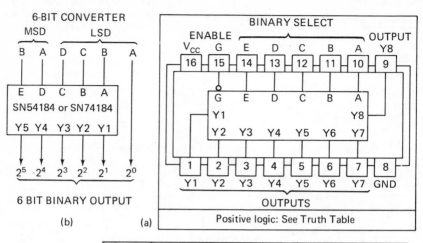

6 BIT BINARY OUTPUT

(b)   (a)

Positive logic: See Truth Table

TRUTH TABLE
BCD-TO-BINARY
CONVERTER

(c)

BCD WORDS	INPUTS (See Note A)						OUTPUTS (See Note B)				
	E	D	C	B	A	G	Y5	Y4	Y3	Y2	Y1
0·1	L	L	L	L	L	L	L	L	L	L	L
2·3	L	L	L	L	H	L	L	L	L	L	H
4·5	L	L	L	H	L	L	L	L	L	H	L
6·7	L	L	L	H	H	L	L	L	L	H	H
8·9	L	L	H	L	L	L	L	L	H	L	L
10·11	L	H	L	L	L	L	L	L	H	L	H
12·13	L	H	L	L	H	L	L	L	H	H	L
14·15	L	H	L	H	L	L	L	L	H	H	H
16·17	L	H	L	H	H	L	L	H	L	L	L
18·19	L	H	H	L	L	L	L	H	L	L	H
20·21	H	L	L	L	L	L	L	H	L	H	L
22·23	H	L	L	L	H	L	L	H	L	H	H
24·25	H	L	L	H	L	L	L	H	H	L	L
26·27	H	L	L	H	H	L	L	H	H	L	H
28·29	H	L	H	L	L	L	L	H	H	H	L
30·31	H	H	L	L	L	L	L	H	H	H	H
32·33	H	H	L	L	H	L	H	L	L	L	L
34·35	H	H	L	H	L	L	H	L	L	L	H
36·37	H	H	L	H	H	L	H	L	L	H	L
38·39	H	H	H	L	L	L	H	L	L	H	H
ANY	X	X	X	X	X	H	H	H	H	H	H

H = high level, L = Low level, X = irrelevant

NOTES: A.   Input conditions other than those shown produce highs at outputs Y1 through Y5.

B.   Outputs Y6, Y7 and Y8 are not used for BCD to binary conversion.

Figure 14-32.   A 6-bit BCD-to-binary converter: (a) symbol; (b) converter cell; (c) truth table. (Courtesy of Signetics Corporation.)

through Y_5 as shown in the truth table. Outputs Y_6, Y_7, and Y_8 are not used. Figure 14-33 illustrates how twenty-eight 74184s may be used to convert six BCD decades to binary.

Figure 14-34 shows a 74185 connected to form a 6-bit binary-to-BCD converter. The truth table shows the input-output functions. The 74185 is another specially programmed 7488. A 16-bit binary-to-BCD converter using 16 74185s is illustrated in Figure 14-35.

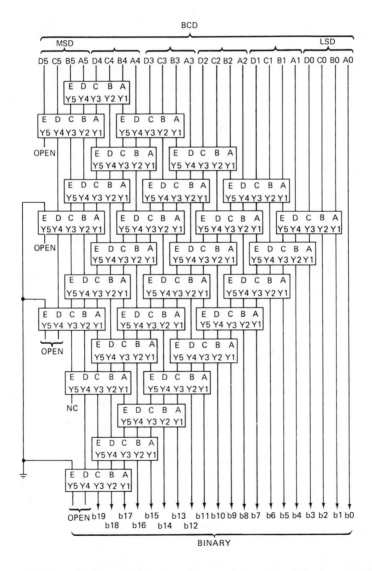

Figure 14-33. Six-decade BCD-to-binary converter. (Courtesy of Texas Instruments.)

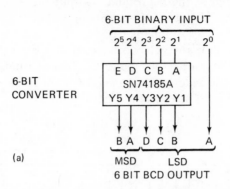

6-BIT BINARY INPUT

6-BIT
CONVERTER

(a)

MSD   LSD
6 BIT BCD OUTPUT

TRUTH TABLE

BINARY WORDS		INPUTS						OUTPUTS							
		BINARY SELECT					ENABLE								
		E	D	C	B	A	G	Y8	Y7	Y6	Y5	Y4	Y3	Y2	Y1
0	1	L	L	L	L	L	L	H	H	L	L	L	L	L	L
2	3	L	L	L	L	H	L	H	H	L	L	L	L	L	H
4	5	L	L	L	H	L	L	H	H	L	L	L	L	H	L
6	7	L	L	L	H	H	L	H	H	L	L	L	L	H	H
8	9	L	L	H	L	L	L	H	H	L	L	L	H	L	L
10	11	L	L	H	L	H	L	H	H	L	L	H	L	L	L
12	13	L	L	H	H	L	L	H	H	L	L	H	L	L	H
14	15	L	L	H	H	H	L	H	H	L	L	H	L	H	L
16	17	L	H	L	L	L	L	H	H	L	L	H	L	H	H
18	19	L	H	L	L	H	L	H	H	L	L	H	H	L	L
20	21	L	H	L	H	L	L	H	H	L	H	L	L	L	L
22	23	L	H	L	H	H	L	H	H	L	H	L	L	L	H
24	25	L	H	H	L	L	L	H	H	L	H	L	L	H	L
26	27	L	H	H	L	H	L	H	H	L	H	L	L	H	H
28	29	L	H	H	H	L	L	H	H	L	H	L	H	L	L
30	31	L	H	H	H	H	L	H	H	L	H	H	L	L	L
32	33	H	L	L	L	L	L	H	H	L	H	L	L	L	H
34	35	H	L	L	L	H	L	H	H	L	H	H	L	H	L
36	37	H	L	L	H	L	L	H	H	L	H	H	L	H	H
38	39	H	L	L	H	H	L	H	H	L	H	H	H	L	L
40	41	H	L	H	L	L	L	H	H	H	L	L	L	L	L
42	43	H	L	H	L	H	L	H	H	H	L	L	L	L	H
44	45	H	L	H	H	L	L	H	H	H	L	L	L	H	L
46	47	H	L	H	H	H	L	H	H	H	L	L	L	H	H
48	49	H	H	L	L	L	L	H	H	H	L	L	H	L	L
50	51	H	H	L	L	H	L	H	H	H	L	H	L	L	L
52	53	H	H	L	H	L	L	H	H	H	L	H	L	L	H
54	55	H	H	L	H	H	L	H	H	H	L	H	L	H	L
56	57	H	H	H	L	L	L	H	H	H	L	H	L	H	H
58	59	H	H	H	L	H	L	H	H	H	L	H	H	L	L
60	61	H	H	H	H	L	L	H	H	H	H	L	L	L	L
62	63	H	H	H	H	H	L	H	H	H	H	L	L	L	H
ALL		X	X	X	X	X	H	H	H	H	H	H	H	H	H

H = high level
L = low level
X = irrelevant

(b)

Figure 14-34.   Binary-to-BCD converter: (a) 6-bit cell; (b) truth table. (Courtesy of Texas Instruments.)

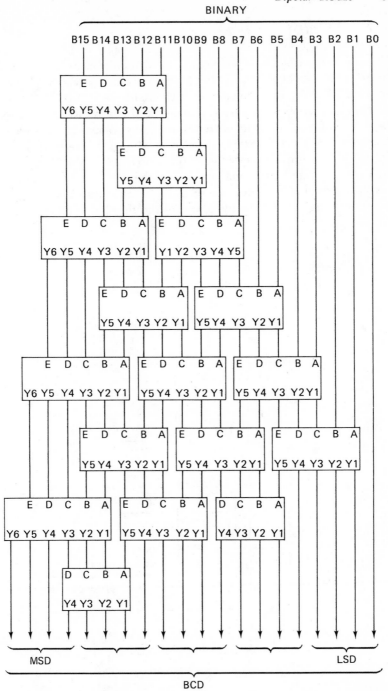

Figure 14-35. A 16-bit binary-to-BCD converter. (Courtesy of Signetics Corporation.)

14.7 MOS ROMs

The application of MOS technology to read-only memories is ideal because of the small geometry of the MOS transistor. Very dense layouts are possible, allowing for memories with thousands of cells.

Figure 14-36 shows a small array of single transistor cells. Notice that some of the transistors have missing gates. When the row-select lines are activated, the transistors in that row with gates are turned "on," resulting in a low resistance between the source and drain. However, those transistors without gates will remain in the "off" condition and a high resistance results. The presence of these "nontransistors" is programmed by eliminating the gate or by adjusting the threshold voltage.

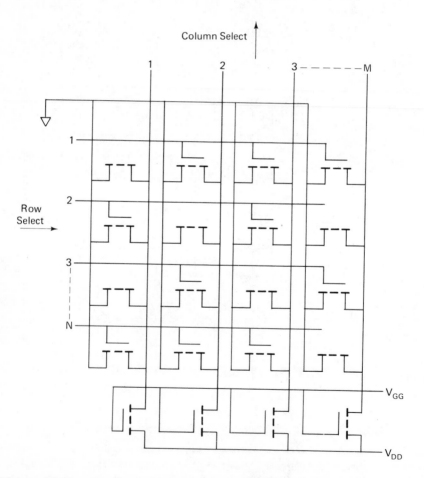

Figure 14-36. Small array of single MOS transistor cells.

The simplest form of the MOS transistor uses a metal gate which has been deposited over the channel area. A given transistor is made inactive by etching away the gate during the manufacturing process. (To program a logical 1 into a given cell, the transistor must remain active.) A more reliable technique is to raise the threshold voltage to a point that exceeds the logic level. This is accomplished by providing a thick layer of oxide between the metal gates and the transistor's channel (Figure 14-37). If a transistor is to become active, the oxide over that transistor is made thinner by etching some of the oxide away. This is accomplished by substituting a single photo-mask used in the MOS fabrication. The transistor with the thick oxide is not active because a greater voltage than that provided by the logic level is required to turn it on (high threshold).

The MOS ROM in Figure 14-38 is an example of those presently available. It is a 1024-bit device which can be organized in 128 words by eight bits or 256 words by four bits. This device may be programmed for a special requirement, or it may be obtained as an off-the-shelf preprogrammed device. For example, it may be ordered to perform the function of converting EBCDIC code to ASCII code. When used as a code converter, the input code is used to address the ROM and its output is the converted code.

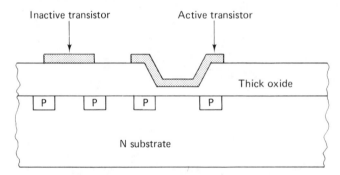

Figure 14-37.   Cross section of a ROM chip showing active and inactive transistors.

14.8   EPROMs

EPROMs are read-only memories which can be erased and reprogrammed many times. Figure 14-39 shows a 1702A erasable and electrically reprogrammable 2048-bit (256 word by 8-bit) read-only memory. It is ideally suited for uses where fast turnaround and program experimentation are important. The 1702A is packaged in a 24 pin dual in-line package

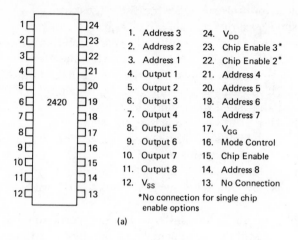

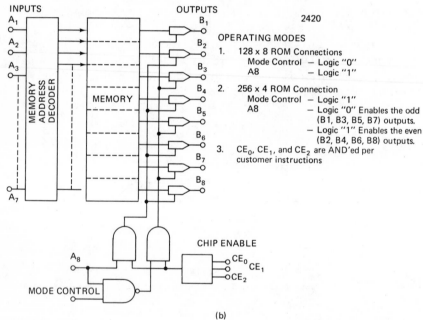

Figure 14-38. A 1024-bit MOS ROM: (a) pin connections; (b) block diagram. (Courtesy of Signetics Corporation.)

with a transparent lid. The transparent lid allows the user to erase the stored program by exposing the chip to ultraviolet light. This device is well suited for use in microprocessor applications. Figure 14-40 shows

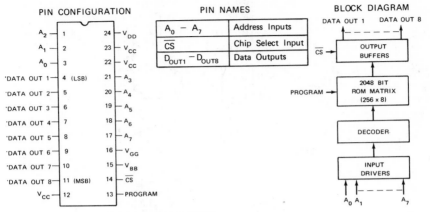

Figure 14-39. 1702A EPROM. (Courtesy of Intel Corp.)

Figure 14-40. PROM programmer. (Courtesy of Intel Corp.)

a PROM programmer with a 1702A plugged into a socket on its front panel. This system has an internal programmer which allows the EPROM to be programmed in a matter of seconds.

14.9 CHARACTER GENERATORS

Another application of the ROM is that of character generation. The ROM referred to in Figure 14-41 is intended for that purpose. You will

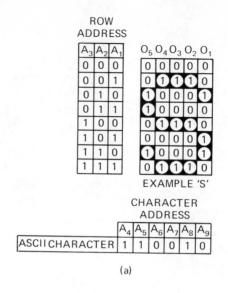

(a)

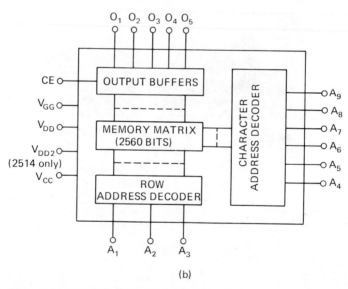

(b)

Figure 14-41. A character generator ROM: (a) character format; (b) block diagram. (Courtesy of Signetics Corporation.)

notice that the character is made up from an array of 35 dots (5 × 7). The block diagram shows that there are five outputs corresponding to the five columns as indicated. The entire character is addressed by the inputs A_4 through A_9. However, it is necessary to address the rows individually with the inputs A_1, A_2, and A_3. The total ASCII character font is shown in Figure 14-42. This device may also be used in a CRT character generator as shown in the block diagram in Figure 14-43. A shift

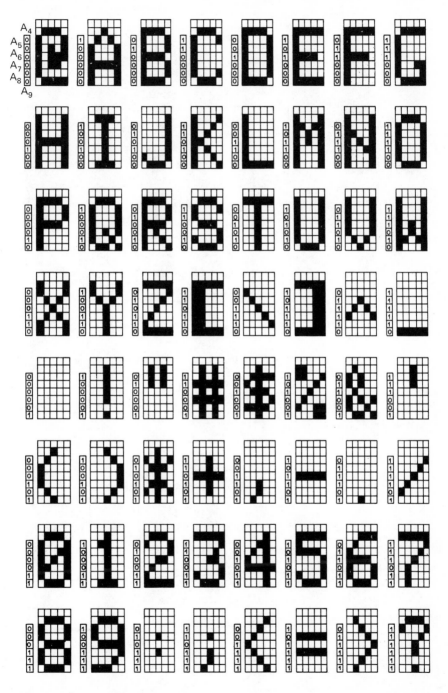

Figure 14-42. ASCII character font (Courtesy of Signetics Corporation.)

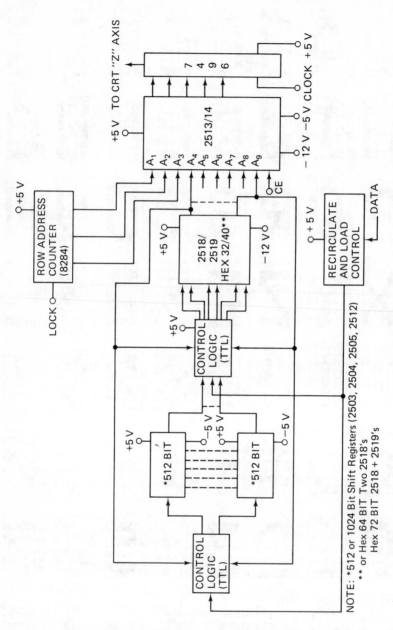

Figure 14-43. CRT character generator. (Courtesy of Signetics Corporation.)

NOTE: *512 or 1024 Bit Shift Registers (2503, 2504, 2505, 2512)
** or Hex 64 BIT Two 2518's
Hex 72 BIT 2518 + 2519's

380

register memory is required to store all of the possible characters to be displayed at a given time on the CRT.

14.10   LOOK-UP TABLES

The ROM may also be used as a look-up table for such data as log or trigonometric functions. Figure 14-44(a) and (b) shows a ROM and its truth table.

It is a 1024-bit monolithic MOS read-only memory that has been programmed to solve for the angle θ whose tangent value x is known; i.e., to obtain the solution to the equation: $\theta = $ arctan x.

Values of x are defined in the look-up table for $0 < x < 1$ with reciprocal angles corresponding from $0° \leqslant \theta \leqslant 45°$. For values of $x \geqslant 1$, the reciprocal of x (i.e., $1/x$) must be entered and the output angle must be complemented to obtain the actual value.

The input is divided into 128 equal parts for x. Thus, the appropriate input address is (128) (x) to the nearest whole integer for obtaining the appropriate ROM address. The input code is the ROM address expressed in binary with A_1 being the least significant bit. For input values greater than unity, the decimal reciprocal is to be taken prior to entry of the binary address.

The output has been normalized for 45°. To obtain the true angular reading, the output should be multiplied by 45°; i.e., $\theta = (\theta_{output}) \times 45°$ where θ_{output} is the decimal equivalent of the output. The output code is the normalized value of the angle θ expressed in binary. The output lines B_1, B_2, . . . B_8 are binary place values $1/2$, $1/4$, . . . $1/256$. To obtain angles between 45° and 89.6° which occur when input values of x are equal to or greater than unity, either complement the output binary code and add a 1, or complement the resultant angular value (i.e., subtract from 90°).

The 8-bit output code has been rounded off. That is, if another bit of even lower significance had been computed for the given arctangent value was a binary "1," it would have carried over into the LSB of the 8-bit code. If it was a binary "0," it would have been dropped.

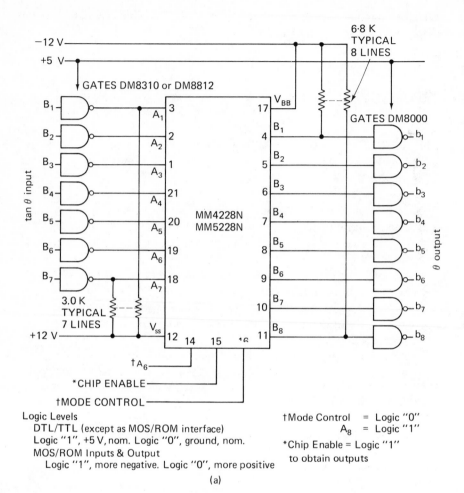

Figure 14-44. A look-up table ROM: (a) device showing input and output connections; (b) truth table. (Courtesy of National Semiconductor.)

EXAMPLE:

Find the angle whose tangent is 0.258.

The input address is 128 × 0.258, or 33 to the nearest integer. Expressed in binary, this is 0100001, and is the actual input code to the converter. The converter will generate the binary value .01010010, whose decimal equivalent is 0.3203125.

Thus, $\theta = 0.320 \times 45° = 14.4°$.

(The above information courtesy of National Semiconductor Corp. "MOS Integrated Circuits" Catalog, October 1970.)

ADDRESS 128(x)	B8	B7	B6	B5	B4	B3	B2	B1
0	0	0	0	0	0	0	0	0
1	0	1	0	0	0	0	0	0
2	1	0	1	0	0	0	0	0
3	1	1	1	0	0	0	0	0
4	0	1	0	1	0	0	0	0
5	0	0	1	1	0	0	0	0
6	1	1	1	1	0	0	0	0
7	0	1	0	0	1	0	0	0
8	0	0	1	0	1	0	0	0
9	1	1	1	0	1	0	0	0
10	1	0	0	1	1	0	0	0
11	0	0	1	1	1	0	0	0
12	0	1	1	1	1	0	0	0
13	1	0	0	0	0	1	0	0
14	1	1	0	0	0	1	0	0
15	0	1	1	0	0	1	0	0
16	0	0	0	1	0	1	0	0
17	1	1	0	1	0	1	0	0
18	1	0	1	1	0	1	0	0
19	0	0	0	0	1	1	0	0
20	0	1	0	0	1	1	0	0
21	1	0	1	0	1	1	0	0
22	1	1	1	0	1	1	0	0
23	0	1	0	1	1	1	0	0
24	0	0	1	1	1	1	0	0
25	1	1	1	1	1	1	0	0
26	1	0	0	0	0	0	1	0
27	0	0	1	0	0	0	1	0
28	0	1	1	0	0	0	1	0
29	0	0	0	1	0	0	1	0
30	1	1	0	1	0	0	1	0
31	1	0	1	1	0	0	1	0
32	0	0	0	0	1	0	1	0
33	0	1	0	0	1	0	1	0
34	0	0	1	0	1	0	1	0
35	1	1	1	0	1	0	1	0
36	1	0	0	1	1	0	1	0
37	0	1	0	1	1	0	1	0
38	0	1	1	1	1	0	1	0
39	0	0	0	0	0	1	1	0
40	0	1	0	0	0	1	1	0
41	1	0	1	0	0	1	1	0
42	1	1	1	0	0	1	1	0
43	1	0	0	1	0	1	1	0
44	0	0	1	1	0	1	1	0
45	0	1	1	1	0	1	1	0
46	0	0	0	0	1	1	1	0
47	0	1	0	0	1	1	1	0
48	1	0	1	0	1	1	1	0
49	1	1	1	0	1	1	1	0
50	1	0	0	1	1	1	1	0
51	1	1	0	1	1	1	1	0
52	0	1	1	1	1	1	1	0
53	0	0	0	0	0	0	0	1
54	0	1	0	0	0	0	0	1
55	0	0	1	0	0	0	0	1
56	0	1	1	0	0	0	0	1
57	0	0	0	1	0	0	0	1
58	0	1	0	1	0	0	0	1
59	1	0	1	1	0	0	0	1
60	1	1	1	1	0	0	0	1
61	1	0	0	0	1	0	0	1
62	1	1	0	0	1	0	0	1
63	1	0	1	0	1	0	0	1
64	1	1	1	0	1	0	0	1
65	1	0	0	1	1	0	0	1
66	1	1	0	1	1	0	0	1
67	1	0	1	1	1	0	0	1
68	1	1	1	1	1	0	0	1
69	1	0	0	0	0	1	0	1
70	1	1	0	0	0	1	0	1
71	1	0	1	0	0	1	0	1
72	1	1	1	0	0	1	0	1
73	1	0	0	1	0	1	0	1
74	1	1	0	1	0	1	0	1
75	1	0	1	1	0	1	0	1
76	0	1	1	1	0	1	0	1
77	0	0	0	0	1	1	0	1
78	0	1	0	0	1	1	0	1
79	0	0	1	0	1	1	0	1
80	0	1	1	0	1	1	0	1
81	0	0	0	1	1	1	0	1
82	1	0	0	1	1	1	0	1
83	1	1	0	1	1	1	0	1
84	1	0	1	1	1	1	0	1
85	1	1	1	1	1	1	0	1
86	1	0	0	0	0	0	1	1
87	0	1	0	0	0	0	1	1
88	0	0	1	0	0	0	1	1
89	0	1	1	0	0	0	1	1
90	1	1	1	0	0	0	1	1
91	1	0	0	1	0	0	1	1
92	1	1	0	1	0	0	1	1
93	1	0	1	1	0	0	1	1
94	0	1	1	1	0	0	1	1
95	0	0	0	0	1	0	1	1
96	1	0	0	0	1	0	1	1
97	1	1	0	0	1	0	1	1
98	1	0	1	0	1	0	1	1
99	0	1	1	0	1	0	1	1
100	0	0	0	1	1	0	1	1
101	1	0	0	1	1	0	1	1
102	1	1	0	1	1	0	1	1
103	1	0	1	1	1	0	1	1
104	0	1	1	1	1	0	1	1
105	0	0	0	0	0	1	1	1
106	1	0	0	0	0	1	1	1
107	1	1	0	0	0	1	1	1
108	0	0	1	0	0	1	1	1
109	0	1	1	0	0	1	1	1
110	1	1	1	0	0	1	1	1
111	1	0	0	1	0	1	1	1
112	0	1	0	1	0	1	1	1
113	1	1	0	1	0	1	1	1
114	1	0	1	1	0	1	1	1
115	0	1	1	1	0	1	1	1
116	0	0	0	0	1	1	1	1
117	1	0	0	0	1	1	1	1
118	1	1	0	0	1	1	1	1
119	0	1	0	0	1	1	1	1
120	1	0	1	0	1	1	1	1
121	1	1	1	0	1	1	1	1
122	0	0	0	1	1	1	1	1
123	0	1	0	1	1	1	1	1
124	1	1	0	1	1	1	1	1
125	0	0	1	1	1	1	1	1
126	1	0	1	1	1	1	1	1
127	0	1	1	1	1	1	1	1

NOTE: "1" MORE NEGATIVE OUTPUT "0" MORE POSITIVE OUTPUT

MM5228N

(b)

Figure 14-44(b).

14.11   MICRO-INSTRUCTIONS

A very important application of ROMs is that of microprogramming. Microprogramming is the technique of providing "hard-wired" instructions (micro-instructions) to perform various tasks such as subroutines. The micro-instructions are permanently stored and are addressed by a counter. The instruction counter, which is sometimes presettable (Chapter 12), keeps track of the steps in the routine. The output of the ROM provides the logic levels required (mode controls, gating instructions, etc.) by the various circuits used to perform the routine. Figure 14-45

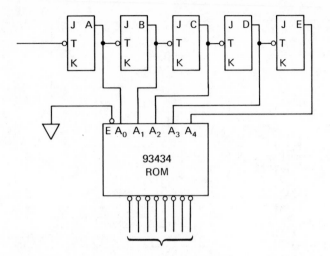

Figure 14-45.   Micro-instruction generator using a ROM.

shows a binary counter with outputs connected to a 32-word by 8-bit ROM. Notice that there are 32 steps possible in a subroutine and that eight outputs are available to at least eight different circuits. Of course, the appropriate instructions must first be "programmed" into the ROM.

A more sophisticated micro-instruction generator can be conditional. That is, the micro-instructions generated are dependent not only on the instruction address but also on certain conditions that exist. An example is depicted in Figure 14-46. Two 2048-bit ROMs are used so that 16 outputs are provided. Although only 32 program steps are shown in this scheme, the number of addresses may be increased by further expansion. The unique characteristic of this circuit, however, is that for each step in the program, eight possible output combinations can be produced. Let's assume that we are at step 8 in the routine (address 01000). Depending on the conditions existing (condition 000 through 111), one of 8 alternate combinations of micro-instructions is produced! If a presettable program counter is used, alternate micro-instructions can cause the counter to go to a prescribed address. This means that branching is achieved, and the solution of problems using an iterative process is possible. This scheme may be used whenever it is desirable to make decisions and to alter the program as those decisions dictate.

A very interesting application of ROMs and microprograms is the HP-80 pocket calculator (Figure 14-47). It is a preprogrammed calculator designed specifically for financial and business applications. Over 30 hard-wired programs are implemented in the HP-80. These programs essentially replace all of the commonly used financial tables such as compound interest, annuities, bonds, and so on. The user need only enter the parameters of the problem and press a key for his answer.

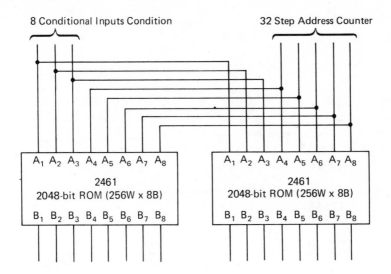

Figure 14-46.   Conditional micro-instruction generator.

Figure 14-47.   The HP-80 has a 10-digit display, plus sign and 2-digit exponent; it weighs nine ounces and operates from rechargeable batteries or the AC line. (Courtesy of Hewlett-Packard.)

An example of HP-80 programming is shown in Figure 14-48. The flow chart shows the algorithm used to solve for the interest i in the equation:

The equation is solved by an iterative technique. The program listing shown is the portion of the program that implements the iteration sequence, that is, the part between A and B in the flow chart.

Seven ROMs are used in the HP-80 and are mounted on a hybrid circuit to form a single assembly, as shown in Figure 14-49.

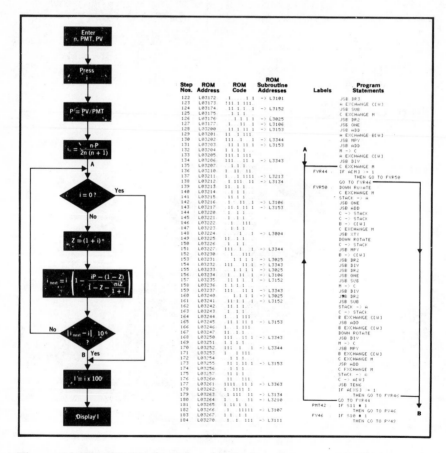

Figure 14-48.   HP-80 algorithm for finding interest rate given present value, payment, and number of periods. Also shown is a partial listing of the program that implements the algorithm. (Courtesy of Hewlett-Packard.)

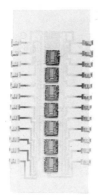

Figure 14-49. HP-80 programs are stored in seven read-only-memory chips mounted on a hybrid circuit. (Courtesy of Hewlett-Packard.)

QUESTIONS AND PROBLEMS—CHAPTER 14

1. What is the purpose of the x and y lines in a magnetic core?

2. Describe how data is written into magnetic cores.

3. Describe how data is read from magnetic cores.

4. What is meant by *destructive readout?*

5. What is the purpose of the *inhibit* line?

6. Why is it necessary to have a memory data register for each plane in a three-dimensional array?

7. What is the primary advantage of core memory over semiconductor memory?

8. What is the primary advantage of bipolar memory over MOS memory?

9. Unaddressed cells do not affect the sense lines in bipolar memory. Why?

10. What is the purpose of the $\overline{\text{STROBE I}}$ line in Figure 14-15?

11. What is the primary advantage of MOS memory over bipolar memory?

12. Differentiate between static and dynamic memory.

13. Explain the "C" in Figure 14-24(a).

14. How are dynamic cells "refreshed"?

15. Describe two ways to program a bipolar ROM.

16. What is a PROM?

17. What is the primary advantage of MOS ROMs over bipolar ROMs?

18. Give four examples of look-up tables.

19. What is meant by *microprogramming*?

20. What are "hard-wired" instructions?

15

Programming

In the previous chapters, the discussion emphasized computer hardware. With more extensive use of minicomputers and the advent of the microprocessor, it has become increasingly important for the technician to understand programming concepts.

A microprocessor is a complete CPU on a single silicon chip. Microprocessors may be substituted for much of the hardware previously used for digital applications. One application of a microprocessor is in a microcomputer. A microcomputer is a complete system using a microprocessor and incorporating RAM, ROM, and I/O devices. For example, Figure 15-2 is Intel's Intellec 8 microcomputer development system. It is used in the development and testing of programs for microprocessor and microcomputer systems.

The basic concept of programming microcomputers is the same as in larger computers. Because of this, a microprocessor will be used to describe programming concepts in this and the following chapter.

The topics covered in this chapter are:

15.1   Programming Concepts

15.2   Flowcharts

Figure 15-1.   PDP minicomputer.

Figure 15-2.   ®Intellec 8 microcomputer development system. (Courtesy Intel Corp.)

15.1   PROGRAMMING CONCEPTS

In order to illustrate how a CPU functions, we will use a "CPU demonstrator." Our CPU is very limited in its operations and its storage capabilities. However, the concepts are consistent with those of a practical CPU. Our CPU demonstrator is word-oriented with eight bits stored at each address.

A block diagram of the CPU demonstrator is shown in Figure 15-3. The memory contains 32 addresses, each of which contains eight bits. The arithmetic logic unit (ALU) performs the arithmetic operations and makes certain logic decisions. The *accumulator* is used to store the results of each arithmetic operation. Most arithmetic operations are performed

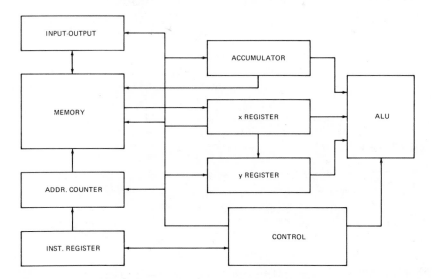

Figure 15-3.   Block diagram of a demonstration CPU.

on the data stored in the accumulator and the data designated by the operand. In order to perform operations on new data, the contents of the accumulator must first be stored in memory so that the new data word may be entered. Therefore, many of the steps in the program are for loading the accumulator ("clear and add") and storing data in memory ("store").

The *control* provides information to set up various functions of the CPU. Control may tell the ALU to add, subtract, multiply, etc., as directed by the op-code. Another function of control is to select the address designated by the operand.

Data stored at a given address is usually referred to as a *data word*. The data word shown in Figure 15-4 is eight bits in length. Our CPU will use the two's complement method for storing negative numbers. Therefore the MSB is the sign bit. The reason the two's complement is stored is that the arithmetic process is simplified. Negative numbers are two's complemented before they are stored and before they are printed out. However, the CPU handles all negative numbers in the two's complement form.

Other information stored in the memory provides instructions to the CPU for processing the data. This information is organized into *instruction words*. For convenience, the instruction word is the same length as the data word. This is done so that an address may be used to store either an instruction word or a data word.

Figure 15-5 illustrates an instruction word of eight bits. The instruction word is composed of two parts: the operation code (op-code) and

Sign bit

b_8 ... b_2 b_1

1	0	0	0	0	1	0	1

Figure 15-4.   Data word representing the negative number −123.

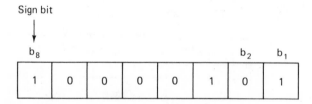

Op-Code   Operand

b_8 b_7 b_6 b_5 b_4 b_3 b_2 b_1

0	1	0	0	1	0	0	1

Figure 15-5.   An 8-bit instruction word. The op-code provides the ADD instruction and the operand indicates address 9.

the operand. The *op-code* designates the operation to be performed by the CPU. The *operand* tells the CPU where to get the data to perform the designated operation.

When a programmer writes a program, he is essentially writing a series of instruction words telling the computer what to do. These words are usually stored in sequence. This is because the CPU "reads" and "executes" the designated operation in the prescribed order of the program.

The instruction word example (Figure 15-5) provides three bits for the op-code. Thus, we are limited to eight operations (000) to (111). The operand contains five bits, limiting the total number of addresses to 32. The op-codes and their designations are listed in Figure 15-6.

Using our CPU demonstrator, let's illustrate the solution of the following problem:

$$A = x + y - 5$$

The variables x and y are obtained from some input unit, such as a card reader, and stored in the memory of the CPU. These variables may be changed after each solution by reading new cards. The address where each variable is stored is selected by the programmer. The number 5 in the equation is a constant and does not change when a new card is read.

Each step of the program is stored at an address selected by the programmer. These addresses are sequential, starting at zero (00000). It

Op-code	Mneumonic Code	Operation	Description
000	HLT	Halt	Stops CPU at end of program
001	CLA	Clear and add	Clears accumulator and adds number stored in operand address.
010	ADD	Add	Adds number at operand address to number in accumulator.
011	SUB	Subtract	Subtracts number at operand address from number in accumulator.
100	MUL	Multiply	Multiplies number at operand address by number stored in accumulator.
101	STO	Store	Stores contents of accumulator in address designated by operand.
110	BRM	Branch if minus	Tells CPU to take next instruction designated by operand address if the content of the accumulator is negative.
111	PNT	Print	Tells CPU to print data stored at address designated by operand.

Figure 15-6.   Op-code designations for the instruction word in Figure 15-5.

is important to note that these addresses are not part of the program but rather serve to designate where the program is stored.

The CPU starts its process by reading the contents of address 00000 (see Figure 15-7). It interprets the op-code (001 = clear and add) and sets up circuits required to perform this operation. It also selects a new address to read as designated by the operand (01000 = address 8). This completes the *instruction phase* of the first step of the program.

Next, the accumulator is cleared, and the contents of address 8 (which is x) are added. In performing this operation, the *execution phase* of the first step is completed. The CPU will now read the instruction contained in the next address of the program (00001). This process is continued until the CPU is instructed to halt. Each step of the program has two parts: the instruction phase and the execution phase.

Address	MN Code	Instruction or Data Word	Explanation
00000	CLA 8	001 01000	Clear and add to the accumulator the content of address 8 which is the value of x.
00001	ADD 9	010 01001	Add to the accumulator the content of address 9 which is the value of y. The content of the accumulator is now x + y.
00010	SUB 10	011 01010	Subtract from the accumulator the content of address 10 which is the number 5. (x + y − 5)
00011	STR 11	101 01011	Store the content of the accumulator (x + y − 5) in address 11
00100	PNT 11	111 01011	Print the content of address 11
00101	HLT	000 00000	Halt. Stops CPU operations
01000		x	DATA
01001		y	DATA
01010		5	Constant

Figure 15-7.   Program for solving the problem $A = x + y - 5$.

15.2   FLOWCHARTS

A diagram commonly used to assist the programmer is called a *flowchart*. The flowchart is also used by engineers and technicians to provide a bird's-eye view of the logic or solution to a given problem. It is made up of several standardized symbols connected by straight lines. Some of the most commonly used symbols are shown in Figure 15-8.

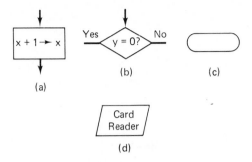

Figure 15-8.   Some of the more commonly used flowchart symbols: (a)
process operation; (b) decision; (c) terminal; (d) input/
output.

The process operation symbol [rectangle in Figure 15-8(a)] tells
us that some specific action is to be taken. The action is indicated by a
statement inside the rectangle. The statement may be written in plain language and expressed algebraically, or other special symbology may be
used. For example, the statement: $x + 1 \rightarrow x$, may be interpreted as:
"One is added to the contents of the x register and the sum is placed
back into the x register." A line with an arrow indicates the input and output of the process operation symbol. By convention, the input should be
either to the top or left side of the symbol, and the output should be from
the bottom or right side. The arrows indicate the direction of logic flow,
or the sequence of operations.

The diamond-shaped symbol [Figure 15-8(b)] indicates that a decision of some sort has to be made. This symbol will always have two
or three outputs and only one input. For example, the symbol in Figure
15-8(b) asks for a decision: "Are the contents of the y register equal to
zero?" Since the answer can only be "yes" or "no," two outputs are indicated. In reading the diagram, however, only one path must be followed,
depending on the decision made.

The terminal symbol in Figure 15-8(c) is used to indicate the start
and the end of a flowchart. Obviously, it is connected with only one line.
Other symbols are used to indicate more specific operations such as the
input/output symbol in Figure 15-8(d). The input/output symbol may
be used to indicate a typewriter, printer, card reader, or other input/
output device.

Figure 15-9 is a flowchart for the program outlined in Figure 15-7.
Notice that all of the operations in the program are in one sequence and
the solution to the problem is concisely diagrammed. Because the operations are in sequence, this is called a *linear program*.

Let's consider a program that determines which of three numbers
is the largest. Figure 15-10 is the flowchart and Figure 15-11 shows the

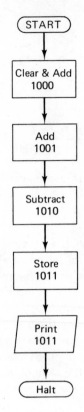

Figure 15-9.   Flowchart for program in Figure 15-7.

program sequence. Note that this program involves decision-making blocks and branching results.

Flowcharts may take many forms. However, care should be taken when constructing a flowchart to represent the problem or sequence in as precise a manner as possible.

15.3 MICRO-INSTRUCTIONS

The op-codes listed in Figure 15-6 designate operations which are considerably different in complexity. For example, the "clear and add" operation requires only that data be cleared from the accumulator and that new data be transferred from another register (this could be thought of as a two-step operation, whereas the multiply operation requires many steps). When the programmer uses the instruction "multiply," he is actually asking the computer to use a set of instructions called *micro-instructions*. Micro-instructions are permanently stored within the CPU. They

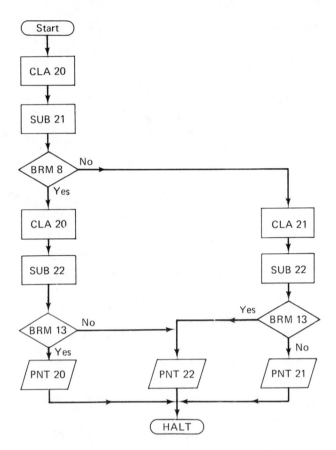

Figure 15-10.   Flowchart of program to determine which of three numbers is largest.

may actually be a number of logic circuits wired so that the various steps are properly controlled. For this reason, micro-instructions are also stored in read-only memory (ROM). Because of the advances in manufacturing techniques of large-scale integrated circuits (LSI), ROMs are widely used today.

The op-code "multiply" causes the CPU to use the micro-instructions for multiplying. These instructions are addressed by the micro-instruction address register (MIAR). The output of this register selects an address in the ROM where the first micro-instruction is stored, as illustrated in the block diagram of Figure 15-12. The first step of the multiply subroutine is then performed, and the MIAR is set to the next address.

Address	MN Code	Instruction or Data Word	Explanation
00000	CLA 20	001 10100	Clear and add the content of address 20, which is the value of x.
00001	SUB 21	011 10101	Subtract contents of address 21 (y) from contents of address 20 (x).
00010	BRM 8	110 01000	Branch to address 8 if minus. If not minus, continue.
00011	CLA 20	001 10100	Clear and add the content of address 20, which is the value of x.
00100	SUB 22	011 10110	Subtract content of address 22 from contents of address 20.
00101	BRM 13	110 001101	Branch to address 13 if minus. If not minus, continue.
00110	PNT 20	111 10100	Print content of address 20.
00111	HLT	000 00000	Halt.
01000	CLA 21	001 10101	Clear and add content of address 21, which is the value of y.
01001	SUB 22	011 10110	Subtract the content of address 22 from contents of address 21.
01010	BRM 13	110 001101	Branch to address 13 if minus. If not minus, continue.
01011	PNT 21	111 10101	Print content of address 21.
01100	HLT	000 00000	Halt.
01101	PNT 22	111 10110	Print content of address 22.
01110	HLT	000 00000	Halt.
10100	x		DATA in address 20.
10101	y		DATA in address 21.
10110	z		DATA in address 22.

Figure 15-11.   Program sequence for flowchart in Figure 15-10.

15.4 MULTIPLICATION SUBROUTINE

Let's follow the sequence of multiplying two numbers. Figure 15-13 shows the flowchart for the multiplication subroutine. In this example, binary numbers 1001 and 101 are multiplied. The op-code "multiply" is detected,

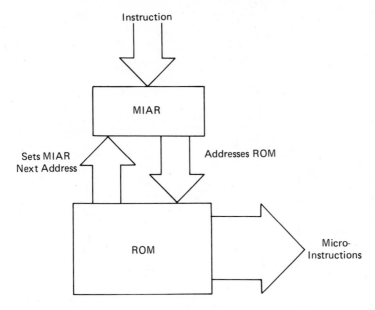

Figure 15-12. Block diagram for micro-instructions.

causing the MIAR to be set to address 001. (From this point until the completion of the multiplication subroutine the CPU is under the control of the micro-instructions.) Notice that in the multiplication subroutine program, Figure 15-14, address 001 causes the multiplier (101) to be loaded into the y register. When this is completed, the MIAR is set to 010. This address stores the micro-instruction, which causes the multiplicand to be transferred to the x register. It also stores the next address (011). You will notice in the flowchart that the next step (address 011) is to make the decision, "Is the content of the LSB position of the y register (y_1) equal to 1?" If the answer is "yes," as it is in this case, the MIAR is set to 100. However, if the answer is "no," the MIAR would be set to 101, as indicated in the flowchart. The micro-instruction at address 100 causes the contents of the x register to be added to the contents of the accumulator and to continue to address 101. Address 101 stores the instruction, which causes the data in the y register to shift one bit position to the right and selects the next address (110). The micro-instruction at address 110 selects one of two possible addresses. Address 000 becomes the next address if the contents of the y register is 0. If not, the next address is 111. This process continues until address 000 is finally selected ending the multiplication subroutine.

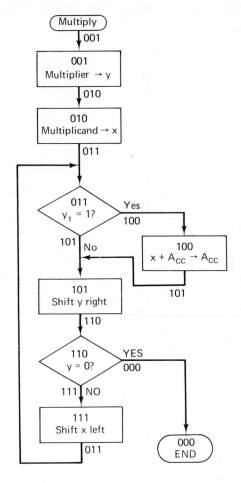

Figure 15-13.   Flowchart for multiplication subroutine.

15.5   *DIVISION SUBROUTINE*

Only eight op-codes were listed in Figure 15-6. However, many more are available to the programmer by the computer. At least one of these codes is for the division subroutine.

The division process is essentially a series of trial subtractions followed by a left shift of the dividend. If the trial subtraction is successful, a 1 is placed in the MSB position of the quotient, and the dividend is shifted to the left. If the subtraction is not successful, a 0 is placed in the quotient, and the dividend is shifted left.

INSTRUCTION ADDRESS	MICRO INSTRUCTION	x REG	y REG	ACCUM.	NEXT ADDRESS
001	Multiplier to y reg.	00000000	00000101	00000000	010
010	Multiplicand to x reg.	00001001	00000101	00000000	011
011	y_1 = 1? Yes	00001001	00000101	00000000	100
100	Add x to Acc.	00001001	00000101	00001001	101
101	Shift y right	00001001	00000010	00001001	110
110	y = 0? No	00001001	00000010	00001001	111
111	Shift x left	00010010	00000010	00001001	011
011	y_1 = 1? No	00010010	00000010	00001001	101
101	Shift y right	00010010	00000001	00001001	110
110	y = 0? No	00010010	00000001	00001001	111
111	Shift x left	00100100	00000001	00001001	011
011	y_1 = 1? Yes	00100100	00000001	00001001	100
100	Add x to Acc.	00100100	00000001	00101101	101
101	Shift y right	00100100	00000000	00101101	110
110	y = 0? Yes	00100100	00000000	00101101	000
000	End				

Figure 15-14. Multiplication program chart.

The flowchart for the division subroutine is shown in Figure 15-15. This routine requires 16 micro-instructions. They are addressed 0000 through 1111. To demonstrate the routine, let's divide 45 (101101) by 5 (101). The program chart for this division appears in Figure 15-16. The first step (address 0001) is to transfer the dividend to the accumulator. The second step is to transfer the divisor (101) to the x register. The divisor is left-justified by the micro-instructions stored in addresses 0011 and 0100. The dividend is also left-justified by the micro-instructions stored in addresses 0101 and 0110. Keep in mind that the MSB positions (x_8 and A_8) are reserved for the sign bits.

The next step in the sequence (address 0111) is the first trial subtraction. If the trial subtraction is successful, a positive number will result and the sign bit (A_8) will equal 0. This decision is shown in the flowchart by the decision block having the address 1000 ($A_8 = 1$?). In this case, the decision is "no" ($A_8 = 0$). Therefore, a 1 is placed in the LSB position of the y register. (When the division process has been completed, this bit becomes the MSB of the quotient.)

The decision at address 1100 ($A_1//x_1$?) means: "Does the LSB of the dividend in the accumulator occupy the same position as the LSB of the divisor in the x register?" If the decision is "yes," the division routine is completed by shifting the y register to the left. If the decision is "no,"

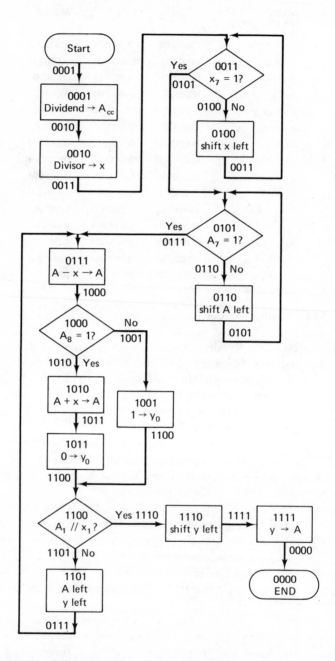

Figure 15-15. Flowchart for division subroutine.

INSTRUCTION ADDRESS	INSTRUCTION	x REG	y REG	ACCUM	NEXT ADDRESS
0001	Dividend to Acc.	00000000	00000000	00101101	0010
0010	Divisor to x	00000101	00000000	00101101	0011
0011	$x_7 = 1$? No	00000101	00000000	00101101	0100
0100	Shift x left	00001010	00000000	00101101	0011
0011	$x_7 = 1$? No	00001010	00000000	00101101	0100
0100	Shift x left	00010100	00000000	00101101	0011
0011	$x_7 = 1$? No	00010100	00000000	00101101	0100
0100	Shift x left	00101000	00000000	00101101	0011
0011	$x_7 = 1$? No	00101000	00000000	00101101	0100
0100	Shift x left	01010000	00000000	00101101	0011
0011	$x_7 = 1$? Yes	01010000	00000000	00101101	0101
0101	$A_7 = 1$? No	01010000	00000000	00101101	0110
0110	Shift Acc left	01010000	00000000	01011010	0101
0101	$A_7 = 1$? Yes	01010000	00000000	01011010	0111
0111	Acc − x to Acc	01010000	00000000	00001010	1000
1000	Acc sign = 1? No	01010000	00000000	00001010	1001
1001	Place "1" in y_0	01010000	000000001	00001010	1100
1100	$A_1 // x_1$? No	01010000	000000001	00001010	1101
1101	Shift Acc and y left	01010000	00000001	00010100	0111
0111	Acc − x to Acc	01010000	00000001	11000100	1000
1000	Acc sign = 1? Yes	01010000	00000001	11000100	1010
1010	Acc + x to Acc	01010000	00000001	00010100	1011
1011	Place "0" in y_0	01010000	000000010	00010100	1100
1100	$A_1 // x_1$? No	01010000	000000010	00010100	1101
1101	Shift Acc and y left	01010000	00000010	00101000	0111
0111	Acc − x to Acc	01010000	00000010	11011000	1000
1000	Acc sign = 1? Yes	01010000	00000010	11011000	1010
1010	Acc + x to Acc	01010000	00000010	00101000	1011
1011	Place "0" in y_0	01010000	000000100	00101000	1100
1100	$A_1 // x_1$? No	01010000	000000100	00101000	1101
1101	Shift Acc and y left	01010000	00000100	01010000	0111
0111	Acc − x to Acc	01010000	00000100	00000000	1000
1000	Acc sign = 1? No	01010000	00000100	00000000	1001
1001	Place "1" in y_0	01010000	000001001	00000000	1100
1100	$A_1 // x_1$? Yes	01010000	000001001	00000000	1110
1110	Shift y left	01010000	00001001	00000000	1111
1111	Transfer y to Acc	01010000	00000000	00001001	0000
0000	End				

Figure 15-16. Division program chart.

both the accumulator and the y register are left-shifted, and the next trial subtraction takes place.

In our example, this trial subtraction produces a negative number in the accumulator and is detected at address 1000. This occurs because a larger number was subtracted from a smaller number. Therefore, the divisor is added back into the accumulator (address 1010) and a 0 is entered into the LSB position of the y register (address 1011). This process continues until the position of the LSB of both the divisor and the dividend agree ($A_1//x_1$?). The y register shifts left once more, and then its contents (quotient) are transferred to the accumulator, thus completing the division subroutine. You will recall that the result of all arithmetic operations remains in the accumulator.

Of course, a typical ROM will have hundreds or thousands of addresses; therefore, only one ROM and one micro-instruction address register are required to provide micro-instructions. Micro-instructions are in the form of *logic levels,* which are used to control the various circuits that perform the operations. One should recognize that a carefully prescribed sequence of events occurs. Therefore, accurate timing circuits are required. (Timing circuits were discussed in Chapter 12. ROMs and micro-instruction generators were covered in Chapter 14.)

15.6 PROGRAMMING MICROCOMPUTERS

The basic concepts for programming microcomputers are the same as for other types of computers. Many microprocessors have very powerful instruction sets, such as Intel's 8080A, shown in Figure 15-17. This is a complete CPU fabricated on a single silicon chip. The instruction set (list of processor instructions) is found in Appendix 1.

Because the architecture (kinds of circuits, number of registers, etc.) of each processor is unique, each has its own set of instructions. The programmer, therefore, must have thorough knowledge of the architecture of the processor he intends to program. This is especially true with respect to the registers and their use. Figure 15-18 shows a block diagram of the Intel 8080A microprocessor.

The accumulator (A register) is part of the ALU. Most mathematical or logic operations are performed between the contents of the accumulator and one of the other registers. The results of these operations always appear in the A register.

The B, C, D, E, H, and L registers are general-purpose types that can be used as either single registers (8-bit), or register pairs (16-bit). The contents of these registers can be modified by using appropriate instructions.

Figure 15-17. 8080A microprocessor chip. (Courtesy Intel Corp.)

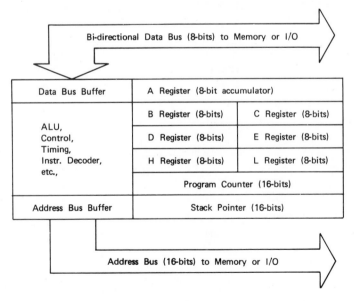

Figure 15-18. Block diagram of 8080A microprocessor. (Courtesy Intel Corp.)

The program counter maintains the memory address of the current program instruction and is incremented automatically as each instruction is executed.

The stack pointer maintains the address of the next available stack location in memory. It can be initialized to use any portion of the read-write memory as a stack. The stack pointer is decremented when data is "pushed" into the stack and incremented when data is "popped" off the stack (i.e., the stack grows "downward"). The stack may be used to retain information such as return addresses, when a variety of subroutines are incorporated in the program. Another important function of the stack is to retain the status of a program when it is interrupted by an I/O device that is requesting immediate service (interrupt).

The 8080A has been designed to greatly simplify system design. Separate 16-line address and 8-line bidirectional data busses are used to allow direct interface to memories and I/O ports. Control signals, which require no decoding, are provided directly by the processor. All busses, including control, are TTL compatible.

15.7   DATA AND INSTRUCTION FORMATS

Data in the 8080A is stored in the form of binary bit integers. All data transfers to the data bus will be in the same format.

$$\boxed{D_7 \quad D_6 \quad D_5 \quad D_4 \quad D_3 \quad D_2 \quad D_1 \quad D_0} \quad \text{Data Word}$$

The program instructions may be one, two, or three bytes in length. Multiple-byte instructions must be stored in successive words in program memory. The instruction formats then depend on the particular operations executed.

One-byte instructions are shown as follows:

$$\boxed{D_7 \longrightarrow D_0} \quad \text{Op-Code}$$

Typical one-byte instructions are: register to register, memory reference, arithmetic or logic, rotate return, push, pop, and enable or disable interrupt instructions.

Two-byte instructions are shown below:

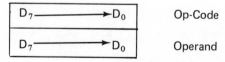

Op-Code

Operand

Typical two-byte instructions are used for immediate mode (immediate data in operand) or for I/O.

Three-byte instructions appear as follows:

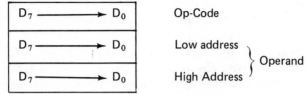

Three-byte instructions are typically: jump, call subroutine, direct load, and store instructions.

15.8   MACHINE LANGUAGE PROGRAMMING

Let's write a program for the 8080A that will: (1) add the data stored at two successive hexadecimal memory addresses 2D and 2E; (2) store the result in memory address 2F; and (3) output the result to port 04. Refer to Appendix 1 for a description of the instruction set.

Address	Mnemonic		Instruction Code (binary)
10	LXI	H,2DH	00100001,00101101,00000000
13	MOV	A,M	01111110
14	INX	H	00100011
15	ADD	M	10000110
16	STA	2FH	00110010,00101111,00000000
19	OUT	04	11010011,00000100
1B	HLT		01110110
	.		
	.		
2D	DATA1		00110100
2E	DATA2		01010110
2F	RESULT		10001010

The program starts at hexadecimal address 10 (decimal 16). The first instruction is LXIH, 2DH, which means, "Load the H and L register pair with the 16-bit number contained in the second and third byte." In this case the H and L registers are used to point to the address in memory containing the first data word. Notice that the second and third bytes are transposed. This is because the processor acts upon the least significant byte first. LXI means, "Load immediate data." The X indicates a 16-bit word follows. The first H, therefore, implies the H and L pair. The second H (in 2DH) indicates the immediate data is hexadecimal.

The second instruction appears at address 13, because the first instruction occupies three addresses. Then this instruction causes data to be moved from the memory location, pointed to by the H and L registers, to the A register.

The third instruction causes the H and L registers to be incremented by one. H and L now point to the second data word.

The next instruction (ADD M) causes the second data word to be added to the first, with the result appearing in the A register. The M indicates a memory location which is pointed to by the H and L registers.

The instruction stored at addresses 16, 17, and 18 can be interpreted as: "Store the data in the accumulator at hexadecimal address 2F." Once again, the second and third bytes are transposed.

At address 19 the instruction is, "Output the contents of the accumulator to port number 4." This causes a binary 4 to appear on the address bus, which selects output port 4. During the same instruction time, data from the accumulator is transferred by way of the data bus. The last instruction in the program simply causes the processor to halt.

The following listing shows all instructions and data for the above program in hexadecimal. Hexadecimal provides a simple means of recording binary data. Notice that one byte equals two hexadecimal digits. Refer to Appendix 2 for the instruction set using hexadecimal notation.

Address	Mnemonic		Instruction Code (Hexadecimal)
1Ø	LXI	H,2DH	21 2D 00
13	MOV	A,M	7E
14	INX	H	23
15	ADD	M	86
16	STA	2FH	32 2F 00
19	OUT	4	D3 04
1B	HLT		76
	.		
	.		
2D	DATA1		34
2E	DATA2		56
2F	RESULT		8A

The binary instruction code listing is called a *machine language program* and can be entered into the microcomputer by way of bit switches on the front panel. The memory address is first selected by activating the direct memory access switch. Address 10 is selected by activating the appropriate switches and pressing the load memory switch. Data can now be entered into this address by setting the appropriate data switches. The increment address switch is pressed, allowing access to the next address. The second byte may now be entered. This process is continued until the entire program is entered.

15.9   ASSEMBLER LANGUAGE PROGRAMMING

In the above example, it was necessary for the programmer to look up the machine language instruction for each step in the program. In addition, it was necessary for him to enter the program bit by bit. As the program becomes more lengthy, the above procedures become tedious and prone to errors.

Manufacturers of minicomputers and microprocessors have developed sophisticated programs which aid in the programming of their systems. These are called *assembler programs*. The mnemonic codes in the previous example are instructions in the 8080A assembler language.

Assembler language programs are written as a sequence of instructions which are converted to executable machine language instructions by the assembler. (Keep in mind that the assembler is a program.)

As illustrated in Figure 15-19, the assembler language program is written by a programmer and is called a *source program*. The assembler converts the source program into an equivalent *object program,* which is the machine language program that can be loaded into memory and executed.

Assembler language instructions must adhere to a fixed set of rules. An instruction has three separate and distinct parts or fields. The first is the label field. It is a name used to reference the instruction address. The next is the code field and it specifies the operation to be performed. The operand field provides any address or data information needed by the code field. Specific rules must be followed in separating the various fields. Some assemblers require a certain number of spaces between fields. Others require use of special characters such as colons, semicolons, or commas. These may be required so that the assembler can

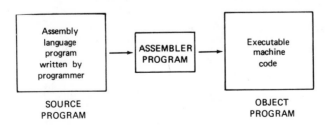

Figure 15-19.   Assembler program converts assembly language source program to object program.

distinguish between the various fields. Usually, an instruction is terminated by a carriage return character.

Let's write the previous program in a form that can be interpreted by the assembler.

Label Field	Code Field	Operand Field
	ORG	10H
DATA:	DS	2
RESULT:	DS	1
	LXI	H, DATA
	MOV	A, M
	INX	H
	ADD	M
	STA	RESULT
	OUT	04
	HLT	
	END	

Note that some new instructions appear in the program. For example, ORG 10H is an instruction to the assembler to tell it to "start assembling the program at address 10." The H designates that this is a hexadecimal number. The 8080 assembler can also recognize octal numbers, as well as binary and decimal. An O or Q following the number indicates octal. B indicates binary, and decimal is indicated when no letter appears.

The instruction DS 2 (define storage) causes the assembler to assign two memory locations for data as indicated in the label field. When the assembler finds the label DATA in the operand field, it will substitute two addresses for the data. The next instruction causes the assembler to substitute one address for the storage of the result.

The programmer may use any combination of letters he chooses for the label. However, the assembler will interpret only the first five characters. The last instruction in the program must be END. This indicates to the assembler that the end of the program has been reached and that the generation of the object program should now begin.

The instructions ORG, DS, and END are examples of pseudo-instructions. These instructions are directed to the assembler but do not cause any object code to be generated.

Before the program can be assembled, it must be stored on a medium such as punched tape. The 8080 assembler is referred to as a *three-pass program*. This means that the source program must be entered three times in order to obtain an object program tape. The procedure is as follows:

1. The assembler program is loaded.
2. The assembler asks for the source program to be loaded. When the source program tape is read in, the assembler processes the first pass, which sets up the symbol tables for the program.
3. The assembler asks for the source tape to be read in a second time. As the tape is read in again, the assembler begins processing the second pass, creating a listing of the program.
4. The assembler asks for the source tape a third time. When the punched tape is read in again, the assembler causes a hexadecimal format, machine-coded program to be outputted on the punch device. This is the object program tape.

Figure 15-20 is the listing of the program. You will notice that two new columns are added to the listing. The first column shows the memory location of the program. The second column contains the actual machine language program in hexadecimal format. Note that the assembler has assigned addresses 10 and 11 for the data. It has also assigned address 12 for the result. The first instruction is in addresses 13, 14, and 15 (three bytes). It should be clear to the student that the instruction MOV A,M has been converted by the assembler to 7E. Assembler languages are used extensively for programming mini- and microcomputers.

```
0000                      ORG      10H
0010          DATA:       DS       2
0012          RESULT:     DS       1
0013 211000               LXI      H,DATA
0016 7E                   MOV      A,M
0017 23                   INX      H
0018 86                   ADD      M
0019 321200               STA      RESULT
001C D304                 OUT      04
001E 76                   HLT
0000                      END
```

Figure 15-20.   Assembler program listing.

15.10   COMPILER LANGUAGES

Some program languages have been developed to assist the programmer in writing programs more rapidly and with less possibilities for error.

These languages are generally written for a specific application. For example, FORTRAN is used for mathematical applications, and the instructions are similar to mathematical expressions. The following instruction is in FORTRAN:

$$NUM = A*B/C$$

This means that the variable NUM is to take the value of A times B, divided by C.

For business applications, COBOL is generally used. The instructions are very similar to plain language. The following is an example of a COBOL instruction:

IF R IS LESS THAN 9 THEN ADD 1 TO R THEN GO TO P2.

When the program is complete, it is entered into the computer, and an object program is generated. The program which performs this task is called the *compiler,* and the language is called the *compiler language.* The program that converts assembler to machine language is called the *assembler.* One must realize that a single instruction from one of these languages generates many machine language words.

Although the compiler language is more efficient for the programmer, it requires huge memory assignments and many operations to perform.

The following is a comparison of compiler, assembler, and machine language programs:

Machine Language	Assembler Language	Compiler Language
0110001100100001	LDA A	D = A+B+C
0100001100100010	ADA B	
0100001100100011	ADA C	
0111001100100100	STA D	

Compiler languages are said to be problem-oriented due to the nature of their instructions. On the other hand, the assembler language is closely related to machine language in that, generally, one machine language instruction is generated from one assembler language instruction. Therefore, the assembler language is referred to as a machine-oriented language. Because compiler languages require huge amounts of memory, they are generally used in computers and large minicomputer systems.

Although it may require a great deal of effort, the assembler language may be optimized by the programmer. This means that a minimum time is required to execute the program, making the assembler language program more ideally suited to real-time applications than compiler language programs.

QUESTIONS AND PROBLEMS—CHAPTER 15

1. Differentiate between: (a) minicomputers; (b) microcomputers; (c) microprocessors.

2. Define each of the following: (a) accumulator; (b) A.L.U.; (c) data word; (d) op code; (e) operand; (f) C.P.U.

3. What is the purpose of a flowchart?

4. What are micro-instructions?

5. Information stored in a memory address may be either instructions or _____.

6. Information stored in the program counter is used to _____.

7. Two sources of information for the "control" block of the CPU are the op-code and the _____.

8. What are the two kinds of information stored in the memory?

9. What are the two clock phases that we discussed?

10. Micro-instructions have control over _____.

11. What are the two parts of the instruction word called?

12. In what kind of language is the object program?

13. The programming language which is most similar to the object program is called _____.

14. A language commonly used in business applications is _____.

15. The language best suited for real time applications is _____.

16. A language commonly used for mathematical applications is _____.

17. In which register of the 8080 microprocessor do the results of all mathematical and logic operations appear?

18. The number of bits for the 8080 data word is _____.

19. For the 8080, the minimum number of bytes in an instruction is one. What is the maximum number of bytes in an instruction?

20. The assembler language instruction MOV A,M means what to the 8080 microprocessor?

21. When writing an assembler language program, what is the label field used for?

22. What are pseudo-instructions?

23. What is meant by a three-pass assembler program?

24. What is the primary advantage of compiler languages?

25. Name two disadvantages of compiler languages.

16

Microprocessors and Microcomputers

Digital systems, from their beginning, have become increasingly efficient and reliable, and are used in ever greater numbers of applications. With the development of the minicomputer, the ability to process data has become a permanent part of various systems. Minicomputers, however, are limited to use in larger systems because of their cost. Smaller systems have relied upon the use of devices such as logic gates, counters, registers, and flip-flops. These "hard-wired" systems are costly to develop; thus their use has been restricted to large-volume applications.

The advent of the microprocessor has lowered development costs significantly, as well as providing the data-processing ability of the minicomputer. Most phases of the products development have been simplified by encoding suitable sequences of instructions, called *programs*, and storing them in the systems ROM. The microprocessor, fabricated on a single chip, performs all of the systems functions by reading the instructions and executing them in sequence. Data is transferred to and from the microcomputer via its I/O ports. Recall that a microcomputer is a complete system incorporating memory, I/O ports, and a microprocessor.

The topics covered in this chapter are:

16.1 Microprocessors

16.2 A Microprocessor Application

16.1 *MICROPROCESSORS*

Figure 16-1 is a photograph of the SC/MP microprocessor kit manufactured by National Semiconductor Corporation. The purpose of the kit
is to develop basic system concepts and familiarization with its instruction
set. The kit uses National's single chip microprocessor (SC/MP). A block
diagram of the microprocessor showing its architecture is found in Figure
16-2. The instruction set for the SC/MP is found in Appendix 3. The
processor is designed for inexpensive systems and requires minimal use
of external components.

Figure 16-3 shows a block diagram of the SC/MP kit. Note that two
chips are used to provide 256 bytes of RAM, which allows enough
memory to write short programs. Typical applications of this processor
may not require more than this. However, SC/MP allows for addressing
a total of 65K (65,536) bytes of memory. Also notice that 512 bytes of
preprogrammed ROM are provided. The ROM contains a debug program and teletype input/output routines called the *system monitor*.

Although the SC/MP kit is intended for experimental purposes,
SC/MP can be used in a large number of applications, including appliances, games, terminals, process controllers, test systems, and small business machines. Figure 16-4 is a block diagram of a typical SC/MP
configuration, with its supporting circuits.

Figure 16-1.   SC/MP microcomputer kit. (Courtesy of National Semiconductor Corp.)

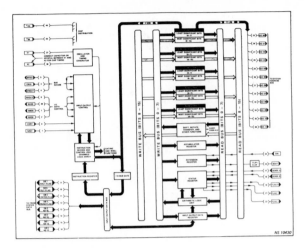

Figure 16-2. Functional block diagram of SC/MP. (Courtesy of National Semiconductor Corp.)

A widely used microprocessor is Motorola's M6800. A functional block diagram is shown in Figure 16-5. The instruction set is shown in Appendix 4. Figure 16-6 illustrates how the M6800 may be expanded into a complete operational system.

Figure 16-7 shows a functional block diagram of Intel's 8080 CPU. Like the M6800 and the SC/MP, it can address 65K of memory and operates on 8-bit bytes. When connected to two additional chips, the 8080 forms the standard interface shown in Figure 16-8.

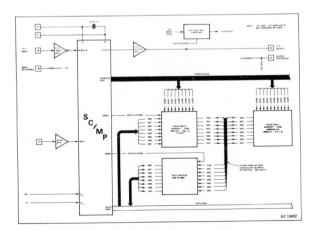

Figure 16-3. SC/MP kit block diagram. (Courtesy National Semiconductor Corp.)

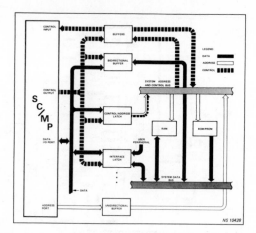

Figure 16-4.   Typical configuration of SC/MP. (Courtesy of National Semiconductor Corp.)

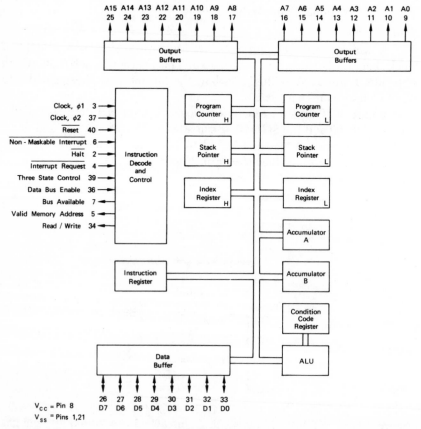

Figure 16-5.   Motorola M6800 functional block diagram. (Courtesy Motorola Semiconductor Products, Inc.)

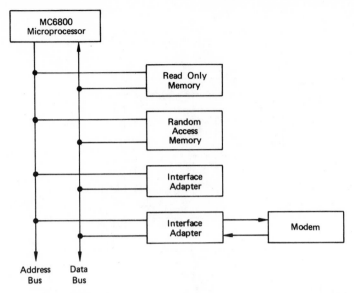

Figure 16-6. M6800 system block diagram. (Courtesy Motorola Semiconductor Products, Inc.)

You will notice that three busses are formed. The address bus is
used to select memory locations or I/O devices. The data bus allows transfer of data or instructions between the CPU and memory or I/O devices.
The third bus is the control bus; it is used to select the direction of data
to or from memory or I/O ports.

From the standard 8080 interface, a complete microcomputer system can be developed. Figure 16-9 illustrates how this may be accomplished.

16.2   A MICROPROCESSOR APPLICATION

In order to develop an understanding of the microprocessor, let's look at
a specific application. The particular application we have chosen does not
utilize all of the 8080's capabilities. However, it may be considered a
typical application.

Let's assume that a manufacturer of some product needs to weigh
and count packages as they come off the line. The packages may contain
any number of products from machine screws to jelly beans. Each package is to be weighed, and if its weight is less than a minimum value (45
grams), or greater than a maximum value (50 grams), it is rejected. If
not, the package is to be labeled with its weight, and the total weight
shall be accumulated. In addition, an accurate count of the labeled packages will be maintained.

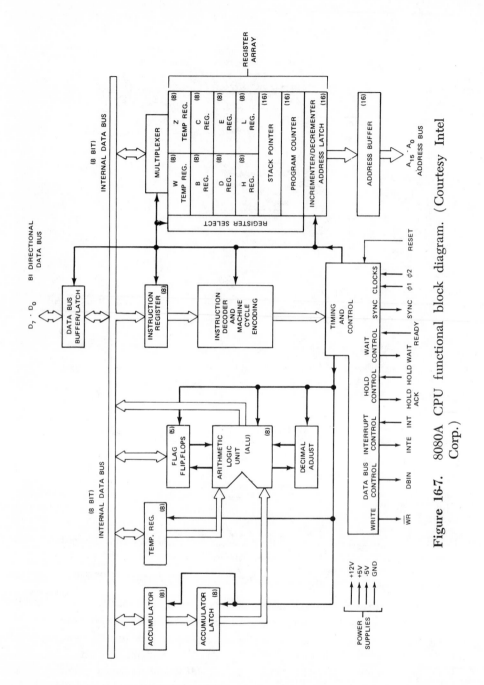

Figure 16-7. 8080A CPU functional block diagram. (Courtesy Intel Corp.)

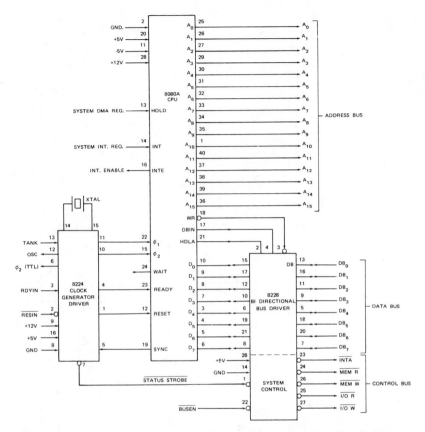

Figure 16-8.   8080A CPU standard interface. (Courtesy Intel Corp.)

Because the instruction set for the 8080 was discussed in the previous chapter, the 8080 will be used for this application. First let's take a closer look at the 8080. Figure 16-8 shows the 8080 with its associated clock generator and control system. The 8080 is a dynamic device, which means that its internal logic circuitry requires a clock with two clock pulses, ϕ_1 and ϕ_2. A single chip package available for this function is the 8224, which is shown, along with its block diagram, in Figure 16-10. The 8224 is controlled by a crystal that is selected by the designer to meet a variety of speed requirements. Notice that it has certain other inputs and outputs that are used for control.

You will recall that each instruction for the 8080 requires one to three bytes. In addition, each instruction requires from one to five machine cycles for fetching and execution, and every machine cycle requires from three to five states for its completion (T_1 through T_5).

Each state has a duration of one clock period. If, for example, the

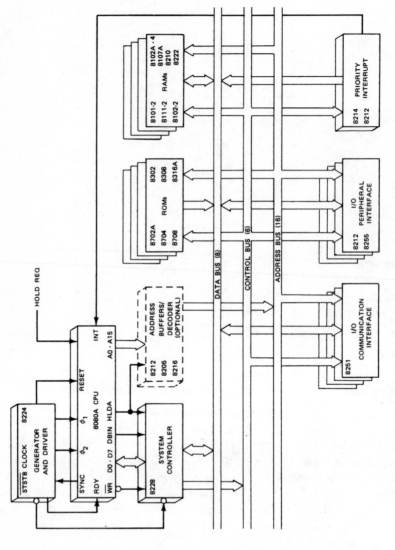

Figure 16-9. 8080A microcomputer system. (Courtesy Intel Corp.)

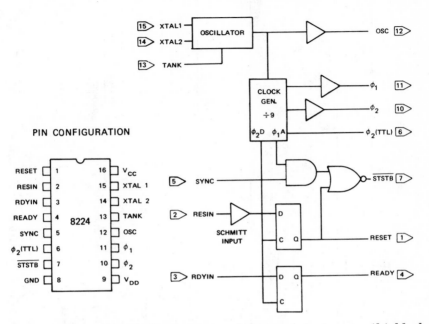

Figure 16-10. 8224 clock generator and driver: (a) pin-outs; (b) block diagram. (Courtesy Intel Corp.)

crystal frequency is 2 MHz, the clock period will be 0.5 μsec. There are three other states: WAIT, HOLD, and HALT. These states may last from one to a finite number of clock periods, as determined by external signals. Refer to the timing diagram in Figure 16-11.

The following is a sequence of events in a machine cycle:

T_1—A memory address or an I/O device number is placed on the address bus (A_{15}–A_0). Status information is placed on the data bus (D_7–D_0).

T_2—The CPU samples the READY and HOLD inputs and checks for HALT instruction.

T_W—This is an optional time period. Processor enters WAIT state, if READY is low, or if HALT instruction has been executed.

T_3—During a read cycle, data is inputted to the processor on the data bus from RAM, ROM, or an I/O device. During a WRITE or OUTPUT cycle, data is transferred from the CPU to RAM or an I/O device by way of the data bus.

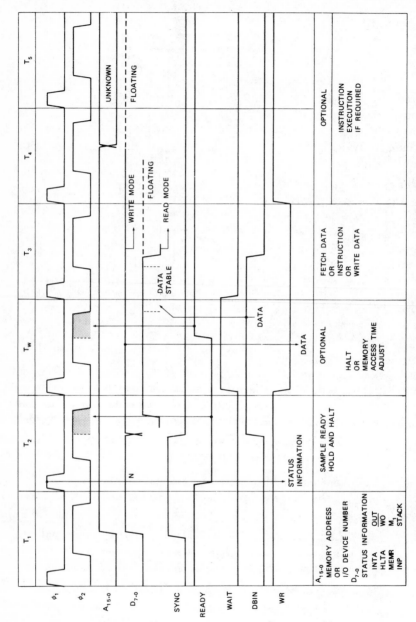

Figure 16-11. Basic 8080A timing cycle. (Courtesy Intel Corp.)

T$_4$, T$_5$—States T$_4$ and T$_5$ are available if the execution of a particular instruction requires them. If not, the CPU may skip one or both of them. T$_4$ and T$_5$ are used only for internal processor operations.

At T$_1$ when status information appears on the data bus, it must be stored (latched) for the duration of the machine cycle. The 8228 system controller and bus driver perform this function, as shown in Figure 16-12. The status information is then decoded by the gating array to provide a control bus. The control bus goes to memory and I/O devices. If the $\overline{\text{MEM R}}$ line is active, memory is activated to read. If the $\overline{\text{MEM W}}$ line is active, memory is activated to write. If $\overline{\text{I/O R}}$ is active, the I/O devices are activated to input data. When $\overline{\text{I/O W}}$ is active, I/O devices are activated to output data. In addition, the 8228 performs the function of a bidirectional bus driver to buffer the 8080 from memory and I/O devices. The bidirectional bus driver is controlled by signals from the gating array to maintain proper data flow or force its outputs to the high impedance state (3-state) for direct memory access.

Now, back to our application problem. Figure 16-13 is a diagram showing one possible implementation of the problem. The data and address busses are extensions of the diagram in Figure 16-8. Notice that RAM and ROM appear between the address and data busses. The ROM is used to store the program. It is a 1702A erasable read-only memory, discussed in Chapter 14. The RAM is used to store the total weight and package count. We are allocating six decimal digits for the package count, eight digits for the accumulated weight, and four digits for the minimum and maximum weights. Because two digits can be stored at each memory

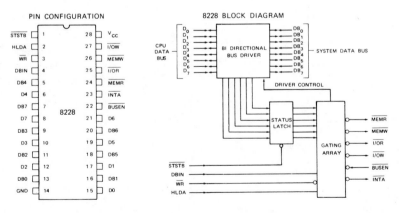

Figure 16-12.   8228 system controller and bus driver: (a) pin-outs; (b) block diagram. (Courtesy Intel Corp.)

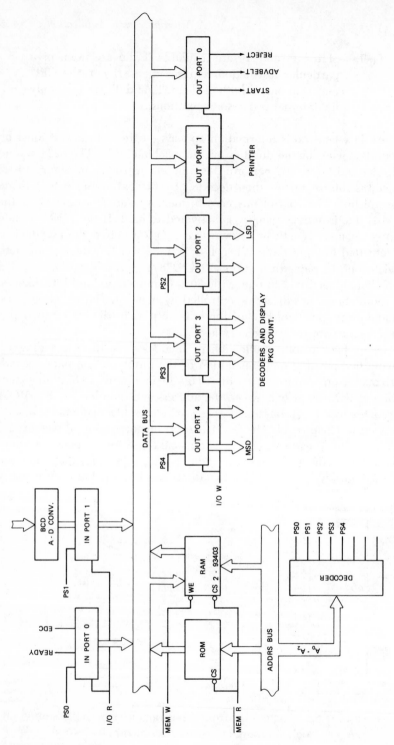

Figure 16-13. Block diagram of implementation of problem example.

address, nine memory locations are required. Two 93403 16W × 4-bit bipolar RAMs will provide sixteen 8-bit addresses. This is suitable for our application. The 93403 is discussed in Chapter 14.

Packages are rejected when their weight is less than the minimum or greater than the maximum values. The processor accomplishes this by outputting a "1" (REJCT) on output port 0. Since the minimum and maximum values are stored in RAM, they can be changed at will.

Figure 16-14 is the flowchart representing the procedure the processor must follow. Block numbers 1 and 2 form a wait loop. The processor is waiting for a package to be in position to be weighed. When the package is in position, the processor outputs a START pulse (block 3). The processor then goes into another wait loop (blocks 4 and 5) waiting for an indication that the analog-to-BCD converter has completed its conversion process. At block 6, the processor inputs data representing the weight of the package. Next, the processor decides if the weight is less than 45 grams (block 7) or greater than 50 grams (block 8). If not, the weight is printed on the package (block 9).

Next, the total weight is accumulated and stored (block 10). In block 11, the package count is accumulated and restored to memory. Blocks 12 and 13 indicate the output of the package count and the total weight, respectively, to LED displays. If the weight of the package is less than 45 or greater than 50 grams, the REJCT pulse is outputted. The last block (15) indicates that a bit has been outputted, which advances the belt so that the next package can be weighed.

Figure 16-15 is the program listing for our application. The first three lines assign memory locations 0100 through 0109 for storage of the total weight, package count, and minimum and maximum package weight.

Input port 1 is used to input the weight of each package. This is an 8-bit byte that has been converted by an analog-to-BCD converter located at the weighing device. The inputting of this data is under the control of the processor.

The procedure is as follows:

1. The processor waits for a "1" on the READY input of input port 0.

2. When the READY appears, the processor outputs a "1" on output port 0 (START). This causes the analog-to-BCD converter to start its conversion process.

3. When the analog-to-BCD conversion is complete, the converter sends an end-of-conversion bit (EOC) to input port 0.

4. The processor then inputs the weight through input port 1.

This procedure is referred to as "handshaking." Notice that the data is inputted on port 1, while the handshaking information uses port 0.

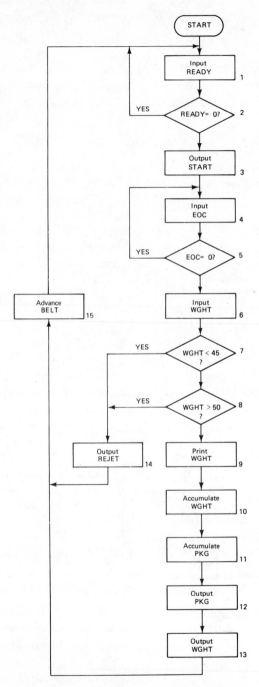

Figure 16-14.   Flowchart of problem example.

```
0100              WGHT    EQU     100H        ; 4 BYTES FOR TOTAL WEIGHT
0104              PKG     EQU     104H        ; 3 BYTES FOR PACKAGE COUNT
0107              MINW    EQU     107H        ; 2 BYTES FOR MIN AND MAX WEIGH
0000                      ORG     10H

                  ; BLOCK 1
0010 DB00         BEGIN:  IN      0           ; INPUT READY
                  ; BLOCK 2
0012 E601                 ANI     01          ; MASK FOR BIT ZERO (READY)
0014 C21000                JNZ     BEGIN       ; WAIT LOOP
                  ; BLOCK 3
0017 3E01                 MVI     A,01
0019 D300                 OUT     0           ; OUTPUT START PULSE
                  ; BLOCK 4
001B DB00         BACK:   IN      0           ; EOC PULSE
                  ; BLOCK 5
001D E602                 ANI     02          ; MASK FOR EOC BIT (BIT 1)
001F CA1B00                JZ      BACK        ; LOOP
                  ; BLOCK 6
0022 210701               LXI     H,MINW      ; MINIMUM WEIGHT VALUE
0025 DB01                 IN      01          ; INPUT WEIGHT
0027 4F                   MOV     C,A         ; MOVE TO C REG
                  ; BLOCK 7
0028 BE                   CMP     M           ; COMPARE WITH MINIMUM WEIGHT
0029 D28500               JNC     REJCT       ; REJECT - TOO LIGHT
                  ; BLOCK 8
002C 23                   INX     H           ; POINTER TO MAXIMUM WEIGHT
002D 79                   MOV     A,C         ; RETURN WEIGHT TO A REG
002E BE                   CMP     M           ; COMPARE WITH MAXIMUM WEIGHT
002F DA8500               JC      REJCT       ; REJECT - TOO HEAVY
                  ; BLOCK 9
0032 79                   MOV     A,C         ; RETURN WEIGHT TO A REG
0033 D301                 OUT     01          ; OUTPUT TO PRINTER
                  ; BLOCK 10
0035 0604                 MVI     B,4         ; 4 BYTES TO B REG
0037 210001               LXI     H,WGHT      ; 4 BYTES = 8 DIGITS
003A AF                   XRA     A           ; CLEAR CARRY
003B 79                   MOV     A,C         ; RETURN WEIGHT TO A REG
003C 8E         LOOP:     ADC     M           ; BINARY ADD
003D 27                   DAA                 ; DECIMAL ADJUST
003E 77                   MOV     M,A         ; STORE
003F 23                   INX     H           ; POINT TO NEXT BYTE
0040 05                   DCR     B           ; DECREMENT B REG
0041 C23C00               JNZ     LOOP        ; IF NOT LAST BYTE, JUMP
                  ; BLOCK 11
0044 210401               LXI     H,PKG       ; PACKAGE COUNT POINTER
0047 7E                   MOV     A,M         ; MOVE PKG COUNT TO A REG
0048 3C                   INR     A           ; INCREMENT COUNT
0049 27                   DAA                 ; DECIMAL ADJUST
004A 77                   MOV     M,A         ; STORE BYTE
004B D25B00               JNC     POUT        ; NO CARRY
004E 23                   INX     H           ; IF CARRY THEN INCREMENT NEXT
004F 7E                   MOV     A,M         ; BYTE
0050 3C                   INR     A
0051 27                   DAA                 ; DECIMAL ADJUST
0052 77                   MOV     M,A         ; STORE BYTE
0053 D25B00               JNC     POUT        ; NO CARRY
0056 23                   INX     H           ; IF CARRY THEN INCREMENT NEXT
0057 7E                   MOV     A,M         ; BYTE
0058 3C                   INR     A
0059 27                   DAA                 ; DECIMAL ADJUST
005A 77                   MOV     M,A         ; STORE BYTE
                  ; BLOCK 12
005B 210401     POUT:     LXI     H,PKG       ; BEGIN PACKAGE COUNT OUTPUT
005E 7E                   MOV     A,M         ; ROUTINE
005F D302                 OUT     02          ; LEAST SIG BYTE
0061 23                   INX     H
0062 7E                   MOV     A,M
0063 D303                 OUT     03          ; NEXT SIG BYTE
0065 23                   INX     H
0066 7E                   MOV     A,M
0067 D304                 OUT     04          ; MOST SIG BYTE
0069 C31000               JMP     BEGIN       ; END OUTPUT ROUTINE
                  ; BLOCK 13
006C 210001               LXI     H,WGHT      ; OUTPUT WGHT ROUTINE
006F 7E                   MOV     A,M
0070 320010               STA     1000H       ; A12 FOR PORT SELECT
0073 23                   INX     H
0074 7E                   MOV     A,M
0075 320020               STA     2000H       ; A13
0078 23                   INX     H
0079 7E                   MOV     A,M
007A 320040               STA     4000H       ; A14
007D 23                   INX     H
007E 7E                   MOV     A,M
007F 320080               STA     8000H       ; A15
0082 C38C00               JMP     BADV        ; JUMP TO BELT ADV ROUTINE
                  ; BLOCK 14
0085 3E02        REJCT:   MVI     A,02H       ; OUTPUT REJECT PULSE
0087 D300                 OUT     0
0089 C31000               JMP     BEGIN
                  ; BLOCK 15
008C 3E08        BADV:    MVI     A,08        ; OUTPUT BELT ADV BIT
008E D300                 OUT     0
0090 C31000               JMP     BEGIN
0000                      END
```

Figure 16-15.   Program listing of problem example.

Handshaking is required because the processor operates at much higher speeds than the mechanical weighing equipment.

The weight of each package is outputted to the printer via output port 1. No handshaking is required here because it is assumed that the package is in position for printing.

After the package has been printed, the belt is advanced and the procedure to weigh the next package begins. The processor controls this by outputting a "1" (BADV) through output port 0.

Output ports 2, 3, and 4 are used to output data representing the package count to decoders and LED displays.

The program begins at address 0010 and ends at address 0093 (decimal 16 through 146). There are 256 locations in ROM, and only 130 are used for this program. The first 16 addresses (0000 through 000F) are reserved for possible program modifications or additions.

The block numbers in the program listing refer to the block numbers in the flowchart (Figure 16-14). Words following a semicolon in any line are not part of the program but are comments used to define and explain program steps.

Addresses 0010 through 0034 are instructions for inputting the weight, determining if it's acceptable, and outputting it to the printer.

At address 0029, the program branches (JUMP) to address 0085 if the weight is less than the minimum value. If the weight is too great, the program branches at address 002F to address 0085. You will notice that at address 0085, the program outputs a reject (REJCT) pulse, which causes the reject mechanism to be activated.

A routine begins at address 0035, which accumulates the total weight of the packages and stores it in address location 0100. A similar program starting at address 0044 maintains a continuous record of the package count at address 0104. Another routine beginning at address 005B outputs the current package count to a LED display.

Address 006C initiates a routine that outputs the current accumulated weight to another LED display. Finally, at address 008C there appears a routine (BADV), which activates the belt advance mechanism. You will notice that if the package has not been rejected, the BADV routine is entered by way of a jump instruction at address 0082. If the package has been rejected, the BADV routine is entered as a continuation of the REJCT routine.

A unique procedure, called *memory mapped I/O*, is used for outputting the accumulated weight. The range of addresses used by ROM and RAM is 0000 through 010F. The last 16 addresses are RAM. Nine address lines (A_0–A_8) are required for addressing. This leaves seven lines (A_9–A_{15}) unused for memory addressing. This situation is typical for small system applications. It is possible to use these address lines for directly selecting I/O ports. Since the total weight requires four bytes, address lines A_{12}–A_{15} were chosen to select four output ports. Figure 16-16 shows the Hewlett-Packard 5082-7300 numeric indicator, a device ideally suited for this application. It combines (a) a 4-bit BCD latch, (b) a seven-segment decoder, and (c) a seven-segment display, all in one package.

Figure 16-17 shows the addressing scheme. Eight of the indicators are required for the display. Data is latched into two devices at a time by activating the latch enable (LE) input with an address line. The two least significant bytes are enabled by A_{12}. This address line is activated by selecting address 1000. The next two digits are enabled by A_{13}, which is activated by selecting address 2000. Address 4000 activates A_{14}, and address 8000 activates A_{15}.

These I/O ports are selected in the program as if they were memory locations. For example, the instruction "STA 1000H" means "Store the contents of the accumulator at address 1000." Since this is an output port rather than a memory address, data is actually outputted. This method of selecting I/O ports is known as *memory mapped I/O*.

Address lines A_9 through A_{11} could have been used for memory mapping output port 02, 03, and 04 (package count output). However, for demonstration purposes, more than one method was chosen for addressing the output ports.

Presently, the applications for microprocessors may seem rather large. In the future, however, microprocessors will be extensively used in applications limited only by man's imagination. In particular, these microprocessors will control and maximize the efficient use of energy.

Figure 16-16.   Hewlett-Packard 5082-7300 numeric indicator. (Courtesy Hewlett-Packard.)

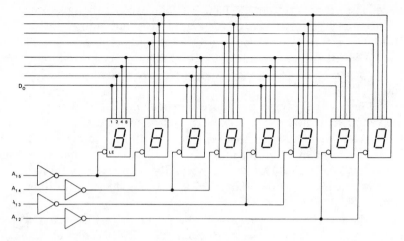

Figure 16-17. Memory mapped I/O ports.

QUESTIONS AND PROBLEMS—CHAPTER 16

1. In the 8080, for what are the H and L registers used?

2. For the 8080, how many bits are stored in the program counter?

3. How many memory locations is the 8080 capable of addressing?

4. For the 8080, what is the purpose of the control bus?

5. What is meant by "handshaking"?

6. In Figure 16-14, a wait loop is indicated by block numbers 1 and 2. What is the processor waiting for?

7. For the microprocessor application problem, what are the names of the inputs and outputs of port zero?

8. For the 8080 microprocessor, the assembler language instruction ADD B means to add the contents of the B register to the accumulator (A reg.). What is the machine language instruction?

9. For the SC/MP, what is the machine language instruction for ADD?

10. What is memory mapping as related to I/Os?

11. In the applications problem covered in this chapter, what is the memory mapped address for the least significant byte of the display. (See Figure 16-17.)

12. What address should be used to address the most significant byte of the memory mapped display? (See Figure 16-17.)

13. Referring to the 8080 instruction set (Appendixes 1 and 2), what is the instruction for moving data from memory to the accumulator?

14. Referring to the 8080 instruction set (Appendixes 1 and 2), what is the instruction for adding the contents of memory to the contents of the accumulator?

15. Recall that the H and L register pair is used to point to a memory location. The mnemonic code for loading the H and L register pair is LXI H, address. Write a short program which will add the contents of memory location 0100_{16} to 0101_{16} and store the result in address 0102_{16}.

16. Write a short program for the 8080 which will add two 16-bit numbers. The first number is stored in hex addresses 0100 and 0101. The second is stored in hex addresses 0110 and 0111. Store the result in hex addresses 0120 and 0121. Numbers larger than 8 bits stored in memory should be arranged such that the least significant byte appears first.

Appendixes

8080 MICROPROCESSOR

DATA AND INSTRUCTION FORMATS

Data in the 8080 is stored in the form of 8-bit binary integers. All data transfers to the system data bus will be in the same format.

$$D_7\ D_6\ D_5\ D_4\ D_3\ D_2\ D_1\ D_0$$

DATA WORD

The program instructions may be one, two, or three bytes in length. Multiple byte instructions must be stored in succesive words in program memory. The instruction formats then depend on the particular operation executed.

One Byte Instructions

| D_7 | D_6 | D_5 | D_4 | D_3 | D_2 | D_1 | D_0 | OP CODE |

TYPICAL INSTRUCTIONS

Register to register, memory reference, arithmetic or logical, rotate return, push, pop, enable or disable Interrupt instructions

Two Byte Instructions

| D_7 | D_6 | D_5 | D_4 | D_3 | D_2 | D_1 | D_0 | OP CODE |
| D_7 | D_6 | D_5 | D_4 | D_3 | D_2 | D_1 | D_0 | OPERAND |

Immediate mode or I/O instructions

Three Byte Instructions

D_7	D_6	D_5	D_4	D_3	D_2	D_1	D_0	OP CODE
D_7	D_6	D_5	D_4	D_3	D_2	D_1	D_0	LOW ADDRESS OR OPERAND 1
D_7	D_6	D_5	D_4	D_3	D_2	D_1	D_0	HIGH ADDRESS OR OPERAND 2

Jump, call or direct load and store instructions

For the 8080 a logic "1" is defined as a high level and a logic "0" is defined as a low level.

INSTRUCTION SET

Summary of Processor Instructions

Mnemonic	Description	D7	D6	D5	D4	D3	D2	D1	D0	Clock[2] Cycles
MOV r1,r2	Move register to register	0	1	D	D	D	S	S	S	5
MOV M,r	Move register to memory	0	1	1	1	0	S	S	S	7
MOV r,M	Move memory to register	0	1	D	D	D	1	1	0	7
HLT	Halt	0	1	1	1	0	1	1	0	7
MVI r	Move immediate register	0	0	D	D	D	1	1	0	7
MVI M	Move immediate memory	0	0	1	1	0	1	1	0	10
INR r	Increment register	0	0	D	D	D	1	0	0	5
DCR r	Decrement register	0	0	D	D	D	1	0	1	5
INR M	Increment memory	0	0	1	1	0	1	0	0	10
DCR M	Decrement memory	0	0	1	1	0	1	0	1	10
ADD r	Add register to A	1	0	0	0	0	S	S	S	4
ADC r	Add register to A with carry	1	0	0	0	1	S	S	S	4
SUB r	Subtract register from A	1	0	0	1	0	S	S	S	4
SBB r	Subtract register from A with borrow	1	0	0	1	1	S	S	S	4
ANA r	And register with A	1	0	1	0	0	S	S	S	4
XRA r	Exclusive Or register with A	1	0	1	0	1	S	S	S	4
ORA r	Or register with A	1	0	1	1	0	S	S	S	4
CMP r	Compare register with A	1	0	1	1	1	S	S	S	4
ADD M	Add memory to A	1	0	0	0	0	1	1	0	7
ADC M	Add memory to A with carry	1	0	0	0	1	1	1	0	7
SUB M	Subtract memory from A	1	0	0	1	0	1	1	0	7
SBB M	Subtract memory from A with borrow	1	0	0	1	1	1	1	0	7
ANA M	And memory with A	1	0	1	0	0	1	1	0	7
XRA M	Exclusive Or memory with A	1	0	1	0	1	1	1	0	7
ORA M	Or memory with A	1	0	1	1	0	1	1	0	7
CMP M	Compare memory with A	1	0	1	1	1	1	1	0	7
ADI	Add immediate to A	1	1	0	0	0	1	1	0	7
ACI	Add immediate to A with carry	1	1	0	0	1	1	1	0	7
SUI	Subtract immediate from A	1	1	0	1	0	1	1	0	7
SBI	Subtract immediate from A with borrow	1	1	0	1	1	1	1	0	7
ANI	And immediate with A	1	1	1	0	0	1	1	0	7
XRI	Exclusive Or immediate with A	1	1	1	0	1	1	1	0	7
ORI	Or immediate with A	1	1	1	1	0	1	1	0	7
CPI	Compare immediate with A	1	1	1	1	1	1	1	0	7
RLC	Rotate A left	0	0	0	0	0	1	1	1	4
RRC	Rotate A right	0	0	0	0	1	1	1	1	4
RAL	Rotate A left through carry	0	0	0	1	0	1	1	1	4
RAR	Rotate A right through carry	0	0	0	1	1	1	1	1	4
JMP	Jump unconditional	1	1	0	0	0	0	1	1	10
JC	Jump on carry	1	1	0	1	1	0	1	0	10
JNC	Jump on no carry	1	1	0	1	0	0	1	0	10
JZ	Jump on zero	1	1	0	0	1	0	1	0	10
JNZ	Jump on no zero	1	1	0	0	0	0	1	0	10
JP	Jump on positive	1	1	1	1	0	0	1	0	10
JM	Jump on minus	1	1	1	1	1	0	1	0	10
JPE	Jump on parity even	1	1	1	0	1	0	1	0	10
JPO	Jump on parity odd	1	1	1	0	0	0	1	0	10

Appendix 1. 8080 instruction set. (Courtesy Intel Corp.)

437

8080 MICROPROCESSOR

Mnemonic	Description	Instruction Code [1]								Clock [2]
		D_7	D_6	D_5	D_4	D_3	D_2	D_1	D_0	Cycles
CALL	Call unconditional	1	1	0	0	1	1	0	1	17
CC	Call on carry	1	1	0	1	1	1	0	0	11/17
CNC	Call on no carry	1	1	0	1	0	1	0	0	11/17
CZ	Call on zero	1	1	0	0	1	1	0	0	11/17
CNZ	Call on no zero	1	1	0	0	0	1	0	0	11/17
CP	Call on positive	1	1	1	1	0	1	0	0	11/17
CM	Call on minus	1	1	1	1	1	1	0	0	11/17
CPE	Call on parity even	1	1	1	0	1	1	0	0	11/17
CPO	Call on parity odd	1	1	1	0	0	1	0	0	11/17
RET	Return	1	1	0	0	1	0	0	1	10
RC	Return on carry	1	1	0	1	1	0	0	0	5/11
RNC	Return on no carry	1	1	0	1	0	0	0	0	5/11
RZ	Return on zero	1	1	0	0	1	0	0	0	5/11
RNZ	Return on no zero	1	1	0	0	0	0	0	0	5/11
RP	Return on positive	1	1	1	1	0	0	0	0	5/11
RM	Return on minus	1	1	1	1	1	0	0	0	5/11
RPE	Return on parity even	1	1	1	0	1	0	0	0	5/11
RPO	Return on parity odd	1	1	1	0	0	0	0	0	5/11
RST	Restart	1	1	A	A	A	1	1	1	11
IN	Input	1	1	0	1	1	0	1	1	10
OUT	Output	1	1	0	1	0	0	1	1	10
LXI B	Load immediate register Pair B & C	0	0	0	0	0	0	0	1	10
LXI D	Load immediate register Pair D & E	0	0	0	1	0	0	0	1	10
LXI H	Load immediate register Pair H & L	0	0	1	0	0	0	0	1	10
LXI SP	Load immediate stack pointer	0	0	1	1	0	0	0	1	10
PUSH B	Push register Pair B & C on stack	1	1	0	0	0	1	0	1	11
PUSH D	Push register Pair D & E on stack	1	1	0	1	0	1	0	1	11
PUSH H	Push register Pair H & L on stack	1	1	1	0	0	1	0	1	11
PUSH PSW	Push A and Flags on stack	1	1	1	1	0	1	0	1	11
POP B	Pop register pair B & C off stack	1	1	0	0	0	0	0	1	10
POP D	Pop register pair D & E off stack	1	1	0	1	0	0	0	1	10

										Cycles
POP H	Pop register pair H & L off stack	1	1	1	0	0	0	0	1	10
POP PSW	Pop A and Flags off stack	1	1	1	1	0	0	0	1	10
STA	Store A direct	0	0	1	1	0	0	1	0	13
LDA	Load A direct	0	0	1	1	1	0	1	0	13
XCHG	Exchange D & E, H & L Registers	1	1	1	0	1	0	1	1	4
XTHL	Exchange top of stack, H & L	1	1	1	0	0	0	1	1	18
SPHL	H & L to stack pointer	1	1	1	1	1	0	0	1	5
PCHL	H & L to program counter	1	1	1	0	1	0	0	1	5
DAD B	Add B & C to H & L	0	0	0	0	1	0	0	1	10
DAD D	Add D & E to H & L	0	0	0	1	1	0	0	1	10
DAD H	Add H & L to H & L	0	0	1	0	1	0	0	1	10
DAD SP	Add stack pointer to H & L	0	0	1	1	1	0	0	1	10
STAX B	Store A indirect	0	0	0	0	0	0	1	0	7
STAX D	Store A indirect	0	0	0	1	0	0	1	0	7
LDAX B	Load A indirect	0	0	0	0	1	0	1	0	7
LDAX D	Load A indirect	0	0	0	1	1	0	1	0	7
INX B	Increment B & C registers	0	0	0	0	0	0	1	1	5
INX D	Increment D & E registers	0	0	0	1	0	0	1	1	5
INX H	Increment H & L registers	0	0	1	0	0	0	1	1	5
INX SP	Increment stack pointer	0	0	1	1	0	0	1	1	5
DCX B	Decrement B & C	0	0	0	0	1	0	1	1	5
DCX D	Decrement D & E	0	0	0	1	1	0	1	1	5
DCX H	Decrement H & L	0	0	1	0	1	0	1	1	5
DCX SP	Decrement stack pointer	0	0	1	1	1	0	1	1	5
CMA	Compliment A	0	0	1	0	1	1	1	1	4
STC	Set carry	0	0	1	1	0	1	1	1	4
CMC	Compliment carry	0	0	1	1	1	1	1	1	4
DAA	Decimal adjust A	0	0	1	0	0	1	1	1	4
SHLD	Store H & L direct	0	0	1	0	0	0	1	0	16
LHLD	Load H & L direct	0	0	1	0	1	0	1	0	16
EI	Enable Interrupts	1	1	1	1	1	0	1	1	4
DI	Disable interrupt	1	1	1	1	0	0	1	1	4
NOP	No-operation	0	0	0	0	0	0	0	0	4

NOTES: 1. DDD or SSS – 000 B – 001 C – 010 D – 011 E – 100 H – 101 L – 101 Memory – 111 A.

2. Two possible cycle times (5/11) indicate instruction cycles dependent on condition flags.

Appendix 1. (Continued)

JUMP		CALL		RETURN		RESTART			ROTATE⁺	
C3	JMP	CD	CALL	C9	RET	C7	RST	0	07	RLC
C2	JNZ	C4	CNZ	C0	RNZ	CF	RST	1	0F	RRC
CA	JZ	CC	CZ	C8	RZ	D7	RST	2	17	RAL
D2	JNC	D4	CNC	D0	RNC	DF	RST	3	1F	RAR
DA	JC	DC	CC	D8	RC	E7	RST	4		
E2	JPO	E4	CPO	E0	RPO	EF	RST	5		
EA	JPE	EC	CPE	E8	RPE	F7	RST	6		
F2	JP	F4	CP	F0	RP	FF	RST	7		
FA	JM	FC	CM	F8	RM					
E9	PCHL									

JUMP / CALL / RETURN column: } Adr

CONTROL

00	NOP
76	HLT
F3	DI
FB	EI

MOVE IMMEDIATE		Acc IMMEDIATE		LOAD IMMEDIATE		STACK OPS		
06	MVI B,	C6	ADI	01	LXI B,	C5	PUSH B	
0E	MVI C,	CE	ACI	11	LXI D,	D5	PUSH D	
16	MVI D,	D6	SUI	21	LXI H,	E5	PUSH H	
1E	MVI E,	DE	SBI	31	LXI SP,	F5	PUSH PSW	
26	MVI H,	E6	ANI					
2E	MVI L,	EE	XRI			C1	POP B	
36	MVI M,	F6	ORI			D1	POP D	
3E	MVI A,	FE	CPI			E1	POP H	
						F1	POP PSW*	

MVI column: } D8   ADI column: } D8   LXI column: } D16

DOUBLE ADD

09	DAD	B
19	DAD	D
29	DAD	H
39	DAD	SP

E3	XTHL
F9	SPHL

MOVE

40	MOV	B,B
41	MOV	B,C
42	MOV	B,D
43	MOV	B,E
44	MOV	B,H
45	MOV	B,L
46	MOV	B,M
47	MOV	B,A
48	MOV	C,B
49	MOV	C,C
4A	MOV	C,D
4B	MOV	C,E
4C	MOV	C,H
4D	MOV	C,L
4E	MOV	C,M
4F	MOV	C,A
50	MOV	D,B
51	MOV	D,C
52	MOV	D,D
53	MOV	D,E
54	MOV	D,H
55	MOV	D,L
56	MOV	D,M
57	MOV	D,A

INCREMENT		DECREMENT		LOAD STORE			SPECIALS	
04	INR B	05	DCR B					
0C	INR C	0D	DCR C					
14	INR D	15	DCR D			EB	XCHG	
1C	INR E	1D	DCR E	0A	LDAX	B	27	DAA*
24	INR H	25	DCR H	1A	LDAX	D	2F	CMA
2C	INR L	2D	DCR L	2A	LHLD	Adr	37	STC⁺
34	INR M	35	DCR M	3A	LDA	Adr	3F	CMC⁺
3C	INR A	3D	DCR A					
03	INX B	0B	DCX B	02	STAX	B		
13	INX D	1B	DCX D	12	STAX	D	INPUT/OUTPUT	
23	INX H	2B	DCX H	22	SHLD	Adr	D3	OUT
33	INX SP	3B	DCX SP	32	STA	Adr	DB	IN

OUT / IN column: } D8

D8 = constant, or logical/arithmetic expression that evaluates to an 8 bit data quantity.

* = all Flags (C, Z, S, P, AC) affected

D16 = constant, or logical/arithmetic expression that evaluates to a 16 bit data quantity.

⁺ = only CARRY affected

Appendix 2.   8080 instruction set using hex notation. (Courtesy Intel Corp.)

MOVE (cont)			ACCUMULATOR*		
58	MOV	E,B	80	ADD	B
59	MOV	E,C	81	ADD	C
5A	MOV	E,D	82	ADD	D
5B	MOV	E,E	83	ADD	E
5C	MOV	E,H	84	ADD	H
5D	MOV	E,L	85	ADD	L
5E	MOV	E,M	86	ADD	M
5F	MOV	E,A	87	ADD	A
60	MOV	H,B	88	ADC	B
61	MOV	H,C	89	ADC	C
62	MOV	H,D	8A	ADC	D
63	MOV	H,E	8B	ADC	E
64	MOV	H,H	8C	ADC	H
65	MOV	H,L	8D	ADC	L
66	MOV	H,M	8E	ADC	M
67	MOV	H,A	8F	ADC	A
68	MOV	L,B	90	SUB	B
69	MOV	L,C	91	SUB	C
6A	MOV	L,D	92	SUB	D
6B	MOV	L,E	93	SUB	E
6C	MOV	L,H	94	SUB	H
6D	MOV	L,L	95	SUB	L
6E	MOV	L,M	96	SUB	M
6F	MOV	L,A	97	SUB	A
70	MOV	M,B	98	SBB	B
71	MOV	M,C	99	SBB	C
72	MOV	M,D	9A	SBB	D
73	MOV	M,E	9B	SBB	E
74	MOV	M,H	9C	SBB	H
75	MOV	M,L	9D	SBB	L
--------------			9E	SBB	M
77	MOV	M,A	9F	SBB	A
78	MOV	A,B	A0	ANA	B
79	MOV	A,C	A1	ANA	C
7A	MOV	A,D	A2	ANA	D
7B	MOV	A,E	A3	ANA	E
7C	MOV	A,H	A4	ANA	H
7D	MOV	A,L	A5	ANA	L
7E	MOV	A,M	A6	ANA	M
7F	MOV	A,A	A7	ANA	A

A8	XRA	B			
A9	XRA	C			
AA	XRA	D			
AB	XRA	E			
AC	XRA	H			
AD	XRA	L			
AE	XRA	M			
AF	XRA	A			
B0	ORA	B			
B1	ORA	C			
B2	ORA	D			
B3	ORA	E			
B4	ORA	H			
B5	ORA	L			
B6	ORA	M			
B7	ORA	A			
B8	CMP	B			
B9	CMP	C			
BA	CMP	D			
BB	CMP	E			
BC	CMP	H			
BD	CMP	L			
BE	CMP	M			
BF	CMP	A			

PSEUDO INSTRUCTION

ORG	Adr
END	
EQU	D16
SET	D16
DS	D16
DB	D8 [ ]
DW	D16 [ ]
IF	D16
ENDIF	
MACRO	[ ]
ENDM	

CONSTANT DEFINITION

0BDH	Hex
1AH	
105D	Decimal
105	
72O	Octal
72Q	
11011B	Binary
00110B	
'TEST'	ASCII
'A' 'B'	

OPERATORS

(.)
* , / , MOD,SHL,SHR
+ , −
NOT
AND
OR,XOR

STANDARD SETS

A	SET	7
B	SET	0
C	SET	1
D	SET	2
E	SET	3
H	SET	4
L	SET	5
M	SET	6
SP	SET	6
PSW	SET	6

FLAG BYTE STACK FORMAT

7	6	5	4	3	2	1	0
S	Z	0	A C	0	P	1	C

Adr = 16 bit address

** = all Flags except CARRY affected;
    (exception: INX & DCX affect no Flags)

Appendix 2. (Continued)

Memory Reference Instructions

Description	Op Code	Source Statement	Object Format	Operation	Micro-cycles	Page
			7 6 5 4 3 2 1 0 | 7 6 5 4 3 2 1 0 (m ptr) (disp)			
Load	LD	disp	1 1 0 0 0 m ptr | disp	$(AC) \leftarrow (EA)$	18	5-5
Store	ST	disp(ptr)	1 1 0 0 1	$(EA) \leftarrow (AC)$	18	5-6
AND	AND	@disp(ptr)	1 1 0 1 0	$(AC) \leftarrow (AC) \wedge (EA)$	18	5-6
OR	OR		1 1 0 1 1	$(AC) \leftarrow (AC) \vee (EA)$	18	5-6
Exclusive-OR	XOR		1 1 1 0 0	$(AC) \leftarrow (AC) \triangledown (EA)$	18	5-6
Decimal Add	DAD		1 1 1 0 1	$(AC) \leftarrow (AC)_{10} + (EA)_{10} + (CY/L); (CY/L)$	23	5-7
Add	ADD		1 1 1 1 0	$(AC) \leftarrow (AC) + (EA) + (CY/L); (CY/L), (OV)$	19	5-7
Complement and Add	CAD		1 1 1 1 1	$(AC) \leftarrow (AC) + \sim (EA) + (CY/L); (CY/L), (OV)$	20	5-8

Memory Increment/Decrement Instructions

Description	Op Code	Source Statement	Object Format	Operation	Micro-cycles	Page
			7 6 5 4 3 2 1 0 | 7 6 5 4 3 2 1 0 (ptr) (disp)			
Increment and Load	ILD	disp	1 0 1 0 1 0 ptr | disp	$(AC), (EA) \leftarrow (EA) + 1$	22	5-8
Decrement and Load	DLD	disp(ptr)	1 0 1 1 0 0	$(AC), (EA) \leftarrow (EA) - 1$	22	5-9

Immediate Instructions

Description	Op Code	Source Statement	Object Format	Operation	Micro-cycles	Page
			7 6 5 4 3 2 1 0 | 7 6 5 4 3 2 1 0 (data)			
Load Immediate	LDI	data	1 1 0 0 0 1 0 0 | data	$(AC) \leftarrow data$	10	5-9
AND Immediate	ANI		1 1 0 1 0 1 0 0	$(AC) \leftarrow (AC) \wedge data$	10	5-10
OR Immediate	ORI		1 1 0 1 1 1 0 0	$(AC) \leftarrow (AC) \vee data$	10	5-10
Exclusive OR Immediate	XRI		1 1 1 0 0 1 0 0	$(AC) \leftarrow (AC) \triangledown data$	10	5-10
Decimal Add Immediate	DAI		1 1 1 0 1 1 0 0	$(AC) \leftarrow (AC)_{10} + data_{10} + (CY/L); (CY/L)$	15	5-11
Add Immediate	ADI		1 1 1 1 0 1 0 0	$(AC) \leftarrow (AC) + data + (CY/L); (CY/L), (OV)$	11	5-11
Complement and Add Immediate	CAI		1 1 1 1 1 1 0 0	$(AC) \leftarrow (AC) + \sim data + (CY/L); (CY/L), (OV)$	12	5-11

Transfer Instructions

Description	Op Code	Source Statement	Object Format	Operation	Micro-cycles	Page
			7 6 5 4 3 2 1 0 | 7 6 5 4 3 2 1 0 (ptr) (disp)			
Jump	JMP	disp	1 0 0 1 0 0 ptr | disp	$(PC) \leftarrow EA$	11	5-12
Jump If Positive	JP	disp(ptr)	1 0 0 1 0 1	If $(AC) \geq 0$, $(PC) \leftarrow EA$	9,11	5-13
Jump If Zero	JZ		1 0 0 1 1 0	If $(AC) = 0$, $(PC) \leftarrow EA$	9,11	5-13
Jump If Not Zero	JNZ		1 0 0 1 1 1	If $(AC) \neq 0$, $(PC) \leftarrow EA$	9,11	5-13

Double-byte Miscellaneous Instructions

Description	Op Code	Source Statement	Object Format	Operation	Micro-cycles	Page
			7 6 5 4 3 2 1 0 | 7 6 5 4 3 2 1 0 (disp)			
Delay	DLY	disp	1 0 0 0 1 1 1 1 | disp	count AC to -1, delay = $13 + 2 (AC) + 2$ disp $+ 2^9$ disp microcycles	13 to 131593	5-22

Description	Op Code	Source Statement	Object Format	Operation	Micro-cycles	Page
Extension Register Instructions			7 6 5 4 3 2 1 0			
Load AC from Extension	40	LDE	0 1 0 0 0 0 0 0	$(AC) \leftarrow (E)$	6	5-14
Exchange AC and Extension	01	XAE	0 0 0 0 0 0 0 1	$(AC) \leftrightarrow (E)$	7	5-14
AND Extension	50	ANE	0 1 0 1 0 0 0 0	$(AC) \leftarrow (AC) \wedge (E)$	6	5-14
OR Extension	58	ORE	0 1 0 1 1 0 0 0	$(AC) \leftarrow (AC) \vee (E)$	6	5-14
Exclusive-OR Extension	60	XRE	0 1 1 0 0 0 0 0	$(AC) \leftarrow (AC) \triangledown (E)$	6	5-15
Decimal Add Extension	68	DAE	0 1 1 0 1 0 0 0	$(AC) \leftarrow (AC)_{10} + (E)_{10} + (CY/L); (CY/L)$	11	5-15
Add Extension	70	ADE	0 1 1 1 0 0 0 0	$(AC) \leftarrow (AC) + (E) + (CY/L); (CY/L), (OV)$	7	5-15
Complement and Add Extension	78	CAE	0 1 1 1 1 0 0 0	$(AC) \leftarrow (AC) + \sim (E) + (CY/L); (CY/L), (OV)$	8	5-16
Pointer Register Move Instructions			7 6 5 4 3 2 1 0			
Exchange Pointer Low	30	XPAL ⎱	0 0 1 1 0 0 0 0 ptr	$(AC) \leftrightarrow (PTR_{7:0})$	8	5-16
Exchange Pointer High	34	XPAH ⎰ ptr	0 0 1 1 0 1 0 0	$(AC) \leftrightarrow (PTR_{15:8})$	8	5-17
Exchange Pointer with PC	3C	XPPC ⎰	0 0 1 1 1 1 0 0	$(PC) \leftrightarrow (PTR)$	7	5-17
Shift, Rotate, Serial I/O Instructions			7 6 5 4 3 2 1 0			
Serial Input/Output	19	SIO	0 0 0 1 1 0 0 1	$(E_i) \leftarrow (E_{i-1}), \ SIN \rightarrow (E_7), \ (E_0) \rightarrow SOUT$	5	5-17
Shift Right	1C	SR	0 0 0 1 1 1 0 0	$(AC_i) \leftarrow (AC_{i-1}), \ 0 \rightarrow (AC_7)$	5	5-18
Shift Right with Link	1D	SRL	0 0 0 1 1 1 0 1	$(AC_i) \leftarrow (AC_{i-1}), \ (CY/L) \rightarrow (AC_7)$	5	5-18
Rotate Right	1E	RR	0 0 0 1 1 1 1 0	$(AC_i) \leftarrow (AC_{i-1}), \ (AC_0) \rightarrow (AC_7)$	5	5-18
Rotate Right with Link	1F	RRL	0 0 0 1 1 1 1 1	$(AC_i) \leftarrow (AC_{i-1}), \ (AC_0) \rightarrow (CY/L) \rightarrow (AC_7)$	5	5-19
Single-byte Miscellaneous Instructions			7 6 5 4 3 2 1 0			
Halt	00	HALT	0 0 0 0 0 0 0 0	Pulse H-flag	8	5-19
Clear Carry/Link	02	CCL	0 0 0 0 0 0 1 0	$(CY/L) \leftarrow 0$	5	5-20
Set Carry/Link	03	SCL	0 0 0 0 0 0 1 1	$(CY/L) \leftarrow 1$	5	5-20
Disable Interrupt	04	DINT	0 0 0 0 0 1 0 0	$(IE) \leftarrow 0$	6	5-21
Enable Interrupt	05	IEN	0 0 0 0 0 1 0 1	$(IE) \leftarrow 1$	6	5-20
Copy Status to AC	06	CSA	0 0 0 0 0 1 1 0	$(AC) \leftarrow (SR)$	5	5-21
Copy AC to Status	07	CAS	0 0 0 0 0 1 1 1	$(SR) \leftarrow (AC)$	6	5-21
No Operation	08	NOP	0 0 0 0 1 0 0 0	$(PC) \leftarrow (PC) + 1$	5	5-22

Appendix 3. SC/MP instruction set. (Courtesy National Semiconductor Corp.)

ABA	Add Accumulators	CLR	Clear	PUL	Pull Data
ADC	Add with Carry	CLV	Clear Overflow	ROL	Rotate Left
ADD	Add	CMP	Compare	ROR	Rotate Right
AND	Logical And	COM	Complement	RTI	Return from Interrupt
ASL	Arithmetic Shift Left	CPX	Compare Index Register	RTS	Return from Subroutine
ASR	Arithmetic Shift Right	DAA	Decimal Adjust	SBA	Subtract Accumulators
BCC	Branch if Carry Clear	DEC	Decrement	SBC	Subtract with Carry
BCS	Branch if Carry Set	DES	Decrement Stack Pointer	SEC	Set Carry
BEQ	Branch if Equal to Zero	DEX	Decrement Index Register	SEI	Set Interrupt Mask
BGE	Branch if Greater or Equal Zero	EOR	Exclusive OR	SEV	Set Overflow
BGT	Branch if Greater than Zero	INC	Increment	STA	Store Accumulator
BHI	Branch if Higher	INS	Increment Stack Pointer	STS	Store Stack Register
BIT	Bit Test	INX	Increment Index Register	STX	Store Index Register
BLE	Branch if Less or Equal	JMP	Jump	SUB	Subtract
BLS	Branch if Lower or Same	JSR	Jump to Subroutine	SWI	Software Interrupt
BLT	Branch if Less than Zero	LDA	Load Accumulator	TAB	Transfer Accumulators
BMI	Branch if Minus	LDS	Load Stack Pointer	TAP	Transfer Accumulators to Condition Code Reg
BNE	Branch if Not Equal to Zero	LDX	Load Index Register	TBA	Transfer Accumulators
BPL	Branch if Plus	LSR	Logical Shift Right	TPA	Transfer Condition Code Reg. to Accumulator
BRA	Branch Always	NEG	Negate	TST	Test
BSR	Branch to Subroutine	NOP	No Operation	TSX	Transfer Stack Pointer to Index Register
BVC	Branch if Overflow Clear			TXS	Transfer Index Register to Stack Pointer
BVS	Branch if Overflow Set				
CBA	Compare Accumulators	ORA	Inclusive OR Accumulator	WAI	Wait for Interrupt
CLC	Clear Carry	PSH	Push Data		
CLI	Clear Interrupt Mask				

Appendix 4. M6800 instruction set. (Courtesy Motorola Semiconductor Products, Inc.)

Answers to Odd-Numbered Review Questions

CHAPTER 2

1. $(2 \times 5^2) + (0 \times 5) + (3 \times 5^0) = 50 + 0 + 3 = 53$

3. $(4 \times 5^4) + (4 \times 5^3) + (3 \times 5^2) + (0 \times 5) + (2 \times 5^0)$
   $= 2500 + 500 + 75 + 2 = 3077$

5. $(2 \times 8^2) + (7 \times 8) + (2 \times 8^0) = 128 + 56 + 2 = 186$

7. $(7 \times 8^2) + (1 \times 8) + (7 \times 8^0) = 448 + 8 + 7 = 463$

9. $(1 \times 8^3) + (0 \times 8^2) + (7 \times 8) + (0 \times 8^0) = 512 + 0 + 56 = 568$

11. $(13 \times 16^3) + (10 \times 16^2) + (3 \times 16^1) + (7 \times 16^0)$
    $= 53248 + 2560 + 48 + 7 = 55863$

13. $(9 \times 16^4) + (10 \times 16^3) + (8 \times 16^2) + (11 \times 16)$
    $+ (0 \times 16^0) = 589824 + 40960 + 2048 + 176 + 0$
    $= 633,008$

15. $33/5 \ = 6 + 3$
    $6/5 \ = 1 + 1$
    $1/5 \ = 0 + 1 \qquad\qquad 113_5 \qquad\qquad 33_{10} = 113_5$

$$33/8 \ = 4 + 1$$
$$4/8 \ \ = 0 + 4 \qquad\qquad 41_8 \qquad\qquad 33_{10} = 41_8$$

$$33/16 = 2 + 1$$
$$2/16 \ = 0 + 2 \qquad\qquad 21_{16} \qquad\qquad 33_{10} = 21_{16}$$

17.
$$276/5 \ = 55 + 1$$
$$55/5 \ \ = 11 + 0$$
$$11/5 \ \ = 2 + 1$$
$$2/5 \ \ \ = 0 + 2 \qquad\qquad 2101_5 \qquad\qquad 276_{10} = 2101_5$$

$$276/8 \ = 34 + 4$$
$$34/8 \ \ = 4 + 2$$
$$4/8 \ \ \ = 0 + 4 \qquad\qquad 424_8 \qquad\qquad 276_{10} = 424_8$$

$$276/16 = 17 + 4$$
$$17/16 \ = 1 + 1$$
$$1/16 \ \ = 0 + 1 \qquad\qquad 114_{16} \qquad\qquad 276_{10} = 114_{16}$$

19.
$$999/5 \ = 199 + 4$$
$$199/5 \ = 39 + 4$$
$$39/5 \ \ = 7 + 4$$
$$7/5 \ \ \ = 1 + 2$$
$$1/5 \ \ \ = 0 + 1 \qquad\qquad 12444_5 \qquad\qquad 999_{10} = 12444_5$$

$$999/8 \ = 124 + 7$$
$$124/8 \ = 15 + 4$$
$$15/8 \ \ = 1 + 7$$
$$1/8 \ \ \ = 0 + 1 \qquad\qquad 1747_8 \qquad\qquad 999_{10} = 1747_8$$

$$999/16 = 62 + 7$$
$$62/16 \ = 3 + 14$$
$$3/16 \ \ = 0 + 3 \qquad\qquad 3E7_{16} \qquad\qquad 999_{10} = 3E7_{16}$$

21.
$$4701/5 \ = 940 + 1$$
$$940/5 \ \ = 188 + 0$$
$$188/5 \ \ = 37 + 3$$
$$37/5 \ \ \ = 7 + 2$$
$$7/5 \ \ \ \ = 1 + 2$$
$$1/5 \ \ \ \ = 0 + 1 \qquad\qquad 122301_5 \qquad\qquad 4701_{10} = 122301_5$$

$$4701/8 \ = 587 + 5$$
$$587/8 \ = 73 + 3$$
$$73/8 \quad = 9 + 1$$
$$9/8 \quad = 1 + 1$$
$$1/8 \quad = 0 + 1 \qquad\qquad 11135_8 \qquad\qquad 4701_{10} = 11135_8$$

$$4701/16 = 293 + 13$$
$$293/16 \ = 18 + 5$$
$$18/16 \ = 1 + 2$$
$$1/16 \quad = 0 + 1 \qquad\qquad 125D_{16} \qquad\qquad 4701_{10} = 125D_{16}$$

23. $$10010/5 \ = 2002 + 0$$
$$2002/5 \ = 400 + 2$$
$$400/5 \quad = 80 + 0$$
$$80/5 \quad = 16 + 0$$
$$16/5 \quad = 3 + 1$$
$$3/5 \quad = 0 + 3 \qquad\qquad 310020_5 \qquad\qquad 10010_{10} = 310020_5$$

$$10010/8 \ = 1251 + 2$$
$$1251/8 \ = 156 + 3$$
$$156/8 \quad = 19 + 4$$
$$19/8 \quad = 2 + 3$$
$$2/8 \quad = 0 + 2 \qquad\qquad 23432_8 \qquad\qquad 10010_{10} = 23432_8$$

$$10010/16 = 625 + 10$$
$$625/16 \quad = 39 + 1$$
$$39/16 \quad = 2 + 7$$
$$2/16 \quad = 0 + 2 \qquad\qquad 271A_{16} \qquad\qquad 10010_{10} = 271A_{16}$$

25. $$6075/5 \ = 1215 + 0$$
$$1215/5 \ = 243 + 0$$
$$243/5 \quad = 48 + 3$$
$$48/5 \quad = 9 + 3$$
$$9/5 \quad = 1 + 4$$
$$1/5 \quad = 0 + 1 \qquad\qquad 143300_5 \qquad\qquad 6075_{10} = 143300_5$$

$$6075/8 \ = 759 + 3$$
$$759/8 \ = 94 + 7$$
$$94/8 \quad = 11 + 6$$
$$11/8 \quad = 1 + 3$$
$$1/8 \quad = 0 + 1 \qquad\qquad 13673_8 \qquad\qquad 6075_{10} = 13673_8$$

$6075/16 = 379 + 11$
$379/16 = 23 + 11$
$23/16 = 1 + 7$
$1/16 = 0 + 1$ $17BB_{16}$ $6075_{10} = 17BB_{16}$

27. (a) $111_2 = 7$
    (b) $1001_2 = 9$
    (c) $110011_2 = 51$
    (d) $76_{10} = 1001100_2$
    (e) $273_{10} = 100010001_2$
    (f) $7620_{10} = 1110111000100_2$

29. a. $37_8 = 11111_2$
    b. $634_8 = 110011100_2$
    c. $6732_8 = 110111011010_2$
    d. $A6_{16} = 10100110_2$
    e. $7BC_{16} = 11110111100_2$
    f. $A34D_{16} = 1010001101001101_2$
    g. $101011100101_2 = 5345_8$
    h. $10011101101001_2 = 23551_8$
    i. $110111000010_2 = DC2_{16}$
    j. $100111011001001_2 = 4EC9_{16}$

31. (a) 101 .   010   111   011   100   100
       5 .   2   7   3   4 4_8

      0101 .   0101   1101   1100   1000
        5 .   5   D   C 8_{16}

    (b) 011   001   011 .   010   011   001   101   000
       3   1   3 .   2   3   1   5 0_8

      1100   1011 .   0100   1100   1101
       C   B .   4   C D_{16}

    (c) 011   111   110.   000   000   111   001   100
       3   7   6 .   0   0   7   1 4_8

      1111   1110.   0000   0011   1001   1000
       F   E .   0   3   9 8_{16}

    (d)   7   6   3   7 .   4   7 3_{10}
        0111   0110   0011   0111.   0100   0111 0011_{BCD}

(e)   1   7   6   3   9 .   9   9   0   7 3_{10}
     0001 0111 0110 0011 1001. 1001 1001 0000 0111 0011_{BCD}

(f)   D   A   F   E   1   2.   7   C   A   6 3_{16}
     1101 1010 1111 1110 0001 0010. 0111 1100 1010 0110 0011_2

33.   1st Register   2nd Register
     0000001100101110   0101000000000000

35. Sixteen combinations are possible. However 6 are invalid since there are only 10 decimal digits (0–9). The invalid characters are: 1010 (10), 1011 (11), 1100 (12), 1101 (13), 1110 (14) and 1111 (15). In decimal, two digits are necessary to represent these numbers. Ten in BCD is written 0001 0000 instead of 1010, which is invalid.

37. (a) G = C7 = 1100 0111
     (b) Period = 4B = 0100 1011
     (c) $ = 5B = 0101 1011
     (d) 0 = F0 = 1111 0000
     (e) 6% = F6 6C = 1111 0110 0110 1100
     (f) #8 = 7B F8 = 0111 1011 1111 1000

39. Including the parity (c) bit:
     F   1   N   D   b   1   3
     1110110 1111001 0100101 0110100 0010000 0000001 1000011

41. The control characters are identified by the zone bits 000 and 001. These characters are defined in Fig. 2-9.

CHAPTER 3

1. (a) 1

```
        7    6    3
        2    1    4
      ─────────────
       (9)   7    7
        8
      ─────────────
    1   1    7    7
```

$1177_8 = (1 \times 8^3) + (1 \times 8^2) + (7 \times 8) + (7 \times 8^0)$
     $= 639_{10}$

(b)

1	1	1
2	7	7
6	4	1
(9)	(12)	(8)
8	8	8
1	1	4	0

$1140_8 = (1 \times 8^3) + (1 \times 8^2) + (4 \times 8) + (0 \times 8^0)$
$= 608_{10}$

(c)

1	1	1
7	6	3
3	6	7
(11)	(13)	(10)
8	8	8
1	3	5	2

$1352_8 = (1 \times 8^3) + (3 \times 8^2) + (5 \times 8) + (2 \times 8^0)$
$= 746_{10}$

3. (a) 1

4	A	3
D	1	3
(17)	(11)	6
16		
1 1	B	6_{16}

$11B6_{16} = (1 \times 16^3) + (1 \times 16^2) + (11 \times 16) + (6 \times 16^0)$
$= 4534_{16}$

(b) 1   1

A	B	C	D
E	F	0	1
(25)	(26)	(12)	(14)
16	16		
1 9	A	C	E

$19ACE_{16} = (1 \times 16^4) + (9 \times 16^3) + (10 \times 16^2)$
$+ (12 \times 16) + (14 \times 16^0) = 105166_{10}$

(c)

		1	
1	0	9	A
2	D	1	C

3 (13) (11) (22)

                    16

3   D   B   6

$3DB6_{16} = (3 \times 16^3) + (13 \times 16^2) + (11 \times 16) + (6 \times 16^0) = 15798_{16}$

5. (a) 15_8 (b) 77_8 (c) 337_8

7. (a) $7AC_{16}$ (b) $6A1A_{16}$ (c) $CE9F_{16}$

9. (a) 1

	1	0	1
	1	1	0
1	0	1	1

$1011_2 = (1 \times 2^3) + (0 \times 2^2) + (1 \times 2) + (1 \times 2^0)$
$= 11_{10}$

(b) 1

	1	0	1	1
	1	1	0	1
1	1	0	0	0

$11000_2 = (1 \times 2^4) + (1 \times 2^3) + (0 \times 2^2) + (0 \times 2^1)$
$+ (0 \times 2^0) = 24_{10}$

(c) 1

				1	1	
	1	1	0	0	1	1
	1	0	0	0	1	1
1	0	1	0	1	1	0

$1010110_2 = 86_{10}$

11. (a)

	1	
1010	1010	(true form)
0101	1010	(one's complement)
	10100	
	⎿→1	
	0101	(ans.)

(b) 001011   (c) 0001101

13. (a)   1
        1010   1010   (true form)
        0101   1011   (two's complement)
                    ⅄ 0101   (ans.)

   (b)   1
        110010   110010   (true)
        100111   011001   (two's complement)
                    ⅄ 001011

   (c)   1
        1100110   1100110   (true)
        1011001   0100111   (two's complement)
                    ⅄ 0001101

15. (a)   1
              547   547   (true form)
        (—) 279   721   (ten's complement)
                    1 268
                  — 1 000
                      268   (ans.)

   (b)   1
             7613   7613   (true form)
        (—) 2724   7276   (ten's complement)
                    1 4889
                  — 1 0000
                      4889   (ans.)

17. For the following solutions, the two's complement method is used
    for negative numbers and for subtraction. Eight bit positions are
    assumed.

   (a)   —9   11110111   (two's complement of 9)
        (+) +6   (+) 00000110
                —3   ┌─11111101   (two's complement)
                          ↓ 00000011   (ans.)

   (b)   13   00001101   00001101
        (—) —5   (—) 11111011   (+) 00000101
                18   00010010   (ans.)

   (c)   9   00001001
        (+) —4   (+) 11111100   (two's complement of 4)
                 5   00000101   (ans.)

(d) -13 11110011   11110011
    $(-)$ $\underline{+8}$ $(-)$ $\underline{00001000}$ $(+)$ $\underline{11111000}$
        $\overline{-21}$ $\overline{}$11101011   (two's
                                        complement)
                            $\underline{}$ 00010101   (ans.)

(e)   23   00010111
        $\underline{14}$ $\underline{00001110}$
        $\overline{37}$ $\overline{00100101}$

(f) -6 11111010   11111010   (two's
    $(-)$ $\underline{-9}$ $(-)$ $\underline{11110111}$ $(+)$ $\underline{00001001}$ complement
        $\overline{+3}$ of 6)
                            00000011   (true form
                                        of 9)
                                        (ans.)

(g)   15   00001111   (true form)
    $(+)$ $\underline{-12}$ $(+)$ $\underline{11110100}$ (two's complement)
        $\overline{3}$ $\overline{00000011}$ (ans.)

(h) -7 11111001   (two's complement)
    $(+)$ $\underline{-4}$ $(+)$ $\underline{11111100}$ (two's complement)
        $\overline{-11}$ $\overline{}$11110101   (two's complement)
                $\underline{}$ 00001011   (ans.)

19. (a) 361_8 0361
        $(-)$ $\underline{254_8}$ $\underline{7524}$ (eight's complement)
                        $\overline{\text{X } 0105}$ (difference in true form)
        carry is
        dropped——

(b) 3273_8 3273
        $(-)$ $\underline{2426_8}$ $\underline{5352}$ (eight's complement)
                        $\overline{\text{X } 0645}$ (difference in true form)

(c) 3106_8 3106
        $(-)$ $\underline{1727_8}$ $\underline{6051}$ (eight's complement)
                        $\overline{\text{X } 1157}$ (difference in true form)

(d) 2267_8 2267
        $(-)$ $\underline{4365_8}$ $\underline{3413}$ (eight's complement)
                        $\overline{5702}$ (invalid answer)

Because the octal numbers represent binary numbers and the MSB is the sign bit, the 4 in the octal number 4365 tells us that the number is negative and in the eight's complement form. The operation is subtract. Therefore the subtrahend is complemented and added. The answer, 5702, apparently indicates a negative number because the MSD is 4 or greater. However, the answer cannot be negative. Therefore an overflow must have occurred. If you think in binary, there was a carry into the MSB position with no carry out. This condition always indicates an overflow (even though mathematically correct). This answer, then, is invalid if it is to be stored in 16-bit positions because the computer looks at the MSB as a *sign bit.*

(e) 1776_8 1776_8

(−) 7177_8 0601_8 (eight's complement)

$\overline{2577}$ (diff. in true form)

The subtrahend is negative (MSD is 4 or greater). The operation is subtract. Therefore, the subtrahend is complemented and added as before. The answer, however, is positive because the MSD is less than 4.

(f) 4063_8 4063

(−) 7274_8 0504   (eight's complement)

$\overline{4567}$ (difference in eight's complement form)

The answer 4567 apparently indicates a negative number because the MSD is 4 or greater. If we think in binary, the sign bit is 1. Neither rule for overflow exists. Therefore the answer is valid.

21. (a)   4C   004C

(−) 2D   FFD3   (sixteen's complement)

↘ $\overline{001F}$ (difference in true form)

(b)   A27   0A27

(−) 8AB   F755   (sixteen's complement)

↘ $\overline{017C}$ (difference in true form)

(c)   CD27   CD27

(−) 7A4D   85B3   (sixteen's complement)

↘ $\overline{53EA}$ (difference in true form)

```
                    7B4A
(d)      7B4A       4B59       (sixteen's complement)
      (−) B4A7      ‾C6A3      (invalid answer)
```

Because the hexadecimal numbers represent binary numbers, and the MSB is the sign bit, the B in the hexadecimal number B4A7 tells us that the number is negative and in the eight's complement form. The operation is subtract. Therefore the subtrahend is complemented and added. The answer C6A3 apparently indicates a negative number because the MSD is 8 or greater (the sign bit is 1). Thinking in binary, there was a carry into the MSB position with no carry out. This is an overflow condition and the answer is invalid.

```
(e)      927B₁₆      927B
      (−) C41F₁₆     3BE1       (complement subtrahend)
                     ‾CE5C      (difference in sixteen's comple-
                                ment form)
```

The answer CE5C apparently indicates a negative number because the MSD is 8 or greater. If we think in binary, the sign bit is 1. Neither rule for overflow exists. Therefore the answer is valid.

```
(f)      12A6₁₆      12A6
      (−) A4BC₁₆     5B44       (complement subtrahend)
                     ‾6DEA      (difference in true form)
```

The subtrahend is negative (MSD is 8 or greater). The operation is subtract. Therefore the subtrahend is complemented and added as before. The answer, however, is positive because the MSD is less than 8.

```
                    1     1                    (carry)
23. (a)      463    0000  0100  0110  0011
          (+) 642   0000  0110  0100  0010
             ‾1105  0000  1010  1010  0101     (binary sum)
                    0000  0110  0110  0000     (add 6 to invalid sums)
                    0001  0001  0000  0101     (valid BCD sum)
```

		1				
				1		(carry)

(b)   207   0000   0010   0000   0111
(+) 463   0000   0100   0110   0011
───   670   0000   0110   0111   1010   (binary sum)
                      0000   0000   0000   0110   (add 6 to invalid sums)
         ──────────────────────────────────
                      0000   0110   0111   0000   (valid BCD sum)

1   1   (carry)

(c)   843   0000   1000   0100   0011
(−) 656   1001   0011   0100   0011   (nine's complement)
───   187   1010   1011   1000   0110   (binary sum)
                      0110   0110   0000   0000   (add 6 to invalid sums)
         ──────────────────────────────────
             1 0000   0001   1000   0110

└─────────────────────────────►1   (end around carry)
         ──────────────────────────────────
                      0000   0001   1000   0111   (valid BCD sum)

1   1   (carry)

(d)   906   0000   1001   0000   0110
(−) 627   1001   0011   0111   0010   (nine's complement)
───   279   1010   1100   0111   1000   (binary sum)
                      0110   0110   0000   0000
         ──────────────────────────────────
             1 0000   0010   0111   1000

└─────────────────────────────►1   (end around carry)
         ──────────────────────────────────
                      0000   0010   0111   1001   (valid BCD sum)

(e)   1   (carry)

         206   0000   0010   0000   0110
(−) 673   1001   0011   0010   0110   (nine's complement)
─── 467   1001   0101   0011   1100   (binary sum)
                      0000   0000   0000   0110   (add 6 to invalid sums)
         ──────────────────────────────────
             0 1001   0101   0011   0010   (valid BCD sum)

└─────────────────────────────►0   (zero carry indicates the
                                                          sum is negative and in
                                                          (nine's complement
                                                          form)
         ──────────────────────────────────
         − 0000   0100   0110   0111   (nine's complement of
                                                          BCD sum yields valid
                                                          negative diff.)

(f)   463   0000   0100   0110   0011

   (−) 674   1001   0011   0010   0101   (nine's complement)

    − 211   1001   0111   1000   1000

            0000   0000   0000   0000   (no invalid sums)

      0   1001   0111   1000   1000   (valid BCD sum)

                                  → 0   (zero carry indicates the sum is negative and in nine's complement form)

   − 0000   0010   0001   0001   (nine's complement of BCD sum yields valid negative difference)

CHAPTER 4

1. AND, OR, NOT

3. Two or more

5.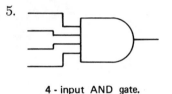

4 - input AND gate.

7.

B	C	D	$\overline{D}$	$\overline{BCD}$
0	0	0	1	0
0	0	1	0	0
0	1	0	1	0
0	1	1	0	0
1	0	0	1	0
1	0	1	0	0
1	1	0	1	1
1	1	1	0	0

9.

A	B	C	ABC	$\overline{ABC}$
0	0	0	0	1
0	0	1	0	1
0	1	0	0	1
0	1	1	0	1
1	0	0	0	1
1	0	1	0	1
1	1	0	0	1
1	1	1	1	0

11.

A	B	Ā	ĀB	(A + ĀB)	(A + B)	(A + ĀB) + (A + B)
0	0	1	0	0	0	0
0	1	1	1	1	1	1
1	0	0	0	1	1	1
1	1	0	0	1	1	1

Equal

13.

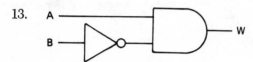

15.

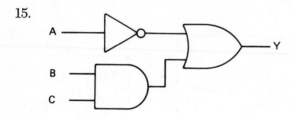

17.

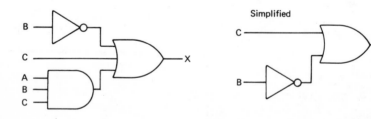

Simplified

19. (a) $\overline{A + B}$ (b) $(A + B) + C = A + B + C$ (c) $AB + \overline{C}$

21. Commutative property—This is a mathematical or logical property which shows that the value of a circuit or mathematical quantity is not affected by the order in which the variables appear.

Distributive property—This is a mathematical property which demonstrates that expressions may be simplified by factoring or multiplying.

DeMorgan's Theorem—DeMorgan's Theorem is used to simplify

Boolean expressions. DeMorgan's Theorem states that the AND operator may be changed to OR or vice-versa if each variable in the expression is complemented and the entire expression is also complemented.

23. 1. $A + AB$
    2. A (15b)

25. 1. $\overline{\overline{AB}}$
    2. $\overline{\overline{A}} + \overline{\overline{B}}$ (14b)
    3. $A + B$ (13a)

27. 1. $A + B + AB + BC$
    2. $A + AB + B + BC$ (6b)
    3. $(A + AB) + (B + BC)$ (7b)
    4. $(A) + (B)$ (15b)
    5. $A + B$ (7b)

29. 1. $XYZ(\overline{A}BD + ABD) + X(\overline{A}X + AB + A\overline{B})$
    2. $XYZ[BD(\overline{A} + A)] + X[\overline{A}X + (AB + A\overline{B})]$ (8a, 7b)
    3. $XYZ[BD(1)] + X[\overline{A}X + A(B + \overline{B})]$ (12b, 8a)
    4. $XYZBD + X(\overline{A}X + A)$ (10a, 12b, 10a)
    5. $XYZBD + X(A + X)$ (16b)
    6. $XYZBD + X$ (15b)
    7. X (15b)

31. 1. $\overline{Y(AB\overline{C} + \overline{A}BC + ABC)}$
    2. $\overline{Y} + (AB\overline{C} + \overline{A}BC + ABC)$ (14b, 13a)
    3. $Y + [AB\overline{C} + C(\overline{A}B + AB)]$ (8a)
    4. $\overline{Y} + AB\overline{C} + C$ (12b, 10a)
    5. $\overline{Y} + AB + C$ (16b)

33.

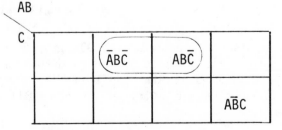

Map $= A\overline{B}C + B\overline{C}$

35.

AB

CD

$\overline{ABCD}$			$\overline{A}B\overline{C}\overline{D}$
$\overline{ABC}D$			$A\overline{B}C\overline{D}$
			$A\overline{B}C\overline{D}$
			$A\overline{B}C\overline{D}$

$$\text{Map} = \overline{A}B\overline{C} + A\overline{B} = \overline{B}(\overline{A}\overline{C} + A)$$

CHAPTER 5

1. For current sourcing, the driving gate must provide the current to activate the input of the gate being driven. For current sinking, the driving gate must provide a sink for the driven gate. In current mode switching, the direction of the current is switched.

3. When V_{in} is greater than V_{BB} (HIGH level), Q_1 is turned on and Q_2 is turned off, allowing the output to go HIGH.

5. RTL gates are current sourcing gates.

7. RTL gates are relatively slow, require high power, and have a low fan-out factor. Their main disadvantage, however, is poor noise immunity.

   DTL gates are somewhat faster than RTL, but possess a higher immunity to noise. Power requirements are less and the fan-out is about 8 compared to 5 for RTL.

   TTL gates are faster than either RTL or DTL gates. Power dissipation is similar to that in DTL gates. Although the noise immunity of TTL gates is not as good as that of DTL gates, it is much better than for RTL gates. Fan-out is higher than in either RTL or DTL gates, about 10.

9. TTL gates operate in the current sinking mode.

11. If A is HIGH, the output will be the complement of B.

13. Current mode switching is used in ECL gates.

15.

TRUTH TABLE

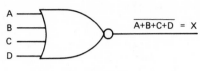

$\overline{A+B+C+D} = X$

NOR Gate

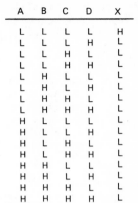

A	B	C	D	X
L	L	L	L	H
L	L	L	H	L
L	L	H	L	L
L	L	H	H	L
L	H	L	L	L
L	H	L	H	L
L	H	H	L	L
L	H	H	H	L
H	L	L	L	L
H	L	L	H	L
H	L	H	L	L
H	L	H	H	L
H	H	L	L	L
H	H	L	H	L
H	H	H	L	L
H	H	H	H	L

$\overline{A}\ \overline{B}\ \overline{C}\ \overline{D} = X$

DeMorgan's Equivalent

17. (a) Q_1 never saturates.

    (b) Q_2 saturates to produce a LOW output.

19. ECL. Because there are no saturating transistors.

21., 23., 25.

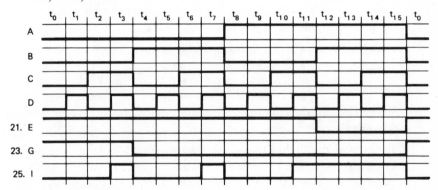

CHAPTER 6

1. Some advantages of MOS over bipolar integrated circuits are that they: (1) are easier to process; (2) require less power; (3) are smaller; (4) have greater number of elements per chip thereby producing LSI circuits.

3. V_{DD}—Drain supply.

5. V_{GG}—Gate supply.

7. (a) 5 V (HIGH): 0 V (LOW)

    (b) -5 V (HIGH): 0 V (LOW)
    (c) -5 V (LOW): -10 V (HIGH)
    (d) -10 V (HIGH): 5 V (LOW)

9.

A	B	$\bar{A}$	$\bar{B}$	AB	$\overline{AB}$	$\overline{A+B}$
0	0	1	1	0	1	1
0	1	1	0	0	1	1
1	0	0	1	0	1	1
1	1	0	0	1	0	0

11. A ratio inverter is an inverter circuit whose output voltage is determined by the ratio of two resistances. For the MOS gate these are the resistances of two transistors. One of the two transistors may be turned off or on by the input voltage. This causes the output voltage to equal V_{CC} times the ratio of the two resistances. Power is dissipated when both transistors are on.

13. The ratioless-powerless inverter dissipates no DC power at all and no DC power supply is required. The output voltage is maintained by pulses originating from the clock pulses.

15. For NMOS or PMOS transistors the threshold voltage is rather high (approximately 3.5 to 4.5 volts) necessitating rather high supply voltages. Bipolar devices have much lower threshold voltages—on the order of 0.5 volts. The effect of the silicon gate process is to lower the threshold voltage of MOS devices to the range 0.5 to 2.5 volts.

Because of the higher threshold voltages for MOS gates, the noise immunity is typically higher than for bipolar transistors. CMOS gates have a noise immunity approaching 50% of the supply voltage.

17.   19.   21.

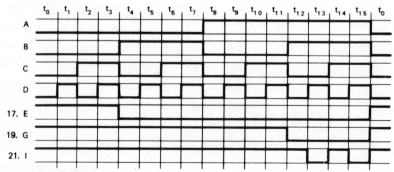

CHAPTER 7

1. (a) Bistable means having two stable states. As related to flip-flops, the two states are SET and RESET.
(b) Astable means having no stable state. An astable flip-flop is a free-running oscillator.
(c) Monostable means having one stable state. A one-shot or single-shot is a monostable multivibrator.

3.

5.

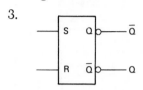

7.

$\overline{S}$	$\overline{R}$	Q	$\overline{Q}$	
L	L	H	H	INVALID
L	H	H	L	SET
H	L	L	H	RESET
H	H	NC	NC	REM.

9. "Race" refers to the possibility of the flip-flop's state changing when input variables change. For example, if the input variables are affected by the state of the flip-flop, it may not be possible to attain the desired state.

11. Once the flip-flop has been triggered, any changes in the input variables will not affect the state of the flip-flop until it has been cocked again.

13. The primary disadvantage of the master-slave RST flip-flop is that it has an invalid input condition.

15.

*To effect the J and K inputs, the flip-flop must be clocked.

J*	K*	S_D	R_D	Q	$\overline{Q}$	
L	L	L	L	NC	NC	Remember*
L	L	L	H	L	H	Direct Reset
L	L	H	L	H	L	Direct Set
L	L	H	H	?	?	INVALID
L	H	L	L	L	H	Clocked Reset*
L	H	L	H	L	H	Direct Reset
L	H	H	L	H	L	Direct Set
L	H	H	H	?	?	INVALID
H	L	L	L	H	L	Clocked Set*
H	L	L	H	L	H	Direct Reset
H	L	H	L	H	L	Direct Set
H	L	H	H	?	?	INVALID
H	H	L	L	Q	Q	Complement*
H	H	L	H	L	H	Direct Reset
H	L	H	L	H	L	Direct Set
H	L	H	H	?	?	INVALID

17. The level indicator on the T input indicates the direction of the transient causing the flip-flop to trigger. For example, if a low level indicator is present, the flip-flop will be triggered when the clock pulse changes from HIGH to LOW.

19. The astable flip-flop is used as a source for clock pulses.

21.,   23.

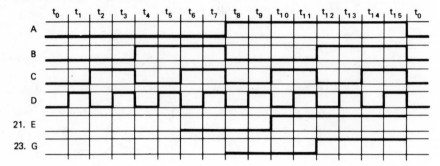

21. E

23. G

CHAPTER 8

1. The basic logic functions AND, OR, and NOT are all inherent in NAND and NOR gates. This is because each gate has a DeMorgan's equivalent and, in addition, each gate inverts.

3. The use of specific circuits to put into effect the desired logic function(s) is called implementation.

5. The low level indicator tells the technician that a low level voltage is required to activate the function described by the symbol. If the low level indicator appears at the output, it tells him that a low level voltage will be present when that function has been activated.

7. (a)

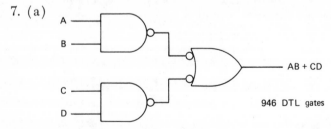

AB + CD

946 DTL gates

(b)

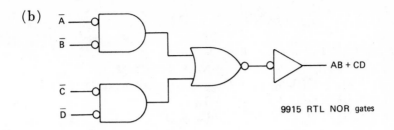

AB + CD

9915 RTL NOR gates

(c)

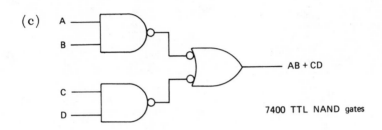

AB + CD

7400 TTL NAND gates

9. Although many may exist, this is one possible solution:

A, B, C, D, E, F, H	Buttons for 15¢ candy.
I, J, K, L	Buttons for 10¢ candy.
D_S	Dime sense
N_S	Nickel sense
A_S through L_S	Candy sense
A_d through L_d	Candy dispense
CH_1	Dime change
CH_2	15¢ change
CH_3	10¢ change
Q	Quarter input
D	Dime input
N	Nickel input
R	Return
$\overline{S}$	Out of 15¢ candy
$\overline{T}$	Out of 10¢ candy

$$A_d = A{\cdot}A_S(D{\cdot}N + N_1{\cdot}N_2{\cdot}N_3 + D_1{\cdot}D_2{\cdot}N_S + Q{\cdot}D_S)$$
$$\text{(Duplicate for } B_d \text{ through } H_d)$$

$$I_d = I{\cdot}I_S(N_1{\cdot}N_2 + D_1 + Q{\cdot}D_S{\cdot}N_S)$$
$$\text{(Duplicate for } J_d \text{ through } L_d)$$

$$\overline{S} = (A \cdot \overline{A_S} + B \cdot \overline{B_S} + C \cdot \overline{C_S} + D \cdot \overline{D_S}$$
$$+ E \cdot \overline{E_S} + F \cdot \overline{F_S} + G \cdot \overline{G_S} + H \cdot \overline{H_S})$$
$$\overline{T} = (I \cdot \overline{I_S} + J \cdot \overline{J_S} + K \cdot \overline{K_S} + L \cdot \overline{L_S})$$
$$R = \overline{S} \cdot Q \cdot \overline{D_S} + \overline{T} \cdot Q \cdot \overline{D_S} + N_S + \overline{S} \cdot D_1 \cdot D_2 \cdot \overline{N_S}$$
$$CH_1 = Q(A_d + B_d + C_d + D_d + E_d + F_d + G_d + H_d)$$
$$CH_2 = Q(I_d + J_d + K_d + L_d)$$
$$CH_3 = D_1 \cdot D_2 (A_d + B_d + C_d + D_d + E_d + F_d + G_d + H_d)$$

11.

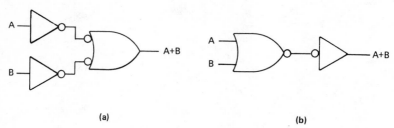

(a)   (b)

13.

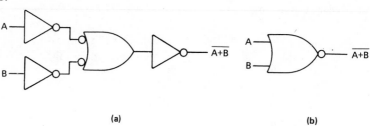

(a)   (b)

15.

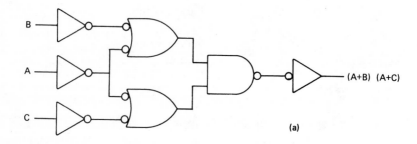

(a)

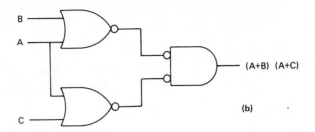

(b)

17.

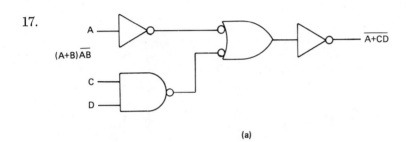

(a)

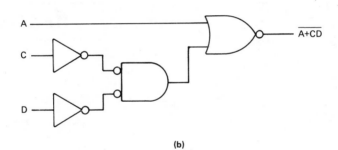

(b)

19.

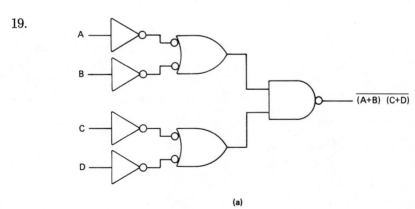

(a)

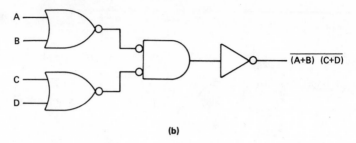

(b)

21.   23.   25.   27.

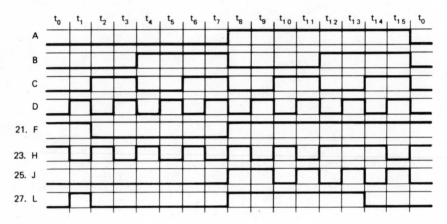

CHAPTER 9

1. Parallel or serial as related to registers refers to the mode of data transfer. When data is entered into all flip-flops of a register simultaneously, such a register is called parallel entry register.

   Parallel registers provide inputs to and outputs from all flip-flops simultaneously. This allows complete data words to be transferred by one clock pulse.

   Serial registers are shift registers. Data is entered and exited one bit at a time. In order to shift data through a shift register, one clock pulse for each bit position of the register is required.

   Parallel registers are used when high speed of data transfer is required. Serial registers are used in slower systems, but their advantage is that fewer circuits are usually required.

3. Any number of flip-flops may be used in a register. However, as the number increases (longer shift registers) limitations are imposed due to the greater time required.

5. Data can be shifted in either direction. The normal shift register shifts data to the right. Data can be shifted left by reversing the order of the connections between flip-flops. Registers can be made to shift either right or left by using gates to switch these connections. (See Fig. 9-10.)

7. Information is transferred to the output pins with a HIGH to LOW transition of the clock pulse.

9. Data in dynamic shift registers must be recirculated (refreshed) because the bits of information are stored in capacitances which need periodic recharging.

11. Because the complement of the input data is gated into the reset inputs of the flip-flops, zeros or ones may be entered. If all zeros are entered, all of the flip-flops are reset.

13. Buffer register.

15. 15 (1111). It was not reset and there is no way to enter zeros.

17. They allow accessability to all bits stored in the register at a single terminal. Recirculating maintains storage of data when it is shifted.

19.

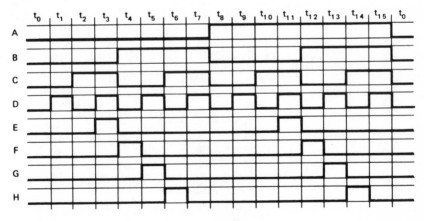

CHAPTER 10

1. An examination of the truth table for the XS-3 code (page 28) shows that the least significant bit is the complement of the 1-bit in the 8421 code. However, further investigation along the same

line reveals no pattern that is easy to implement. If an 8421 weighting is used on the XS-3 code, we see that the code is three (0011) greater than the 8421 code.

The most practical solution is to subtract 0011 from each character of the XS-3 code. This may be accomplished by adding the two's complement of 0011 (1101) to each character to be converted. For example, when the character 4 (0111) is to be converted:

$$
\begin{array}{ll}
0111 & \text{(4 in XS-3 code)} \\
(+)1101 & \\
\hline
0100 & \text{(4 in 8421 code)}
\end{array}
$$

Notice that the carry out of the MSB position is discarded.

Adders are discussed in Chapter 13. The following diagram shows how a 7483 4-bit full adder (page 305) may be used to implement the conversion of the XS-3 code to the 8421 BCD code. It should be apparent that the XS-3 code may be produced by adding 0011 to the 8421 code.

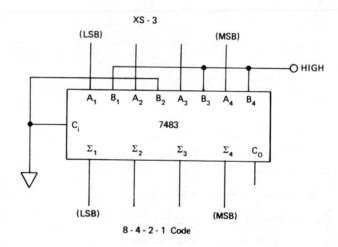

3. A micro-instruction is a logic level used to activate a particular circuit or change its mode.

5. Since there are 9 inputs (1–9), it would be logical that only one input would be activated at a given time. The greatest number of gates that can be activated simultaneously is 3. This occurs when a 7 (0111) is encoded.

7. Multiplex refers to the use of data lines or circuits for multiple purposes. This is accomplished by time sharing.

9.

(a)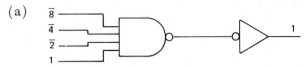

(b)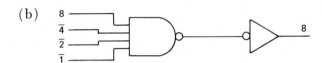

(c)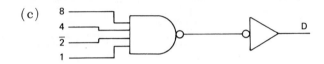

11. Four OR gates are required to encode a decimal 6.

13. A two input AND gate may be used to encode a decimal digit 9.

15. Input I_6 (pin 7) is selected if $S_0 = 0$, $S_1 = 1$, $S_2 = 1$.

17.

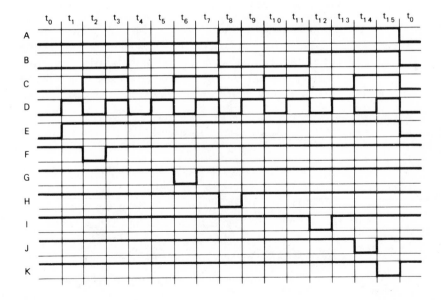

19.

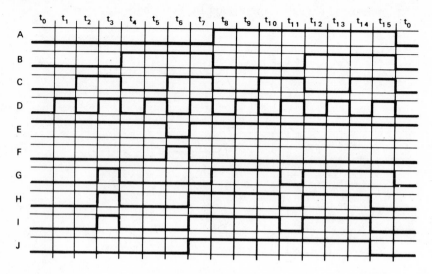

CHAPTER 11

1. Because the J-K flip-flop can be programmed to complement (change state with each input pulse), it is used in counters.

3. The most popular method is illustrated in Fig. 11-5. The two bit flip-flop is prevented from setting on the tenth count by connecting its J input to $\overline{8}$. Because the 8-bit flip-flop must be triggered from the 1-bit flip-flop, it is programmed to complement only when a 2-bit and 4-bit exist. The 8-bit flip-flop, therefore, will reset on the tenth count.

5. The inverter is used to produce an active LOW for the J input of the 8-bit flip-flop.

7. Synchronous counters are often used to solve the "race" problem.

9. The first of the two AND gates is used to detect the presence of a 1-bit AND a 2-bit in order to activate the 4-bit flip-flop to complement. The second AND gate is used in the same manner to detect the presence of an (1-bit AND 2-bit) AND a 4-bit in order to activate the 8-bit flip-flop to complement.

11. CP_2 in Fig. 11-14 is the clock pulse enabling jam entry into the register.

13. The OR gate in Fig. 11-21 is used to allow either a count-up or a count-down mode depending on which AND gate is activated.

15. A feedback shift-counter can be decoded for time durations.

17.

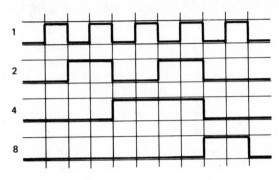

19.

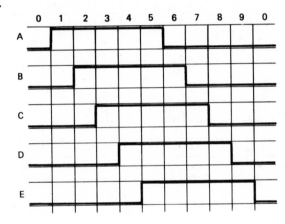

CHAPTER 12

1. Shift pulses are generated by a "clock." The clock is composed of an oscillator, counters, and decoder circuits.

3. The feedback shift counter is easier to decode. Usually, any single time period may be decoded with a 2 input gate. In addition, pulses having a duration greater than one time period are easily decoded.

5. The pulse widths of ϕ_1 and ϕ_2 are determined by the oscillator. Note that ϕ_1 and ϕ_2 are decoded from negative pulse of the oscillator.

7. ϕ and $\bar{\phi}$ identify two distinctly separate times. A bit-time is initiated when ϕ goes HIGH. Because time is required for voltage levels to settle, a second pulse is usually required to be used as a trigger source during a bit-time period. $\bar{\phi}$ goes high in the middle of a time period and is used for this purpose.

9. A function sequencer is used to provide micro-instructions based on conditions that exist as well as time intervals. The micro-instructions are decoded from both the clock and the circuit conditions.

11. The counters in the clock must be reset to zero after the completion of the operation. This can be done by decoding the available conditions. (See Figure 12-25).

13. A left digit shift is accomplished by providing four less than the normal number of clock pulses (32) in one of the clock cycles.

15.

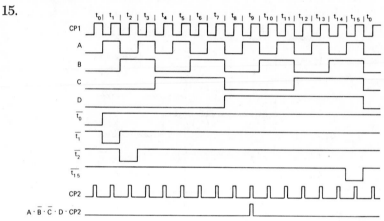

17.

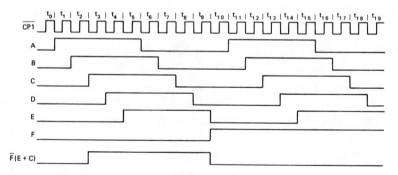

19.

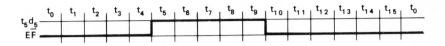

CHAPTER 13

1. (a) $S = x \oplus y$
    $C_0 = xy$
   (b) $D = x \oplus y$
    $B_0 = \overline{x}y$ assuming $x \oplus y$

3. The purpose of the inverter drawn in Fig. 13-4 is to provide the true form (active high) of the output (sum) of the first half adder. If the complements of A and B are not provided, inverters are required at the input to produce A and B.

5. A full adder is required because an input for the carry out of the next lower bit position is necessary. Two half adders and an OR gate produce a full adder.

7.

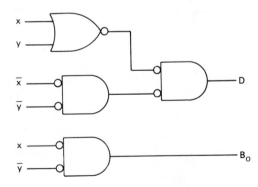

9. $(A = B) = AB + \overline{A}\overline{B}$

11.

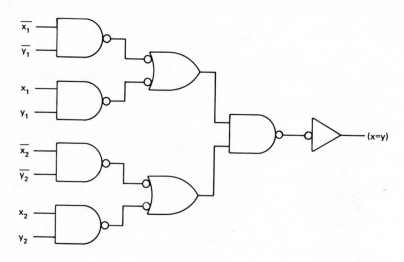

13. Carry-look-ahead circuits are used to overcome the long propagation delay of multi-bit adders.

15. (a)   862   862
        (−) 345   (+) 654   nine's complement
                          1 516
                          └─▸ 1   end around carry
                           517
    (b)   670   670
        (−) 913   (+) 086   nine's complement
                           756   *
                          −243

* If a carry is not produced, the sum is in the nine's complement form and is a negative number.

                    1   1
17. (a)   360   0011 0110 1001 0011
        (+) 741 (+) 0011 1010 0111 0100
                    0111 0001 0000 0111
                    1101 0011 0011 1101
                    0100 0100 0011 0100

In the above problem a carry was produced out of the most significant digit position. This carry is added to zeros in the next position. One must remember that a zero is represented by 0011 in XS-3.

```
                           1
(b)      709      1010  0011  1100
     (+) 119  (+) 0100  0100  1100
                 ─────────────────
                 1110  1000  1000
                 1101  1101  0011
                 ─────────────────
                 1011  0101  1011
```

```
                                                  1
19. (a)     1011  0110  0111          1011  0110  0111
        (−) 1000  1010  0110—1̄s̄→(+)   0111  0101  1001
                                     ─────────────────────
                                   ⌐1 0010  1100  0000
                                   │  0011  1101  0011
                                   └→EAC ───────────────→1
                                     ─────────────────────
                                      0101  1001  0100
                                                  1
    (b)     0110  0110  1001          0110  0110  1001
        (−) 1100  0111  0100—1̄s̄→(+)   0011  1000  1011
                                     ─────────────────────
                                      1001  1111  0100
                                      1101  1101  0011
                                     ─────────────────────
                                      0110  1100  0111
                 (Recomplemented)    −1001  0011  1000
```

21. If the mode is "subtract" and the sign bit equals zero, the output must be recomplemented. The inverter is used to allow a carry of zero to activate the AND function for complementation.

23.

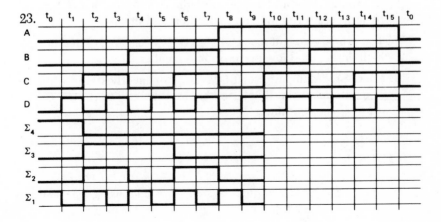

CHAPTER 14

1. The x and y lines in a core memory plane are used to select a particular core and switch its state to a one or zero.

3. Data is read from core memory by switching all cores at a given address to zero. Those cores that were in the one state "switch," causing a current to be induced in the sense line while those cores that were in the zero state induce no current. This data must be stored in a register (MDR) or it will be lost. The read cycle is always followed by a write cycle so that the data may be stored back in the cores.

5. The inhibit line is activated during the write cycle so that zeros may be written.

7. Core memory is non-volatile, that is, it will not lose data when there is no power.

9. In order to select a bipolar memory cell both the x and y lines must be selected. If they are not, the sense line output has a high impedance for that cell because of the reverse bias potential between the emitter and base.

11. High density and low power requirements are the principal advantages of MOS over bipolar devices.

13. The "C" in Fig. 14-24(a) indicates the shunt capacitance which stores the charge representing the state of the cell.

15. Bipolar ROMs may be programmed: (1) in the field by a programming device which selects the addresses in sequence and either blows a resistor element open (a one is stored) or leaves it intact (a zero is stored); (2) during the manufacturing process, where the resistor elements are either connected or not connected by a process of aluminum deposition and etching.

17. The primary advantage of MOS ROMs over bipolar ROMs is their high density and low power requirements.

19. Micro-programming uses ROMs to permanently store instructions. The outputs of the ROM may be used directly to control the various steps in a routine.

CHAPTER 15

1. (a) A complete computer system typically using SSI and MSI devices for the CPU.

(b) A complete computer system using a microprocessor.

(c) A complete CPU fabricated on a single silicon chip.

3. To organize a problem in a logical sequence.

5. Data.

7. Operand.

9. (a) Instruction phase.
   (b) Execution phase.

11. (a) Op-code.
    (b) Operand.

13. Assembler language.

15. Assembler language.

17. "A" register.

19. 3.

21. A name appears in this field to reference a particular instruction.

23. This means the source program must be entered three times before an object code program is produced.

25. (a) They require huge amounts of memory.
    (b) They are not suitable to real time applications.

CHAPTER 16

1. Point to memory locations containing data; may be also used for temporary storage.

3. 65,536.

5. "Handshaking" is a procedure whereby the microprocessor communicates with a peripheral device before data is inputted or outputted. Information to start a process in a peripheral device, or information that data is ready, is included in "handshaking." Handshaking may be used when timing in the peripheral device differs from that in the microprocessor.

7. The inputs are: READY and EOC.
   The outputs are: START, ADVBELT AND REJECT.

9. 11110XXX. The three least significant bits are used to signify a register which contains the address of the operand.

11. Any address that produces a "1" on the A_{12} line will address the least significant byte. However, only hexadecimal 1000 should be

used because the lesser significant bits are used to address memory and the more significant bits are used for output ports.

13. The mnemonic code is MOV A,M and the machine language code is: 0111 1110 (7E).

15.

```
LXI    H, 0100H        ; Load 0100 in H and L
MOV    A,M             ; Move data to Acc.
INX    H               ; Increment H and L
ADD    M               ; Add contents of 0101 to Acc.
INX    H               ; Increment H and L
MOV    M,A             ; Store result in 0102
```

Note that the instruction ADD is the same as ADC *except* that the ADC instruction will add a carry which was produced from any previous addition.

Index

DATE DUE

EXPRESSION	FIRST LEVEL LOGIC	NAND IMPLEMENTATION	NOR IMPLEMENTATION
14. $xy + xz = x(y + z)$ Distributive Property	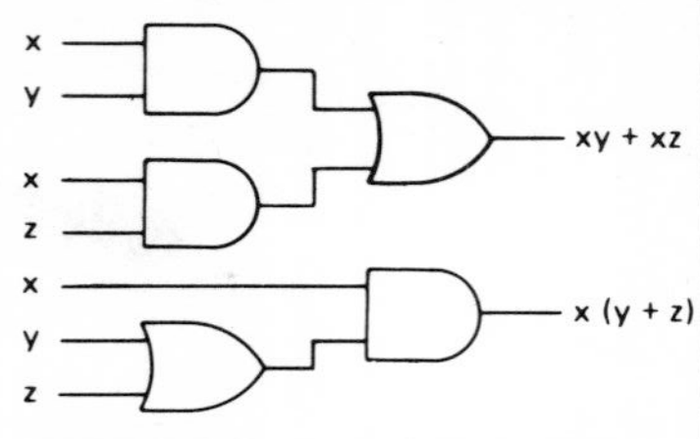	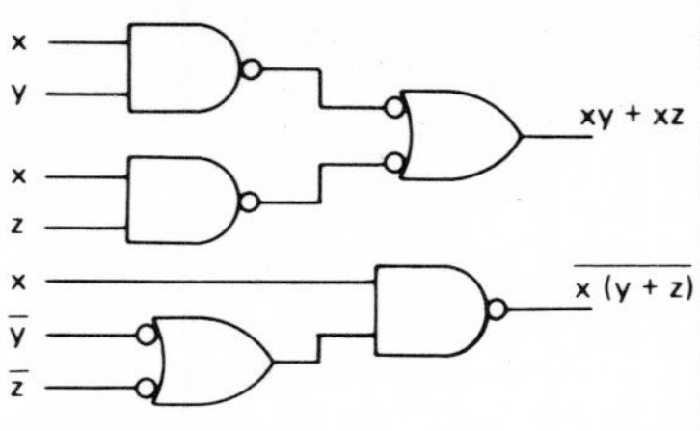	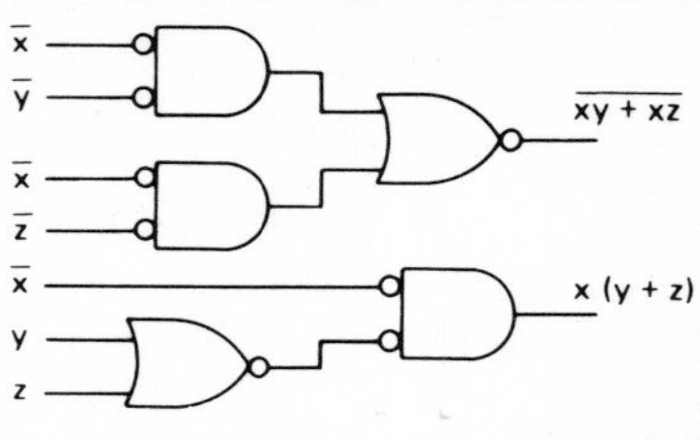
15. $(x + y)(x + z) = x + (y \cdot z)$ Distributive Property	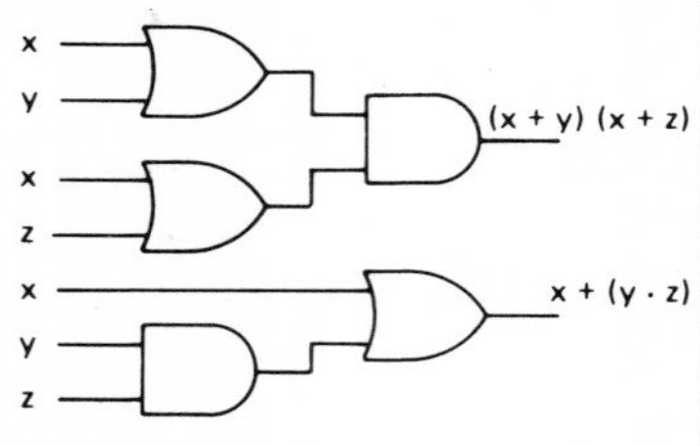	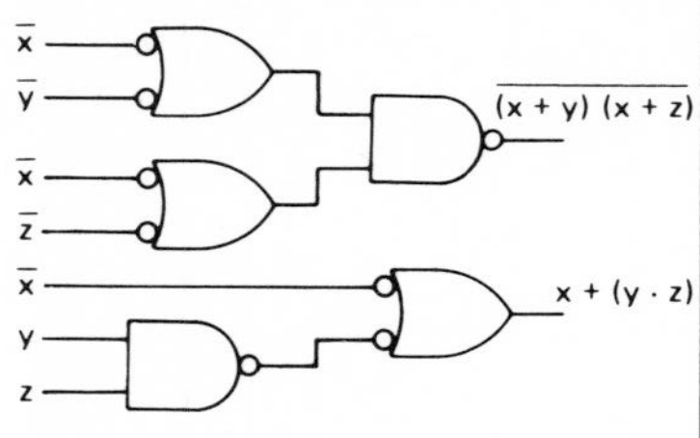	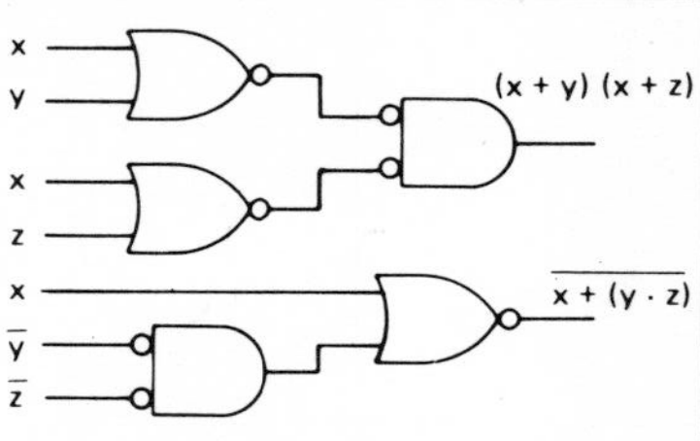
16. $x + (xz) = x$ Absorption Theorem	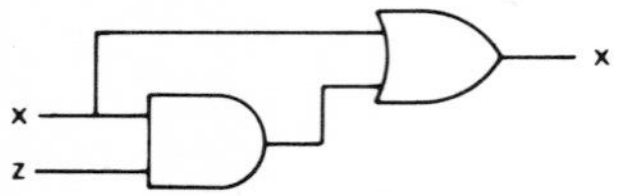	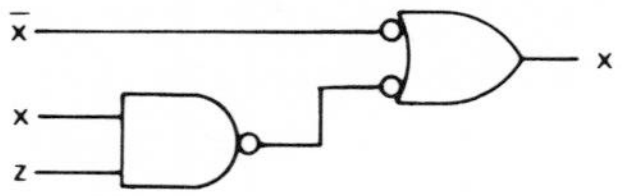	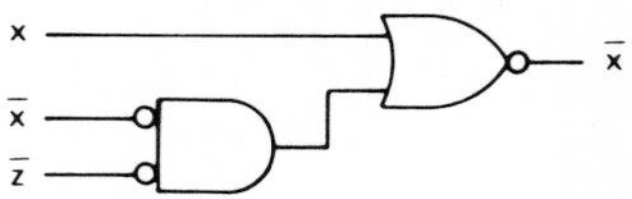
17. $x(x + y) = x$ Absorption Theorem	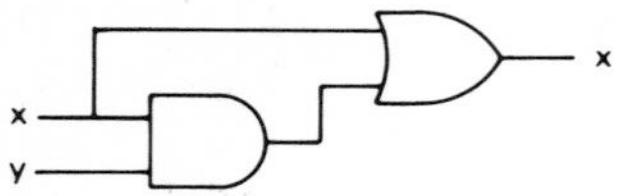	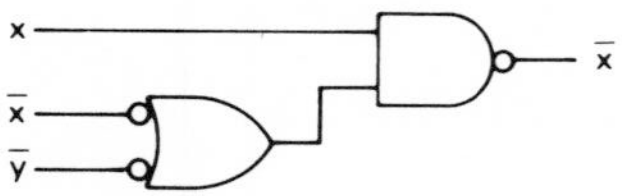	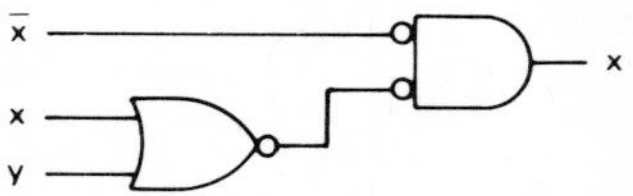
18. $x + \overline{x}y = x + y$	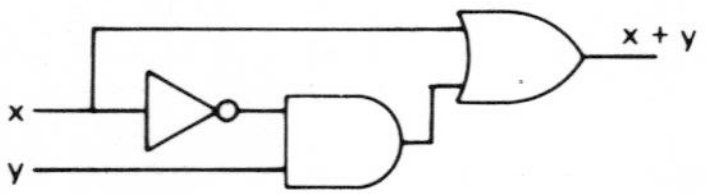	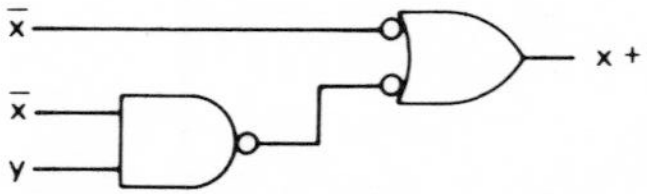	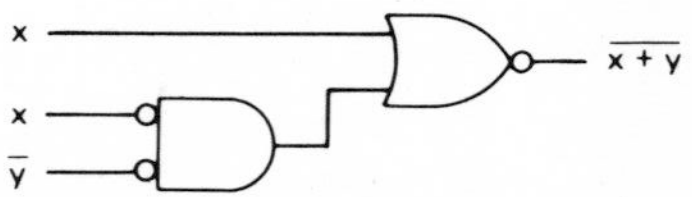

Figure 8-10(c). Implementation of logic expressions